Illustrated BRITISH Firearms PATENTS 1714-1853

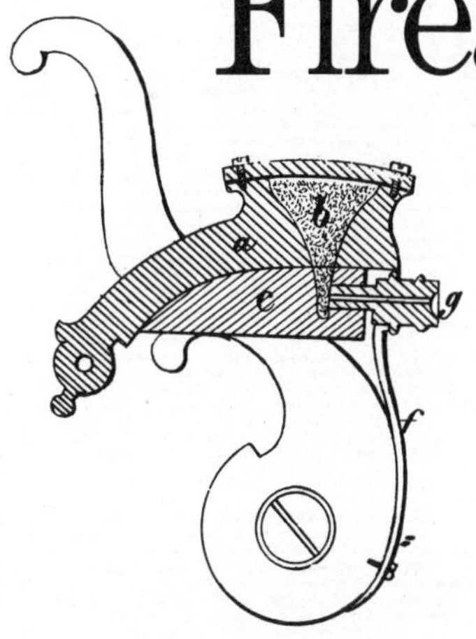

Fig. II.

Fig. 9.

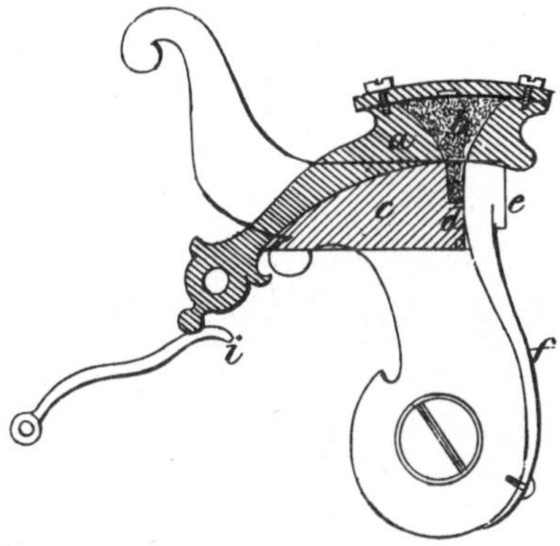

Illustrated BRITISH Firearms PATENTS 1714-1853

Fig. 3.

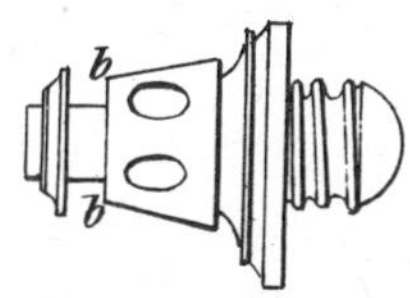

Compiled and Edited by *Stephen V. Grancsay*

and Merrill Lindsay

Fig. 10.

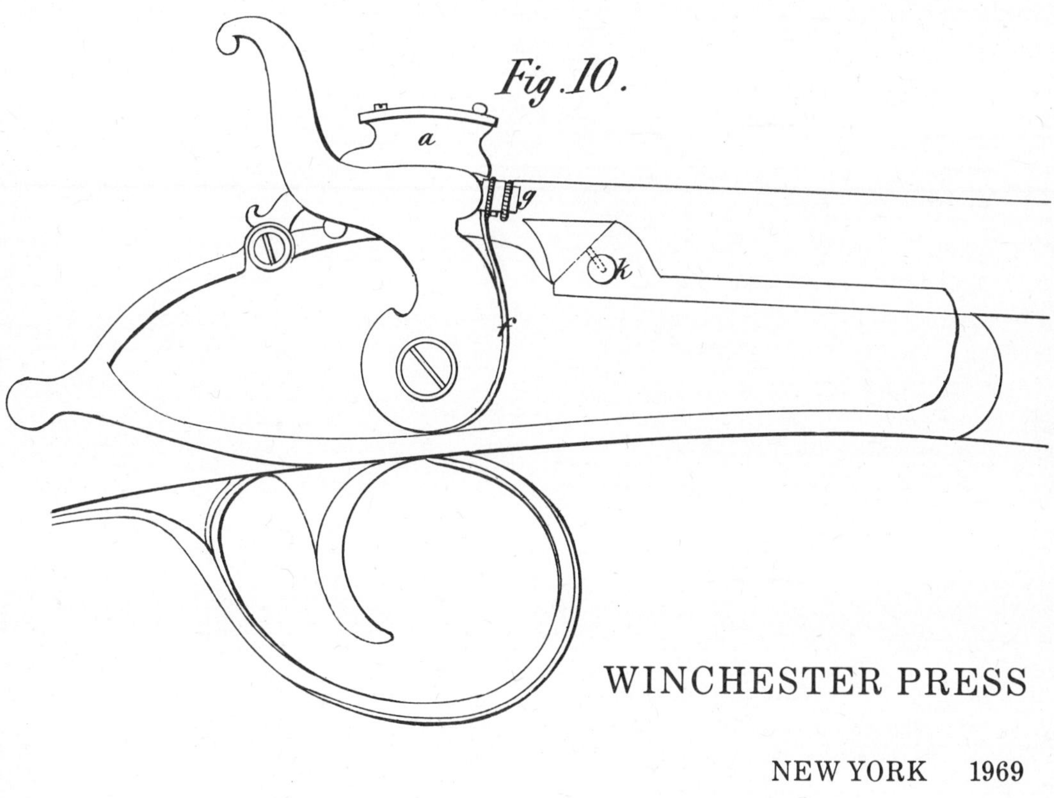

WINCHESTER PRESS

NEW YORK 1969

Library of Congress catalog card number: 77-99750
SBN: 87691-008-8

BRITISH FIREARMS PATENTS from 1714 to 1853

THE ACTUAL PATENT DRAWINGS and descriptions of new gun inventions have been more talked about than seen. Frequently, a serious collector of firearms will state that such and such an invention or improvement was patented in a given year where, as a matter of fact, no such patents exist. The patents that do exist make fascinating reading.

Over a century ago, five large volumes of English Patent Law regarding firearms were published containing a selection from the patent records of all of the firearm patents recorded from the beginning of the patent office to the time of the publication of the volumes in 1853. In the beginning of the patent office, new inventions were described and patents granted on the descriptions, probably accompanied by a demonstration of the new firearm, but no drawings were filed for permanent record. Patents in the early days were in the form of an exclusive royal grant to manufacture or exploit, and did not always deal with new inventions.

The first invention recorded is that of Reynold Hoxon in 1592 for a foolproof powder and shot flask which presumably had some sort of measuring device "by the help whereof any, be he never so unskillful, may charge any such peece with a just proportion of powder and bullett at one instant, without removing his hand or danger of fireing of himself or any of his fellowes." More than seventy years were to pass before we find the next patent having to do with gun design, although there were patents issued for manufacturing gunpowder, digging for saltpeter, making machines for casting bullets and the casting of cannon barrels.

Starting with the 1660's, there was a flurry of activity at the patent office. Between 1661 and 1682, there were three patents granted for repeating small arms. The Marquis of Worcester took out a patent in 1661 for an "invencon to make certyne guns or pistolls which in the tenthe parte of one minute may—be re-charged." No more useful information is supplied except that the barrel of the piece remains fixed. Scholars have surmised that this patent may have covered the Kalthoff repeater. The Kalthoff guns were first built in 1648 in Denmark. A son of the inventor, Caspar Kalthoff, had moved to England where he was known to have built guns for Worcester. Worcester, however, was active in sponsoring all types of mechanical and engineering innovations, so that it is impossible to prove from the wording of the patent what kind of a gun is designated. There was no drawing submitted with the words and no specifications.

Three years later, in 1664, Abraham Hill, a gunmaker, got a patent for a repeater. The words seem to suggest that the patent covers several types of breech loading mechanism, but the text is very unclear. One piece is a "gun or pistoll, the breech whereof rises upon a hindge by a contrivance of a motion from under it." Another gun carries seven or eight charges in the stock of the gun and is operated by thrusting forward the sight. This gun took a square cartridge, as did one of Puckle's guns designed to shoot heathens. I have seen a cartridge wheel lock gun in the Russell Aitken collection which sprung open at the breech when the rear sight was pushed forward. Aitken's gun is German but of approximately the same period. I know of only one surviving gun signed by Abraham Hill in the Frank Bivens collection, it is a Berselli system, loose powder and ball repeater with a side lever. The lock is signed, "Abr^m^ Hill." It has a brass twist-off barrel with a lug for a barrel wrench, brass frame and butt plate. Hill's patent would seem to cover a number of repeating breech loading systems, but the wording is purposely so obscure that one can only guess. There are, of course, no specifications or engineering drawings. In 1682, Charles Cardiff got a patent for what is clearly a gun designed to fire superimposed loads "to discharge twice or thrice or more, severall and distinct shotts in a singell barrell and locke, with one priming, and with double locks oftener, reserving one or more shotts 'till occasion offer... the mistery of maine lyeing in the charge." No specifications or drawings of this gun exist, but we have all seen guns of this type made at this and earlier periods.

If you disregard glass grenadoes, and engines which will destroy "any number of French ships," there was no more activity, at least at the patent office, in the area of gun invention through the rest of the 17th century. It is only in 1718 when we come to the Puckle patent that we begin to find useful specifications and drawings to accompany the patents. It is here that the present volume starts.

From Puckle's wonderful invention, a hand-cranked machine gun which shot round bullets at Christians and square bullets at Turks, we go to Isaac de la Chaumette, the French Huegenot emigré who invented the breech loading gun with a vertical screw at the breech which twisted down to allow the top loading of ball and powder. Unfortunately, the patent rambles on about breech loading cannon and winds up with revolving hospital mattresses. There are no specifications and there were no drawings, or if there were drawings, they have not survived. Volume One of the original set of books covers approximately a century of gun invention including the famous breech loader, an improvement on the Chaumette, which was designed by Major Patrick Ferguson as an answer to the Kentucky rifles used by the American Revolutionaries.

The Ferguson is illustrated in the patent books and actually saw service in the hands of Ferguson's special company of riflemen.

From the leisurely pace of 18th century invention, the number of patents themselves show the tremendous increase of mechanical invention starting in the second decade of the 19th century. Where Volume One of the set covered a hundred years of invention, Volume Five of the series covers but two years. By this time, Americans had gotten into the patent race and both Colt and Sharps are represented. Gun patents inspired by a desire to make money as well as by pure litigatiousness covered every phase of design and application, including many ideas which had been used by gunmakers for centuries.

The most important of the patents of the 150 years between 1714 and the mid-19th century have been selected by Stephen Grancsay from the five volumes in his library. The drawings are reproduced where they appear in the books as well as in full size in the book's accompanying portfolio. At the beginning of Volumes One and Five, there is an index of all firearms patents by the name of the inventor and also in chronological order. These have been reproduced in full. The reader will find a wealth of information in this volume and will be more of an authority on this most important period of firearms history for having had the opportunity to consult the original source materials presented here.

MERRILL LINDSAY

INTRODUCTION

THE HISTORY OF IMPROVEMENTS IN FIREARMS since they were introduced into the military systems of Europe in the fourteenth century shows very slow progress. It was not until the early decades of the nineteenth century that inventions were accelerated. In appraising these inventions one must consider the contemporary scientific discoveries in physics, chemistry and metallurgy; also the available machinery. At the time when the flintlock appeared to have reached the acme of structural excellence in the hands of Nock, Wilkinson, the Mantons, Egg, and other eminent gunsmiths of the past generation, when its beauty of finish and mechanical accuracy were at the highest, its doom was sealed. Some French chemists worked out the secret of fulminating chlorates; a Scotch clergyman applied them to a gun lock; and the reign of flint which had lasted for more than two hundred years was over.

The Society of Arts was the originator of the Great Exhibition of 1851. In the period preceding the Exhibition (in which no flintlocks were shown) three improvements took place in small arms, viz. the application of detonating powder to discharge firearms and subsequently the substitution of a detonating cap, the invention of an effective system for loading at the breech, and thirdly, an improvement in the form of the projectile, especially as related to the rifle. The first use of fulminate ignition in 1807 and the subsequent development of the percussion cap, which did away with the process of priming, were essential steps toward modern breech loading, repeating arms, metallic cartridges, practical revolvers, and automatic weapons.

The complete list of firearms patents from 1718 to 1852 which is included here along with the complete index, includes everything from fundamental inventions to mere trifles. It may be of some interest to note that the various minute changes introduced into the making of firearms suggested to Lieutenant-General Pitt-Rivers the earliest systematic attempt to apply the theory of evolution to the products of human handiwork. Of the numerous patents listed, a selection of fifty was made. Only four of those included here are of the eighteenth century. We may mention two noteworthy items, the Ferguson rifle (1776) and the patent breech of Henry Nock (1787). The improved breech caused the powder to be ignited in the center and thus greatly improved the strength and regularity of the shooting. Among the nineteenth-century noteworthy inventions are: the Forsyth firearms using the first form of percussion ignition; the flintlock breech-loading revolver patented in 1818 by Elisha Haydon Collier (also later fitted with pill-lock and cap lock); and the sequence of Colt revolvers (first patented in England in 1835) that command more attention than any other antique firearm, especially because of the subtle variations of Colt's numerous designs. We should also mention the bolt-action system that was first applied to military arms in the Dreyse needle gun (invented between 1827-1829), adopted by the Prussian army in 1840. The first models were muzzle loaders and then, in 1837, a breech loader with bolt action was made. This was a single-shot breech loader using a paper cartridge with its primer at the base of the bullet. A long needle was driven forward through the powder charge, exploding the primer by compressing it sharply against the base of the bullet. The tremendous effect of this weapon was demonstrated during the war between Austria and Germany in 1866, in which the breech loader-armed Germans quickly overcame the Austrians, still using muzzle-loaders.

The detonating system, which substituted a fulminate for flint as a means of igniting the charge of powder, was patented by the Reverend Alexander Forsyth in 1807. The explosive powder was contained in a small magazine pivoted to the lock, and by turning it on its axis a few grains of the powder were conveyed beneath a striking pin and were ignited on the descent of the hammer. Forsyth's invention gave instantaneous discharge and greatly improved the certainty of fire. The chief disadvantage of the flintlock fowling piece was the delay between the fall of the hammer and ignition of the charge. Following Forsyth, the percussion system went through the various embryonic stages of loose detonating powder, pellets, and tube primers, and emerged as the perfected percussion cap. The simplest of the early percussion locks was the tube lock. In this a small copper tube containing fulminate was partly inserted into the touch-hole, and the uncovered end crushed by the hammer. The copper cap soon succeeded the tube and the system rapidly met with success, and, according to Colonel Hawker, it was in general use by sportsmen all over the world by 1825.

The cap-lock system of ignition was a two-stage invention which consists of the nipple and detonating cap. The nipple was devised for patch-lock firearms soon after 1807. A detonating patch was pressed fast to the striking nose of the hammer and driven on the nipple by the falling hammer. The invention of the percussion cap has been credited to Joshua Shaw, a British landscape painter who established himself in Philadelphia, but there are several other legitimate claimants.

It took several decades to develop the metallic cartridge. A great variety of loose ammunition breech loaders, using percussion-cap ignition but cartridges of paper or other combustible material, appeared before 1850. Johannes Samuel Pauly in 1812 patented a breech-loading gun with a break-open breech. It had an internal firing pin and a center-fire cartridge with a brass head and a wrapped paper body. The significant feature of this cartridge is the soft head that was designed to expand and form a seal against the escape of flame and gas, and this is the principle on which almost all firearms function today. M. Lefaucheux received a French patent in 1835 that covered a practical breech-loading shotgun and its special cartridge. In 1847, Houiller, a Paris gunsmith, patented the "Pin Fire" cartridge, in which a metal cap was placed in a wad at the base of the cartridge, and an anvil in the form of a long wire rested in the cap and projected through both the cartridge and the barrel. The invention of the pin fire, which was the first "gas-tight" cartridge, was an important step towards the general adoption of the breech-loading mechanism.

Long before Christian Sharps patented his design for a breech-loading rifle on September 12, 1848, many men such as Cookson, Ferguson, Hall, and others had invented and made breech-loading arms. However, most of these early types depended upon some form of threaded or rotary breech block which was awkward to operate. The Sharps principle employed a falling block operated by an under lever that served as a trigger guard. Sharps's 1848 rifle was an experimental issue in the British forces but it was withdrawn, as the gas escape with paper cartridges was considered dangerous. Our selection of patents terminates with Christian Sharps's slant breech which was granted in 1853 (No. 712) and which improved the breech seal.

The Rifle Brigade, 95th Regiment, was raised in 1800 and was armed with Ezekiel Baker's Flintlock Rifle. This model was loaded with great difficulty and for a short period a small wooden mallet was supplied to ram home the bullet. No doubt it was thought that the flintlock ("Brown Bess") that had served the British Army so well from 1720 to 1840 had no rival, for it was not until 1834 that the British military authorities made trials that proved to them the great superioriy of the percussion system. When the Rifle Brigade was re-armed in 1838 with the Brunswick rifle, the percussion system was adopted for that rifle and for the troops generally. Thus at last the armed forces had the benefit of a means of ignition for small arms that functioned instantaneously, besides making them far less liable than before to misfire or to become useless in damp or wet weather.

Rifled guns, however, had been used for many years before they were generally adopted for military use. They took too long to load and so were not considered suitable for combat. The invention of the Minié ball permitted rapid loading and insured that the ball would fit the rifle grooves properly when fired. The Minié ball is a cylindro-conical bullet with three grooves filled with tallow to lubricate and a conical cavity in the base. In the original, this was plugged with a sheet-iron cup which the powder gases drove into the bullet, expanding it. The bullet diameter—less than the bore—permitted it to drop down the barrel when loading. The British adopted the Minié Rifled Musquet in 1851. It was superseded by the Pattern 1853 rifled musket (Lovell's Improved Brunswick rifle), also known as "The Enfield." The rifle had won its battle with the muzzle-loader even before the invention of the metallic cartridge.

It would be difficult to ascertain all the gradations of improvement that have taken place in firearms since their first introduction. In fact the origin of the principal mechanisms of firearms antedating the percussion system cannot be ascribed with certainty to any individual, nor to any country, nor to a specific date. It is only in recent years that the invention of the true flintlock has been ascribed to Marin le Bourgeoys of Lisieux, France, early in the seventeenth century. And the name of the inventor of the percussion cap has not yet been convincingly acknowledgd.

A study of extant firearms shows that invention is always a combination of old elements of technology and science. Many of the features of Colt's revolver had appeared prior to his patent. Firearms with a revolving cylinder, containing a number of chambers and firing through a single barrel, had been in use in the seventeenth century. A combination miquelet revolver and sword that belonged to Louis XIII of France (Inventory no. 288) is in the Metropolitan Museum of Art. It was necessary, however, to move the breech by hand and the limitations of the means of ignition hindered their development. Another seventeenth century feature adopted by Colt was the removable shoulder stock of a wheel lock gun, thus enabling it to be converted to a pistol. Elisha Collier anticipated the two principal features of the Colt revolver: the revolving of the cylinder by cocking the hammer is shown in the Collier-Coolidge shotgun; the locking and alignment of the chamber with the barrel at the moment of firing is a distinguishing feature of the Collier flintlock revolver patented in 1818. John H. Hall, who in 1810 invented a successful breech-loading flintlock rifle, preceded Colt in the plan to have military weapons made with uniform parts. The folding trigger, a feature of the Paterson Colt, appears on a pair of pistols made about 1810 by the Paris gunsmith Le Page. However, it was the combination of good features that made the Colt revolver popular. His breech-loading revolver could discharge its five or six rounds as rapidly as the eye can wink.

In order to enable the student to study some of the older inventions, a bibliography of works dealing with technical features of ancient firearms is included. In these publications one will find information on early breech loaders, rifles, multi-shot firearms, revolvers, and other weapons. Merrill Lindsay in his book *One Hundred Great Guns* describes these mechanisms and illustrates them with historical examples.

If one wishes to write the history of English firearms in the first half of the nineteenth century, here is the source material. Since the original publication is rare and not readily available, the present book is indispensable to students of the development of firearms during a critical period of their history. The subsequent progress that had been made in the manufacture of firearms in England was greatly due to the Exhibition of 1851, for in that exhibition the display of firearms in the American section drew attention to the subject, and led to the introduction of machinery into England for their manufacture. The American system included mass manufacture, by power-driven machinery, by machinery especially designed to serve its particular purpose, and by the use of the principle of interchangeable parts. This system was apparently first introduced by H. Blanc, who worked in French Government Arsenals late in the eighteenth century.

STEPHEN V. GRANCSAY
Curator Emeritus, Department of Arms & Armor
The Metropolitan Museum of Art

SELECTIVE BIBLIOGRAPHY OF PUBLICATIONS DEALING WITH FIREARMS HAVING SPECIAL TECHNICAL FEATURES

Anonymous. *Great Britain, Comm. of Patents. Abridgements of specifications relating to fire-arms and other weapons, ammunition, and accoutrements.* London, 1857-80. 2 vols., 12o.

Alm, Josef. "Bössa med Börlas." (Copper primer gun). *Liv Rust Kammeren,* 1941, Vol. 2, 71-75, 2 figs.

Double-barreled copper primer gun invented by Charles Lancaster of London.

Arquebusers de France, Les. Bulletin Bimestriel (Directeur de la Publication: M. Moy). Paris, No. 1, January 1964—No. 31, 1968.

Baxter, D. R. "Early Lever-Action Repeating Rifles." *Guns Review,* Jan. 1965, Vol. V, No. 1.

On June 10, 1641, Peter Kalthoff took out a patent for a gun that fired twenty-one shots with one loading. This referred to the lever action repeating wheel lock and flintlock arms for which this maker and his family were to become renowned.

Baxter, D. R. *Superimposed Load Firearms 1360-1860.* Hong Kong, *South China Morning Post,* 1966. 536 pp. 177 pls. 8o.

List of Makers and Inventors of Superimposed Load Firearms, pp. 341-344.

Bedford, Peter A. "Gold, Platinum and Touch-holes." *Journal of Arms & Armour Society,* March, 1965, Vol. V, No. 1, 27-32.

Bedford, R. & P. A. "A Pair of Egg Pistols Belonging to the Duke of York and a Note on the Tap-Action." *Journal of Arms & Armour Society,* June 1968, Vol. 6, No. 2, 31-37.

Blackmore, Howard L. "Gillmore's Musket 1816." *The Gun Collector,* Jan. 1953, 796-800, 3 figs.

Gillmore's main idea was the use of a cup or wad to hold the ball tight in the barrel and so enable the size to be decreased from the standard 14½ to 18 to the pound.

Blackmore, Howard L. "The Douglas Flintlock 1817." *The Gun Collector,* April 1953, 818-820, 2 figs.

The cock had a separate head containing a double pair of jaws facing opposite directions and held in position by a wing nut. If one flint broke, the wing nut was loosened and the other flint swivelled to the front in a matter of seconds.

Blackmore, Howard L. "The Seven-Barrel Guns." *Journal of Arms & Armour Society,* June 1955, Vol. I, No. 10, 165-182, 4 pls.

Blackmore, Howard L. "The Experimental Firearms of Henry Nock." *Journal of Arms & Armour Society,* 1956, Vol. II, 69-109, 7 pls.

Blackmore, Howard L. "Chamber's Repeating Flintlock." *American Rifleman,* Sept. 1958, 21-23.

Blackmore, Howard L. *British Military Firearms 1650-1850.* London, Herbert Jenkins, 1961. 296 pp. 83 pls. 33 figs. 8o.

Lock, barrel and stock marks, pp. 280-282, 102 figs. Bibliography, pp. 284-286.

Includes Chapters on: Early Rifles and Breechloaders; The Experimental Arms of Henry Nock.

Blackmore, Howard L. *Guns and Rifles of the World.* N. Y., The Viking Press, 1965. 134 pp. 1059 figs. 4o.

Bibliography, pp. 125-126.

Blair, Claude. "A Note on the Early History of the Wheel-lock." *Journal of Arms & Armour Society,* March 1961, Vol. III, No. 9, 221-256, 11 pls.; March 1964, Vol. IV, No. 9, 187-188.

Blair, Claude. *Pistols of the World.* London, B. T. Batsford, 1968. 206 pp. 823 pls. and line drawings. 4o.

Bibliography, pp. 188-196.

Blanch, H. J. *A Century of Guns: A Sketch of the Leading Types of Sporting and Military Small Arms.* London, John Blanch & Son, 1909. 153 pp. 21 pls. 8o.
Gunsmiths in the London Directory, 1812, p. vii.

Bolton, G. *Remarks on the present defective state of fire-arms...with an explanation of a newly invented patent gun-lock.* London, 1795. 113 pp. 8o.
Letters Patent No. 2041 of 1795.

Bordino, S. M. *Fucile di fanteria con esca fulminante.* Turin, Fodratti, 1839. 8o.
A work on the percussion gun. The author was Director of the Royal Arms Factory. French edition, 1841. Government print.

Bossi, Giuliano. *Breve Discorso delli Doppii Archibugi A Routa Inventati per G. (iuliano) B. (ossi) Romano.* Parigi, 1629. pls. 8o.

Bossi, Giuliano. *Breve Trattato d'Alcune Inventioni che Sono State Fatte per Rinforzare e Radoppiare Li Tiri Degli Archibugi e Moschetti,* per Giuliano Bossi Romano. Anversa, Girolamo Verdussen, 1625. 120 pp. fig. 8o.
"A Brief Treatise On Some Inventions Pertaining to Reinforcement and Repeating Fire with Arquebuses and Muskets."

Boudriot, Jean. *Armes a feu françaises, modeles reglementaires, 1717-1836.*
Three albums of text and line drawings made to scale, of French regulation guns and pistols. Paris, 1961-1965.

Buttin, Charles. "Les fusils de Sardaigne." In: *Beiträge zur Geschichte der Handfeuerwaffen.* Festschrift zum 80. Geburtstag von Moritz Thierbach. Dresden, Wilhelm Baensch, 1905. pp. 169-209, 4 pls.

Buttin, Charles. "Pistolets de Louis XIII par Pierre Bergier de Grenoble et P. Chabrioux de Montelimar." *Armes Anciennes,* Geneva, 1957, Vol. II, No. 10, 75-83, pl.
Waterproof two-shot wheel lock pistols. Made by Bergier in Grenoble, c. 1635 for Louis XIII, in the Musée de l'Armée, Paris.

Carpenter, A. C. "A Fifteen Shot Wheellock Repeating Pistol of Circa 1650." *Journal of Arms & Armour Society,* March-June, 1963, Vol. IV, Nos. 5-6, 93-95, 4 pls.

Chesney, F. R. *Observations on the Past and Present State of Fire-Arms, and of the Probable Effects in War of the New Musket.* London, Longman, 1852. 376 pp. 8o.

Christie, James. *A catalogue of an extensive and very valuable assortment of gunlocks, chiefly with gold pans, the workmanship of Mortimer, Nock, Wogden and Manton.* Sold by auction by Mr. Christie, September 10, 1803. London, 1803. 4 pp. 8o.

Cockle, Maurice J. D. *A Bibliography of Military Books up to 1642.* London, 1900.
Reprinted: London, Holland Press, 1957. 268 pp. 4o.

Colt, Samuel. "On the application of machinery to the manufacture of rotating chambered-breech fire-arms, and the peculiarities of those arms." Minutes Proc. Inst. Civil Engineers (with abstracts of discussions), London, 1852, Vol. 11, 30-68, 16 figs.

Cottesloe, Col. Lord. "Notes on the history of the Royal Small Arms Factory, Enfield Lock." Jour. Soc. Army Hist. Research, 1933, Vol. 12, 197-212.
Situated near London; established about 1804 for the manufacture of firearms and continued to the present day.

Cotty, G. H. *Mémoire sur la fabrication des armes portatives de guerre.* Paris, Magimel, 1806. 237 pp. table. 8o.
An exact account of the methods of making various types of firearms. The author is the inventor of the Cotty flintlock.

De Peyster, J. Watts. "The First Breech-Loading Rifle (Ferguson rifle)." *Scribner's Monthly,* 1879-1880, Vol. 19, No. 6 (April), 939-941, 3 figs.

Dillon, Viscount. "On the development of gunlocks from examples in the Tower." *Archaeological Journal,* 1893, Vol. 50, 115-132, 19 figs.

Dorman, G. (Ed.) *Patents for Inventions in the Netherlands.* The Hague, 1942.

Farquharson, G. *Manton v. Parker. Report on the cause... for alleged infringement of plaintiff's patent for an improved gun-lock.* London, 1814. 65 pp. pl. 8o.

Feldhaus, Franz M. *Leonardo der Techniker und Erfinder.* Jena, Eugene Diederichs, 1922. 170 pp. illus. 8o.

Feldhaus, Franz M. "Tesching und Flobert." *Zeitschrift für historische Waffenkunde,* 1918, Vol. 8, 83-84.
Guns of small caliber originated in Teschen prior to 1842. Flobert, a gunsmith in Paris, patented his small-caliber gun in 1849.

Forman, W. H. "Exhibits lock of a snaphaunce musket, and a triple-barreled fire-lock pistol." *Jour. Brit. Arch. Assoc.,* 1859, Vol. 15, 286-287.

G., L. "Le fusil à aiguille inventé en Belgique, vers 1624, par un comte de Hornes." *Messager Sci.,* 1880, Vol. 53, 204-217.

Gamel, J. *Opisanie Tulskago oruzheinago zavoda.* Moskva, 1826. 352 pp. 42 pl. 4o.
Description of the small-arms factory at Tula in its historical and technical aspects.

A translation of pp. 147-263 of this work, by von Bach, is among a collection of memoranda, letters and drawings relating to small-arms of R. Prosser, 1851-1854, in the Patent Office Library, London.

Gessler, Eduard A. "Ein Dreischussgewehr mit Steinschloss, aus der Mitte des 17. Jahrhunderts." *Zeitschrift für historische Waffenkunde,* 1912, Vol. 6, 139-140, fig.
In the Historical Museum, Basel.

Gessler, Eduard A. "Das 'Sturmfässlin,' eine merkwürdige Feuerwaffe." *Zeitschrift für historische Waffenkunde,* 1916, Vol. 7, 224-227, 5 figs.
In the Historical Museum, Basel; first half 17th century.

Giraud, Jean Baptiste. "Pierre Bergier, armurier et horlogier du roi à Grenoble (1633-41)." *Bulletin Archéologique du Comité des Travaux Historique et Scientifique,* Paris, 1900, 506-512.
Waterproof two shot wheel lock pistols. Made by Bergier in Grenoble, c. 1635 for Louis XIII, in the Musée de l'Armée, Paris. Reprint: Lyon, 1901. 24 pp. 8o.

Goodman, J. D. "On the progress of the small arms manufacture." *Jour. Roy. Statist. Soc.,* 1865, Vol. 28, 494-506.
The making of firearms in Birmingham from 1689 to the present.

Grancsay, Stephen V. "Cornel Klett, Hofbüchsenmacher." The story of a rare two-caliber wheel lock. *The Gun Digest,* Chicago, 1955, 9th edition, pp. 47-51, 16 figs.

Guiod, Général. *Armes portatives de guerre et de chasse. Exhibition, London, 1862.* Rapports du Jury française. Vol. IV. Paris, 1862. 8o.

Harrison, G. Charter (Editor). "The J. C. Lowe Collection of Cylinder Guns." *The Gun Collector,* 1950-53, pp. 497-556.

Haven, Charles T. & Belden, Frank A. *A History of the Colt Revolver and the Other Arms Made by Colt's Patent Fire Arms Manufacturing Company from 1836 to 1940.* With a Foreword by Stephen V. Grancsay. N. Y., William Morrow & Co., 1940. 71 pp. illus. 8o.
Reprinted.

Heurteloup, M. *Mémoire sur le Fusil de Guerre.* Paris, Everat, 1836.
See W. Reid, below.

Hoff, Arne. "Dated XVII Century Air-Guns." *The Gun Collector,* 1950-53, pp. 574-576, 5 figs.

Hoff, Arne. "Late Firearms with Snap Matchlock."

Hoff, Arne. "Noch eine Schnellfeuerpistole aus dem 17. Jahrhundert." *Zeitschrift für historische Waffenkunde,* 1939, n.s. Vol. 6, 247, figs.
Seventeenth century German 9 barrel snaphaunce pistol in the Arsenal, Copenhagen.

Hoff, Arne. "Old firearms in which technique lagged behind invention: ingenious devices that have proved their worth in modern times." *Illustrated London News,* July 18, 1936, Vol. 189, 104-105, 6 figs.

Hoff, Arne. "Quelques inventions de la famille Klett à Salzburg." *Armes Anciennes,* Genêve, 1959, Vol. II, No. 6, 133-139.

Reprint of Four Studies on History of Arms (Tojhusmuseest Skrifter 7) Copenhagen, 1963. pp. 9-30, 11 figs.

Hoopes, Thomas T. "Notes on the development of the Baltic flintlock." In: *A Miscellany of Arms and Armor.* Presented by fellow members of the Armor and Arms Club to Bashford Dean in honor of his sixtieth birthday, October 28, 1927. New York, 1927. pp. 45-51, 14 figs.

Hoopes, Thomas T. "Ein Beitrag zum französischen Radschloss." *Zeitschrift für historische Waffenkunde,* 1935, n.s. Vol. 5 (14) 50-53, 15 figs.

Wheel lock pistol of about 1630 inscribed: "Gabriel Gourinal à Perigueux." Stibbert Collection, Florence.

Wheel lock gun inscribed: "Jean Henequin à Metz 1621." National Museum, Munich. It also bears the crowned arms of France and a crowned L, indicating that it very probably belonged at one time to Louis XIII of France.

Hoopes, Thomas T. "The German Back-Striking Lock." *The Gun Collectors Letter,* Feb. 28, 1947, 4 pp. 7 figs. & map.

Hortstein, O. v. "Zwei eigenartige Handfeuerwaffen." *Zeitschrift für historische Waffenkunde,* 1907, Vol. 4, 189-190, illus.

Matchlock pistols.

Hummelberger, Walter & Scharer, Leo. "Die österreichische Militär Repetierwindbüchse und ihr Erfinder Bartholomäus Girardoni." *Zeitschrift für historische Waffen- und Kostümkunde,* 1964, Vol. 6, No. 2, 81-95, 8.figs.; 1965, Vol. 7, No. 1, 24-53, 11 figs.

Jenkins, Paul B. "The story of the primer." *A.R.C.A. Arms Review,* Nov. 1934, Vol. I, 11 pp., 8 figs.

Modern cartridges and their earlier history.

Johnson, Melvin M. & Haven, Charles T. *Ammunition, its history, development and use 1600 to 1943.* N.Y., William Morrow & Co., 1943. 361 pp. illus. 8o.

Chapter Two: Cartridge Development—1600 to the Beginning of Smokeless Powder. pp. 21-68.

Kalmár, Johannes von. "Eine Schnellfeuerpistole des 17. Jahrhunderts." *Zeitschrift für historische Waffenkunde,* 1927, n.s. Vol. 6, 86, figs.

In the Hungarian Historical Museum, Budapest.

Kane, Joseph Nathan. *Walter Hunt, American inventor (1796-1859).* New York, 1935. 75 pp. diagrams. 8o.

Aug. 10, 1848, Wooden Cartridge (enclosing a lead ball designed for the Hunt breech-loading gun).

Aug. 10, 1848, Metallic Cartridge (provided with a percussion point).

Aug. 21, 1849, Piston Breech and Firing Cock Repeating Gun (designed for breech loading, instead of muzzle loading, containing a load of cartridges available for instant use).

Kiener, Ch. H. & Pauilhac, Georges. *L'arquebuse à barillet du Musée de Cluny aux armes de Jean Ernest, Marquis de Brandebourg (1612).*
Paris, 1933. 31 pp. 3 figs. 9 pls. 8o.

Wheel lock revolver gun by a Nuremberg gunsmith, bearing the date 1612 and the initials of the Margrave of Brandenburg.

Review: Michelly, in *Zeitschrift für historische Waffenkunde,* 1934, n.s. Vol. 4 (13), 245.

Koetschau, Karl. "Eine Büchse des Grossen Kurfürsten." Hohenzollern Jahrbuch, 1903, Vol. 7, 92-101, 8 figs. pl.

A self-spanning wheel lock made by Jacob Zimmerman in Berlin, 1646, in the Dresden Historical Museum.

Koetschau, Karl. "Studien aus dem königl. historischen Museum und der königl. Gewehrgalerie zu Dresden." I. Ein Axthammer mit Schiessvorrichtung. II. Die Büchse des Michael Gull vom Jahre 1658. In: *Beiträge zur Geschichte der Handfeuerwaffen.* Festschrift zum 80. Geburtstag von Moritz Thierbach. Dresden, Wilhelm Baensch, 1905. pp. 116-127, ill.

Kronbiegel, C. Otto. "Ueber Nikolaus von Dreyse." Schuss und Waffe, 1910, Vol. 3, 254-256.
Dreyse invented the needle gun in 1828.

Latham, John. "Early breech-loaders." Jour. Royal United Service Inst., 1865, Vol. 9, 88-106, 2 pl. 21 figs.
Reprinted: London, The Arms and Armour Press, 1966.
Starting with the matchlock of Henry VIII dated 1537, in the Tower of London, the author continues a study of the nineteenth century patents up to the time of publication.

Latham, John. *On the progress of breech-loading small arms.* London, Wilkinson Sword Co., Ltd., 1875. 31 pp. illus. 8o.
Reprinted: *Jour. Royal United Service Inst.*, 1875, Vol. 19, 631-653. 2 pls. The author was head of the firm of Wilkinsons', Gun and Sword Makers, Pall Mall.

Lenk, Torsten. *The Flintlock: its origin and development.* London, Holland Press, 1965. 188 pp. 134 pls. 4o.
Sources and Bibliography, pp. 159-166.
Translated by G. A. Urquhart. Edited by J. F. Hayward.

Le Page, H. Fauré. *Etude sur les perfectionnements apportés aux armes et sur les différents systèmes qui ont précedé les modèles en usage actuellement.* Paris, Belin frères, 1909. 78 pp. illus. 8o.

Lindsay, Merrill. *One Hundred Great Guns. An Illustrated History of Firearms.* N.Y., Walker & Co., 1967. 379 pp. Many colored illustrations; marks also illustrated. 4o.
Bibliography, pp. 310-365.

Logan, Herschel C. *Hand Cannon to Automatic—A Pictorial Parade of Hand Arms.* Huntington, West Virginia, Standard Publications, 1944. 8o.
Text and 172 pls. (line drawings).

Logan, Herschel C. "The Chronology and Evolution of Firearms." In: *The Collecting of Guns.* Edited by James E. Serven. Harrisburg, Penn., 1964, pp. 58-68.
The 28 line drawings show ignition systems.

Luschin, A. "Ein altes Repetiergewehr." *Anzeiger German. Museums* (Kunde Deut. Vorzeit), 1872, n.s. Vol. 19, col. 349-350.
Description of a "repeating" gun which the inventor, Balthasar Gerolt von Gegenhofen sent to the Prince, Bishop Otto Friedrich of Laibach, Count of Buchheim, in 1647.

Majendie, Captain V. D. *Military Breech-Loading Small Arms.* London, Arms & Armour Press, 1966. 39 pp. pls. X-XIV.
Reprinted from the *Journal of the Royal United Service Institution,* 1867.

Maynard, George C. "List of magazine guns patented in the United States prior to January 1, 1880." *Magazine of Antique Firearms,* 1912, Vol. 3, 55-61.

Metschl, John. *The Rudolph J. Nunnemacher Collection of Projectile Arms.* Milwaukee, 1928. 2 vols. 113 pls. 8o.
Bulletin Public Museum, City of Milwaukee, 1928, Volume 9.
Includes chapter on "Origins, purposes and rise of projectile arms" by Paul B. Jenkins.
Appendix I. Arms Patents (U.S., 1836-1873), pp. 787-808. Compiled from Stockbridge's work (*q.v.*), with corrections.
Appendix II. American gun makers, 1630-1928, pp. 809-841.

Michelly, D. D. "Kipplaufgewehre in früheren Jahrhunderten." *Zeitschrift für historische Waffenkunde,* 1923, n.s. Vol. 1 (10), 77, figs.
Illustrates a breech loading flintlock gun from the Royal Manufactory in Naples, dated 1786.

Moore, Warren. *Guns, the Development of Firearms, Air Guns and Cartridges Told in Pictures.* N.Y., Grosset & Dunlap, 1963. 104 pp. illus. 8o.

Neumann, E. "Von den allerältesten Repetiergewehren." *Schuss und Waffe*, 1910, Vol. 3, 416-421, 432-436, 454-456, 18 figs.

Patterson, C. Meade. "A Breech-Loading, Self-Spanning Wheel-lock." *The American Arms Collector*, Towson, Maryland; April, 1958, Vol. 2, No. 2, 34-37, 8 figs.

Paulin-Desormeaux, A. O. *Nouveau Manuel Complet de l'Armurier, du Fourbisseeur et de l'Arquebusier, ou Traité Complet et Simplifié de ces Arts.*
Paris, *Encyclopédique de Roret*, 1852. 2 vols. 17 folding plates (1600 figs.)
Earlier edition. Paris, 1827.

Paulsen, J. "En saerpraeget Pistolkonstruktion af en dansk Bossemager." Vaabenhist. Aarboger, 1940-42, Vol. 3, 329-344, 4 figs.
Summary in German, pp. 342-344.
A percussion pistol of special construction by N. S. Jessen, gunsmith of Copenhagen, made between 1853 and 1859.

Peterson, Harold L. (Editor). *Encyclopedia of Firearms.* N.Y., Dutton, 1964. 367 pp. illus. 8o.

R. "A series of patents for fire-arms from the year 1617 downwards preserved in the Patent Office Library, London." *Notes & Queries*, 1866, n.s. Vol. 10, 517.

Reid, (Sir) Alex. John Forsyth. *The Reverend Alexander John Forsyth, M.A., LL.D., Belhelvie, Aberdeenshire, 1768-1843, and his invention of the percussion lock.* Aberdeen, The University Press, 1909. 32 pp. 11 pls. 8o.
This monograph has been reprinted.

Reid, W. "Pauly, Gun Designer." *Jour. Arms & Armour Society*, March & September, 1958, Vol. 2, 181-210, 10 pls., 254-258.

Reid, W. "The Firearms of Baron Heurteloup." *Jour. Arms & Armour Society*, September, 1959, Vol. 3, No. 3, 59-81, portrait and 6 pls.

Renwick, William G. "The Earliest Known Rifle." *The American Rifleman*, March 1953, 15-18, 75.
The single-headed eagle of the design known to have been used by Maximilian I between 1493 and 1508 is painted on the stock. The barrel is of bronze. The lock, apparently a snap matchlock, is unfortunately missing.

Riling, Ray. *Guns and Shooting, a Bibliography*. N.Y., Greenberg, 1951. 434 pp. illus. 8o.

Serven, James E. *Colt Firearms 1836-1954.* Santa Ana, California, 1954. 385 pp. illus. 4o.
Another edition, 1964.

Shaw, Joshua. *A Sketch of History of the Copper Cap, and Percussion Artillery Primer*. Bordentown, N. J., 1847. 7 pp.
Development of the Shaw cap and claims for remuneration for government use.

Sixl, P. "Die ersten mehrläufigen Hand- und Hakenbüchsen." In: *Beiträge zur Geschichte der Handfeuerwaffen. Festschrift zum 80 Geburtstag von Moritz Thierbach.* Dresden, Wilhelm Baensch, 1905. pp. 87-102, 3 pls.

Somerville, J. *Essay on the safety gun.* Edinburgh, 1828. 2 ed. pls. 8o.
(Letters Patent No. 5026 of 1824).

Stern, Walter M. "Gunmaking in Seventeenth-Century London." *Jour. Arms & Armour Society*, March 1954, Vol. 1, No. 5, 58-100.

Stockbridge, V. D. *Digest of Patents Relating to Breech-Loading and Magazine Small Arms (Except Revolvers) Granted in the United States from 1836 to 1873, inclusive. Classified and Arranged According to the Movement of the Principal Parts for Opening and Closing the Breech.* Washington, D.C., 1874. 176 pp. illus.
Reprinted: Norman Flayderman, Greenwich, Conn., 1963.

Stöckel, Johann F. "Ein dreiläufiger Lunten-Drehling im Zeughausmuseum zu Kopenhagen." *Zeitschrift für historische Waffenkunde*, 1937, n.s. Vol. 6, 85-85, figs.

Stöckel, Johann F. *Haandskydevaabens bedommelse.* (Hand Firearms, Their Makers and Marks). Kjobenhavn, Udgivet af Tojhusmuseet 1938-1943. 2 vols. 1076 pp. 6505 figs. of marks in the text; 5 pls. 8o.

Taylerson, A.W.F. "Registered Designs." *Jour Arms & Armour Society,* June 1954, Vol. 1, No. 6, 104-104.

Thierbach, Moritz. *Die geschichtliche Entwickelung der Handfeuerwaffen, bearbeitet nach den in den deutschen Sammlungen nach vorhandenen Originalen.* Dresden, Carl Höckner, 1886. 538 pp. 36 pls. 8o.

Nachträge zur geschichtlichen Entwickelung der Handfeuerwaffen. Dresden, Carl Damm, 1899. 52 pp. 8o.

Reprinted: Akademische Druck- und Verlagsanstalt, Graz, Austria, 1965.

Thierbach, Moritz. "Die ältesten Radschlösser deutscher Sammlungen." *Zeitschrift für historische Waffenkunde,* 1901, Vol. 2, 138-142, 7 figs.

The oldest known existing wheel lock in a German collection is on a gun in the Fürstlichen Sammlung auf Schloss Osterbein bei Gera. It is dated 1541.

Tylden, G. "Some problems of the early Breech-Loaders." *Jour. Army Historical Research,* Vol. 18, 41-43.

Vokes, C. G. "Timing device on a gun." (Letter to the Editor.) *Apollo,* 1949, Vol. 49, 105, fig.

Double-barrel muzzle-loading 10 bore percussion gun with watch mechanism on the lock.

Walcott, Mackenzie. "Early revolvers." *Notes & Queries,* 1856. s. Vol. 1, 311, 382; 1858, two s. Vol. 5, 107, 245, 358, 464.

Cites contemporary references.

Wandrus, Henry. "The Peter Kalthoff Repeating Wheellock Rifle." *The Gun Collector Magazine,* 1947, pp. 81-83, 3 figs.

These Kalthoff rifles were first made in 1645.

Weinitz, Franz. "Joseph Furttenbachs des Aeelteren Erfindung, genannt der Brand-Kugel-Schnapper." *Zeitschrift für historische Waffenkunde,* 1926, n.s. Vol. 2, 38-39, 2 figs.

This invention described in a manuscript sent by Furttenbach to Herzog August den Jüngeren von Braunschweig in 1648, entitled: *Ein Brand Kugel zu demmen.* Preserved in the Library of Wolfenbüttel.

Wettendorfer, Eduard. "A Wheel lock Automatic." *American Rifleman,* March, 1954, Vol. 102, No. 3, 49, 5 figs.

Williamson, Harold F. *Winchester, the Gun that won the West.* Washington, D.C., Combat Forces Press, 1952. 494 pp. illus. 4o.

Bibliography, pp. 409-415.

Winant, Lewis. *Early Percussion Firearms. A History of Early Percussion Firearms Ignition —From Forsyth to Winchester .44/40.* N.Y., William Morrow & Co., 1959. 292 pp. 235 figs. 8o.

Bibliography, pp. 277-283.

Wolff, Eldon G. "A Flint Lock Repeating Musket." *The Gun Collectors Letter,* June 13, 1947, No. 13, 7 pp., 5 figs.

In Milwaukee Public Museum.

Woodcroft, B. *Abridgments of the Specifications Relating to Firearms and other Weapons, Ammunition, and Accoutrements from 1588-1858.* London, published at the Great Seal Patent Office, 1859. 323 pp. 8o.

Second Edition, London, The Holland Press, 1960.

Worcester, Edward Somerset, Earl of. *A Century of Inventions.* London, 1660.

Includes information of multi-firing weapons.

Zernin. "Ein französisches Batterieschloss besonderer Konstruktion." In *Beiträge zur Geschichte der Handfeuerwaffen.* Festschrift zum 80. Geburtstag von Moritz Thierbach. Dresden, 1905. pp. 152-154, 3 figs.

SPECIFICATIONS

OF

PATENTS OF INVENTION

RELATING TO

FIRE-ARMS, PROJECTILES,

&c. &c. &c.

Enrolled under the Act 21 JAMES I.

From 15th May 1718 to 10th September 1852.

TO WHICH ARE PREFIXED,

AN INDEX OF THE NAMES OF THE INVENTORS, AND AN INDEX ARRANGED ACCORDING TO THE SUBJECT-MATTER OF THE INVENTION.

PART I.

LONDON:
PRINTED BY GEORGE EDWARD EYRE AND WILLIAM SPOTTISWOODE
PRINTERS TO THE QUEEN'S MOST EXCELLENT MAJESTY.

1854.

ALPHABETICAL INDEX

OF

PATENTEES OF INVENTIONS

IN

FIRE-ARMS, PROJECTILES,

&c. &c. &c.

Name of Patentee.	Number of Patent.	Date.	Subject-matter of Patent.
A.			
ADAMS, ROBERT - - -	13,527	24th Feb. 1851	Rifles and other fire-arms.
AINGWORTH, BENJAMIN - -	12,732	1st Aug. 1849	Ornamenting iron and other metals for use in the manufacture of gun barrels and other articles.
AITKEN, JOHN - - -	1270	5th Dec. 1780	Loading guns or fire-arms with two or more charges of gunpowder and shot, or balls, and discharging the same successively.
ASPINWALL, THOMAS - -	10,013	16th Jan. 1844	Cannon of wrought-iron or steel, or both combined; machinery used in the making; method of making.
ASPINWALL, THOMAS - -	10,728	23rd June 1845	Ordnance carriage; apparatus for governing the recoil and moving the piece backwards and forwards.
B.			
BARTHELEMEY, PETER - -	2763	18th May 1804	Shield or protection for the human body against sword, bayonet, or pike, also proof against musket balls.
BENTHAM, SAMUEL - -	1838	26th Nov. 1791	Planing wood.
BENTHAM, SAMUEL - -	1951	23rd April 1793	Mode of working wood, metal, and other materials.
BENTLEY, JOSEPH - - -	8024	9th April 1839	Guns, pistols, and other fire-arms.
BENTLEY, JOSEPH - - -	10,280	30th July 1844	Fire arms.
BERGER, CHRISTOPHER - -	1233	7th Sept. 1779	Spiral wedge for elevating, pointing, and lowering cannon and all other pieces of ordnance.
BERNEY, THOMAS FRENCH -	8143	6th July 1839	Cartridges.
BICKFORD, JOHN SOLOMON -	10,928	6th Nov. 1845	Manufacturing the miner's safety fuze.
BICKFORD, WILLIAM - -	6159	6th Sept. 1831	Instrument for igniting gunpowder when used in blasting rocks, and in mining. ("Miner's safety fuze.")
BINGHAM, WILLIAM - -	6166	24th Sept. 1831	Fire-arms of different descriptions.
BISHOP, PETER - - -	11,173	21st April 1846	Manufacture of bayonets.

Name of Patentee	Number of Patent.	Date.	Subject-matter of Patent.
BLIGHT, RICHARD - - -	1218	10th April 1779	Construction of cannons, guns, and gun carriages.
BLYTH, JOHN - - -	11,011	20th Dec. 1845	Diminishing the risk of accidental explosions of gunpowder and other substances liable to explode by contact with fire.
BOCHE, MICHEL - - -	10,683	22nd May 1845	Apparatus for measuring charges of powder and shot.
BODMER, JAMES - - -	3755	23rd Nov. 1813	Loading at the breach, fire-arms with a rifled bore; also a touch hole and moveable sight for fire-arms.
BOLTON, GEORGE - - -	2041	28th Feb. 1795	Gun-lock for muskets, pistols, and other fire-arms.
BOOTHBY, BENJAMIN - -	4705	27th Sept. 1822	Manufacturing cannon shot.
BRADLEY, JOHN - - -	3437	24th April 1811	Manufacturing gun-skelps.
BRADY, JOHN DRUMGOLLE -	14,029	22nd March 1852	Helmets, cartridge-boxes, and other military accoutrements.
BRAMAH, JOSEPH - - -	2652	30th Oct. 1802	Machine for producing straight, smooth, and parallel surfaces, on wood and other materials.
BROCKEDON, WILLIAM - -	9712	25th Apr. 1843	Manufacture of wadding for fire-arms.
BROWNE, HENRY - - -	503	5th Oct. 1728	Making cannon or great guns of iron or brass.
BUNDY, WILLIAM - - -	2932	1st May 1806	Machine for making leaden bullets, and other shot.
BUSH, WILLIAM - - -	8513	20th May 1840	Fire-arms and cartridges.
BYRNE, CHARLES - - -	1003	4th Feb. 1772	Construction of guns and all other fire-arms.

C.

Name of Patentee	Number of Patent.	Date.	Subject-matter of Patent.
CALLOW, EDWARD - - -	13,215	6th Aug. 1850	Muskets, cannon, and other fire-arms; explosive compositions and instruments.
CARBINES, HENRY - - -	10,364	24th Oct. 1844	Fusees, cartridges, and other like explosive instruments.
CARTMELL, THOMAS - -	5033	6th Nov 1824	Cock to be applied to the lock of fire-arms or ordnance, for firing the same by percussion, and whereby the priming is rendered impervious to wind, rain, or damp.
CHANU, ADOLPHE LUDOVIC -	14,065	15th April 1852	Explosive compounds and fusees; also nethods of firing the same.
CHURCH, WILLIAM - - -	14,087	24th April 1852	Fire-arms, ordnance, and projectiles to be used therewith; machinery for the manufacture of parts of such fire-arms, ordnance, and projectiles.
CLANRICARDE, JOHN DE BURGH, [Marquis of.	6139	15th July 1831	Fire-arms, and the projectiles used therewith.
COLES, CHARLES BARWELL -	8205	23rd Aug. 1839	Fixing and carrying fire-arms on horseback.
COLES, WILLIAM - - -	7472	14th Nov. 1837	Gunnery, gun and other carriages, and means of connecting the same.
COLLIER, ELISHA HAYDEN -	4315	24th Nov. 1818	Fire-arms;—also applicable to cannon.
COLT, SAMUEL - - -	6909	22nd Oct. 1835	Fire-arms.
COLT, SAMUEL - - -	12,668	20th June 1849	Fire-arms.

Name of Patentee.	Number of Patent.	Date.	Subject-matter of Patent.
CONGREVE, WILLIAM - -	3134	24th May 1808	Gun carriage, for land or sea service.
CONGREVE, Colonel WILLIAM	3565	11th May 1812	Gun and carronade carriages.
CONGREVE, Sir WILLIAM, Bart.	3937	3rd July 1815	Manufacturing gunpowder.
CONGREVE, Sir WILLIAM -	4853	16th Oct. 1823	Fireworks.
COOK. BENJAMIN - - -	3122	28th Mar. 1808	Making barrels and ramrods for fowling pieces, muskets, pistols, and other fire-arms.
COOK, HENRY - - -	1668	26th Aug. 1788	Elastic sponge for sponging great guns and other fire-arms, and not liable to damage by vermin or water.
COOK, JAMES - - - -	4960	30th May 1824	Making locks for guns, pistols, and other fire-arms.
COOPER, JOSEPH ROCK - -	7610	10th April 1838	Fire-arms.
COOPER, JOSEPH ROCK - -	8347	21st Jan. 1840	Fire-arms, and balls to be used therewith.
COOPER, JOSEPH ROCK - -	12,781	20th Sept. 1849	Fire-arms.
CRUNDWELL, JOHN - -	5845	15th Sept. 1829	Apparatus to be applied to fire-arms in place of locks.

D.

Name of Patentee.	Number of Patent.	Date.	Subject-matter of Patent.
DAVEY, SIMON - - -	14,065	15th April 1852	Explosive compounds and fusees; also methods of firing the same.
DAVEY, THOMAS - - -	10,928	6th Nov. 1845	Manufacturing the miner's safety fuse.
DAVIS, SAMSON - - -	4648	12th Feb. 1822	Lock for guns and other fire-arms.
DAVIS, SAMSON - - -	5055	18th Dec. 1824	Guns and other fire-arms.
DAVIS, SAMSON - - -	6046	6th Dec. 1830	Construction of guns and fire-arms.
DAY, JOHN - - - -	4861	13th Nov. 1823	Percussion gun-locks for various sorts of fire-arms.
DEAKIN, FRANCIS - - -	4724	9th Nov. 1822	Manufacture of holster-cases, cartouch-boxes, and other cases.
DE BARROS, IGNACIO - -	11,880	30th Sept. 1847	Machinery for making lasts for boots and shoes, butts or stocks for fire-arms, and other irregular forms.
DE BERENGER, CHARLES RANDOM, Baron.	4990	27th July 1824	Applying percussion for the purpose of igniting charges in fire-arms generally.
DE BERENGER, CHARLES RANDOM, Baron.	5439	20th Dec. 1826	Powder-flasks, horns, or other utensils for carrying gunpowder, to load therefrom guns, pistols, blunderbusses, and other fire-arms.
DE BERENGER, CHARLES RANDOM, Baron.	5905	27th Feb. 1830	Fire-arms and other weapons of defence.
DE CHATAUVILLARD, LOUIS ALFRED.	12,613	15th May 1849	Fire-arms, cartridges, bullets, bayonets, and ordnance.
DELAVAL, THOMAS - -	846	22nd April 1766	Flux for making glass; making gunpowder from sulphur-stones or brasses found in coal-mines.
DELL, WILLIAM - - -	4491	20th July 1820	Gun barrels.
DELVIGNE, HENRI GUSTAVE -	14,066	17th April 1852	Fire-arms; methods of discharging the same.

Name of Patentee	Number of Patent.	Date.	Subject-matter of Patent.
DEMONDION, AUGUSTUS -	6137	13th July 1831	Guns, muskets, and other fire-arms; cartridges to be used therewith; method of priming; also machinery for making such fire-arms, cartridges, and priming;—applicable to other purposes.
DENNETT, JOHN - - -	7759	2nd Aug. 1838	War rockets, and apparatus for applying the power of rockets for obtaining communication with stranded ships; instrument for pointing mortars for the purpose of throwing shells or firing shot from the same.
DE POGGI, ANTHONY CESARI -	2428	24th July 1800	Constructing and using ordnance, both for sea and land.
DE VILLERS, LOUIS FLORENT [DELANNOIJ.	683	30th May 1753	Making gun carriages of cast iron, to be worked by two men instead of eight as heretofore.
DICKINSON, JOHN - - -	3080	12th Nov. 1807	Cannon cartridge paper.
DICKSON, ISAAC - - -	5726	8th Dec. 1828	Projectile.
DODD, GEORGE - - -	2825	28th Feb. 1805	Royal York gun lock, other gun locks, and and locks of all sorts of fire-arms.
DODDS, ISAAC - - - -	6755	29th Jan. 1835	Machinery for cutting and shaping wood and other materials.
DODDS, ISAAC - - -	6826	30th April 1835	Construction of fire-arms;—partly applicable to the making and using common or other ordnance.
DOWNING, CHARLES - -	5242	15th Aug. 1825	Fowling-pieces and other fire-arms.
DREW, JOSEPH, junior - -	9069	6th Sept. 1841	Machine for rolling and cutting lozenges, also for cutting gun-wads, wafers, and other similar substances.
DUPE, WILLIAM - - -	2252	23rd July 1798	Manufacturing bars of a mixture of iron and steel, for double-barrel gun barrels.
DUPE, WILLIAM - - -	6166	24th Sept. 1841	Fire-arms of different descriptions.

E.

Name of Patentee	Number of Patent.	Date.	Subject-matter of Patent.
EGG, DURS - - - - -	2692	23rd March 1803	Fire-arms.
EGG, DURS - - - - -	3599	25th Sept. 1812	Construction of fire-arms and their locks; also apparatus for trying and loading them.
EGG, JOSEPH - - - -	4727	26th Nov. 1822	Construction of guns and fire-arms, on the self-priming and detonating principle.
EGG, JOSEPH - - - -	6829	9th May 1835	Fire-arms.
ELMSLIE, JAMES AUGUSTUS -	13,377	30th Nov. 1850	Sheathing ships; protecting and confining gunpowder and compounds thereof; materials used for such purposes.

F.

Name of Patentee.	Number of Patent.	Date.	Subject-matter of Patent.
FERGUSON, Captain PATRICK	1139	2nd Dec. 1776	Fire-arms, whereby they may be loaded with more expedition, ease, and safety, and fired with greater certainty, possessing also other advantages.
FORSYTH, ALEXANDER JOHN	3032	11th April 1807	Discharging or giving fire to artillery and other fire-arms; also exploding gunpowder in mines, chambers, cavities, or other places.
FOUCAUD, LEOPOLD - -	6326*	2nd Nov. 1832	Priming percussion locks of guns and pistols. * Note.—*This Patent, though included in the Index purchased from Mr. Woodcroft, does not appear in the Record books of the Great Seal.*
FOX, FRANCIS, junior - -	4427	15th Jan. 1820	Discharge of fire-arms and artillery.
FRIEND, RICHARD - - -	3005	29th Jan. 1807	Making and working gun and carronade carriages.
FRITH, JOHN - - - -	9961	25th Nov. 1843	Manufacture of cannon.

G.

Name of Patentee.	Number of Patent.	Date.	Subject-matter of Patent.
GIFFORD, JOHN WALTER DE [LONGUEVILLE.	14,058	6th April 1852	Fire-arms and projectiles.
GILL, THOMAS - - -	2436	2nd Aug. 1800	Rifling the bores or calibres of cannon, and of musket, carbine, gun, and pistol barrels.
GILLETT, WILLIAM STEDMAN	6554	8th Feb. 1834	Guns and other small arms.
GILLYON, THOMAS AUGUSTUS [GREGORY.	6491	19th Oct. 1833	Ordnance, and the carriages and projectiles to be used therewith.
GODDARD, SAMUEL ASPINWALL	14,087	24th April 1852	Fire-arms, ordnance, and projectiles to be used therewith; machinery for the manufacture of parts of such fire-arms, ordnance, and projectiles.
GOLDEN, WILLIAM - -	9129	2nd Nov. 1841	Fire-arms; bullets and other projectiles to be used therewith.
GOVER, JOHN - - - -	2151	8th Dec. 1796	Carriage for cannon.
GOVER, JOHN - - - -	2803	19th Dec. 1804	Construction of a carriage for all sorts of cannon.
GRIERSON, CHARLES - -	2566	16th Dec. 1801	Breech and lock for single and double barrel guns, pistols, and other fire-arms.

Name of Patentee.	Number of Patent.	Date.	Subject-matter of Patent.
GREEN, JOHN - - - -	1095	8th April 1775	Fire-lock constructed for portability and safety, with the lock so placed as not to obstruct the sight, having the prime secured against the effects of rain, and so contrived that the barrel can be taken from the lock for the purpose of cleaning the same; is also provided wlth a lever that sets the lock in motion, but being removed disengages the action of the gun; the said fire-lock may be used either to a gun, pistol, cannon, or other fire-arm, with one, two, three, or more barrels.
GROVE, JONATHAN - - -	2072	22nd Oct. 1795	Lock for guns or other fire-arms.

H.

Name of Patentee.	Number of Patent.	Date.	Subject-matter of Patent.
HADLEY, GILBERT, - -	577	4th June 1741	Piece of ordnance or cannon, to be charged and discharged eight times in a minute.
HALAHAN, THOMAS - -	5379	22nd June 1826	Machinery for working ordnance.
HALE, JOHN - - - - -	10,849	2nd Oct. 1845	Guns.
HALE, WILLIAM - - -	10,008	11th Jan. 1844	Rockets.
HALL, JOSEPH - - -	2573	16th Jan. 1802	Hammer for guns, pistols, and other fire-arms, which contains the prime and preserves the same from damp.
HARRINGTON, JOSIAH - -	12,543	28th March 1849	Priming fire-arms; apparatus for discharging fire-arms.
HAWKINS, GRIFFIN - -	3447	9th May 1811	Apparatus for the defence of ships against boarding.
HAWKINS, RICHARD FRANCIS -	3028	8th April 1807	Guns and carronade carriages.
HAYCRAFT, JOSEPH - -	2224	23rd March 1798	Gun-carriage.
HAZARD, ERSKINE - -	5402	12th Aug. 1826	Preparing explosive mixtures, and employing them as a moving power for machinery.
HEATON, EMANUEL - -	3794	23rd March 1814	Locks and breeches of fire-arms, by rendering the pans of locks and communication between the priming and loading of fire-arms waterproof.
HEURTELOUP, CHARLES LOUIS [STANISLAS, Baron.	6611	22nd May 1834	Fire-arms.
HEURTELOUP, CHARLES LOUIS [STANISLAS, Baron.	7980	23rd Feb. 1839	Fire-arms, and balls to be used therewith.
HEURTELOUP, CHARLES LOUIS [STANISLAS, Baron.	9084	9th Sept. 1841	Manufacture of continuous priming for certain descriptions of fire-arms; mechanism for applying the same.
HEYWOOD, GEORGE - -	3813	7th June 1814	Turning rolls, or rolling gun and pistol barrels, previous to welding.
HODGES, RICHARD EDWARD -	12,623	29th May 1849	Mechanical purchases;—applicable to projectiles.
HORTON, JOSHUA - - -	5553	11th Oct. 1827	Making hollow cylinders, guns, ordnance, retorts, and various other articles of wrought-iron, of steel, or composed of both those metals.
HOWELLS, THOMAS HOSKINS -	13,403	12th Dec. 1850.	Gun-carriages.
HUNOUT, AUGUSTIN LOUIS -	5155	23rd April 1825	Artillery, musketry, and other fire-arms.
HUNT, THOMAS - - -	14,265	19th Aug. 1852	Fire-arms.

Name of Patentee.	Number of Patent.	Date.	Subject-matter of Patent.

J.

Name of Patentee.	Number of Patent.	Date.	Subject-matter of Patent.
JACKSON, JOHN - - -	4823	29th July 1823	Construction of locks used for the discharge of guns and other fire-arms, on the detonating principle.
JAMES, HENRY - - -	3469	26th July 1811	Manufacture of barrels of fire-arms and artillery.
JENOUR, JOSHUA - - -	5570	28th Nov. 1827	Cartridge or case, and method of enclosing therein shot or other missiles, for loading fire-arms and guns.
JONES, CHARLES - - -	6394	7th March 1833	Percussion-locks applicable to fire-arms.
JONES, CHARLES - - -	6436	12th June 1833	Gun and pistol locks.
JONES, JOHN, junior - -	2911	20th Feb. 1806	Manufacturing barrels for fire-arms.
JONES, JOHN - - -	3267	28th Sept. 1809	Manufacturing skelps for fire-arms.
JONES, JOHN - - - -	3469	26th July 1811	Manufacturing barrels for fire-arms and artillery.
JOVER, WILLIAM - - -	1095	8th April 1775	Fire-lock constructed for portability and safety, with the lock so placed as not to obstruct the sight, having the prime secured against the effects of rain, and so contrived that the barrels can be taken from the lock for the purpose of cleaning the same; is also provided with a lever that sets the lock in motion, but, being removed, disengages the action of the gun; the said fire-lock may be used either to a gun, pistol, cannon, or other fire-arm, with one, two, three, or more barrels.

K.

Name of Patentee.	Number of Patent.	Date.	Subject-matter of Patent.
KEDDIE, THOMAS HAMILTON -	2810	19th Jan. 1805	Cartouch-box.
KENTISH, RICHARD - -	2885	30th Oct. 1805	Armour-waistcoat as a defence against the bayonet, sword, pike, or any pointed instrument, and in many instances may prevent the infliction of a wound from a musket ball.
KLEFT, HENRY, WILLIAM [VANDER	3837	17th Aug. 1814	Constructing a walking-staff to contain a pistol, powder, ball, and screw telescope, pen, ink, paper, pencil, knife, and drawing-utensils.
KUFAHL, GEORGE LEOPOLD [LUDWIG	13,994	3rd March 1852	Fire-arms.

Name of Patentee.	Number of Patent.	Date.	Subject-matter of Patent.
L.			
LACY, JOHN GEORGE - -	6046	6th Dec. 1830	Construction of guns and fire-arms.
LANCASTER, CHARLES WILLIAM.	13,161	3rd July 1850	Manufacture of fire-arms and cannon, and percussion-tubes.
LANCASTER, CHARLES WILLIAM.	13,454	16th Jan. 1851	Manufacture of fire-arms, cannons, and projectiles.
LAWRENCE, DAVID - -	5845	15th Sept. 1829	Apparatus to be applied to fowling-pieces and other fire-arms, in place of locks.
LOAM, MICHAEL - - -	12,406	11th Jan. 1849	Manufacture of fusees.
LOGAN, MICHAEL - - -	2917	13th March 1806	System of marine, fort, and field artillery.
LOGAN, MICHAEL - - -	3196	26th Jan. 1809	Transcendant ordnance or cannon, for marine, fort, or field service.
M.			
MACINTOSH, JOHN - -	14,041	24th March 1852	Ordnance and fire-arms, balls and shells.
MANCEAUX, FRANCOIS JULES -	13,934	29th Jan. 1852	Fire-arms, and instruments and apparatus used in connection therewith.
MANTON, GEORGE HENRY -	5838	2nd Sept. 1829	Locks on fowling-pieces or fire-arms.
MANTON, GEORGE HENRY -	7965	11th Feb. 1839	Fowling-pieces and other fire-arms.
MANTON, GEORGE HENRY -	12,543	28th March 1849	Priming fire-arms; apparatus for discharging fire-arms.
MANTON, JOHN - - -	2178	12th April 1797	Construction of guns and pistols.
MANTON, JOHN - - -	3286	11th Dec. 1809	Lock for guns and pistols.
MANTON, JOHN - - -	3942	21st July 1815	Construction of hammers and pans for the locks of fowling-pieces and fire-arms.
MANTON, JOHN - - -	4577	30th July 1821	Construction of locks of fowling-pieces and fire-arms.
MANTON, JOHN AUGUSTUS -	6572	13th March 1834	Fire-arms.
MANTON, JOSEPH - - -	1865	18th April 1792	Hammer for the locks of fire-arms; breech for double and single-barrelled guns and pistols.
MANTON, JOSEPH - - -	1893	5th July 1792	Triggers for double and single barrelled guns and pistols; wadding for the same.
MANTON, JOSEPH - - -	2722	6th July 1803	Hammer for the locks of fowling-pieces and small-arms.
MANTON, JOSEPH - - -	2986	15th Sept. 1806	Double-barrelled guns.
MANTON, JOSEPH - - -	3558	30th April 1812	Guns and pistols.
MANTON, JOSEPH - - -	3985	29th Feb. 1816	Construction and use of certain of the parts of fire-arms; shoeing horses.
MANTON, JOSEPH - - -	4166	26th Sept. 1817	Locks for fire-arms.
MANTON, JOSEPH - - -	4285	3rd Aug. 1818	Primers of fire-arms; construction of certain of the parts of fire-arms.
MANTON, JOSEPH - - -	5106	26th Feb. 1825	Fire-arms.
MANTON, JOSEPH - - -	5135	25th March 1826	Shot.

Name of Patentee.	Number of Patent.	Date.	Subject-matter of Patent.
MARSHALL, JAMES - -	5511	26th June 1827	Mounting cannons or guns for sea or other service.
MASON, WILLIAM - -	6872	6th Aug. 1835	Making fire-arms and artillery.
MATTHEY, JAMES FREDERICK	2991	4th Dec. 1806	Fire-arms and guns.
MELVILLE, ALEXANDER -	13,215	6th Oct. 1850	Muskets, cannon, and other fire-arms; explosive compositions and instruments.
MIDDLETON, EDWARD - -	14,087	24th April 1852	Fire-arms, ordnance, and projectiles to be used therewith; machinery for manufacture of parts of such fire-arms, ordnance, and projectiles.
MILLS, WILLIAM - - -	5416	18th Oct. 1826	Fire-arms.
MOLLETT, JOHN - - -	11,657	15th April 1847	Fire-arms and cartridges.
MOSER, ABRAHAM ADOLPH -	6196	15th Dec. 1831	Fire-arms.
MOULD, JACOB - - -	5099	19th Feb. 1825	Fire-arms.

N.

Name of Patentee.	Number of Patent.	Date.	Subject-matter of Patent.
NAPIER, DAVID - - -	8385	12th Feb. 1840	Manufacture of projectiles.
NEEDHAM, HENRY - - -	12,432	20th Jan. 1849	Fire-arms.
NEEDHAM, WILLIAM - -	9801	24th June 1843	Fire-arms.
NEWMARCH, BENJAMIN - -	5317	16th Jan. 1826	Exploding fire-arms.
NEWMARCH, BENJAMIN - -	5421	7th Nov. 1826	Fire-arms.
NEWTON, ALFRED VINCENT -	12,048	31st Jan. 1848	Machinery for manufacturing shot and other solid balls.
NEWTON, ALFRED VINCENT -	14,085	22nd April 1852	Priming fire-arms.
NOCK, HENRY - - -	1095	8th April 1775	Firelock constructed for portability and safety, with the lock so placed as not to obstruct the sight, having the prime secured against the effects of rain, and so contrived that the barrel can be taken from the lock for the purpose of cleaning the same; is also provided with a lever that sets the lock in motion, but being removed disengages the action of the gun; the said firelock may be used either to a gun, pistol, cannon, or other fire-arm, with one, two, three, or more barrels.
NOCK, HENRY - - -	1598	25th April 1787	Breeching applicable to guns and other fire-arms.
NOCK, SAMUEL - - -	4054	12th Aug. 1816	Pan of the locks of guns and fire-arms.
NOON, THOMAS - - -	3233	4th May 1809	Guns, pistols, and other fire-arms; applicable to cannon and other large guns.

O.

Name of Patentee.	Number of Patent.	Date.	Subject-matter of Patent.
O'LEARY, CONNOR WILLIAM -	11,864	9th Sept. 1847	Producing power for the discharge of weapons and missiles, and for other purposes.
ORSON, EDWARD FORBES -	5708	18th Sept. 1828	Cartridge for sporting purposes.
OSBORN, HENRY - - -	3590	3rd Aug. 1812	Machine for turning and levelling articles made of iron, preparatory to welding and grinding.

Name of Patentee.	Number of Patent.	Date.	Subject-matter of Patent.
Osborn, Henry - - -	3617	28th Nov. 1812	Welding and making cylinders of iron and steel.
Osborn, Henry - - -	3740	15th Oct. 1813	Making tools for tapering cylinders of iron, steel, or other metal or mixture of metals; also for tapering bars of the same materials.
Osborn, Henry - - -	4105	1st March 1817	Producing cylinders.

P.

Name of Patentee.	Number of Patent.	Date.	Subject-matter of Patent.
Pauly, Jean Samuel - -	3833	4th Aug. 1814	Construction and use of fire-arms.
Pauly, Jean Samuel - -	4026	14th May 1816	Construction and use of fire-arms.
Peckham, John Randell -	2709	10th June 1803	Lock for muskets, fusees, carbines, fowling-pieces, or pistols
Perkins, Jacob - - -	4952	15th May 1824	Throwing shells and other projectiles.
Petitpierre, Jean Henry -	4923	20th March 1824	Machine for making from one piece of leather, without seam or sewing, shoes and slippers, gloves, caps, and hats, cartouche-boxes, sheaths for swords, bayonets, and knives.
Place, Francis - - -	2920	21st March 1806	Locks for muskets, pistols, fowling-pieces, carriage-guns, and every species of fire-arms.
Poole, Moses - - -	7286	19th Jan. 1837	Ordnance and other fire-arms.
Poole, Moses - -	8573	18th July 1840	Fire-arms and apparatus to be used therewith.
Poole, Moses - - -	9119	14th Oct. 1841	Fire-arms.
Poole, Moses - - -	14,052	31st March 1852	Fire-arms.
Pretyman, Robert - -	3722	19th July 1813	Pan touchhole and pan cover of a gun-lock.
Prosser, John - - -	2454	9th Dec. 1800	Waterproof pan and hammer for gun and pistol-locks; also a breech for gun and pistol barrels.
Puckle, James - - -	418	15th May 1718	Portable gun, or machine called a defence.

R.

Name of Patentee.	Number of Patent.	Date.	Subject-matter of Patent.
Raminger, Henry - -	725	29th June 1758	Machine whereby shot used for fowling, and bullets of lead, are made more exactly round and solid than heretofore.
Rawle, William - - -	1173	28th Nov. 1777	Machine for the carriage of soldiers' cartridges.
Redfern, Bartholomew -	6053	17th Dec. 1830	Lock break-off and trigger, for fowling-pieces, muskets, rifles, pistols, and small fire-arms.
Reeves, Charles, junior -	13,991	27th Feb. 1852	Manufacture of bayonets, swords, and other cutting instruments.

Name of Patentee.	Number of Patent.	Date.	Subject-matter of Patent.
Remnant, Stephen - -	765	5th Aug. 1761	Gun-carriages of cast iron.
Richards, George - -	3155	30th July 1808	Single and double cannon, carronades or ordnance, muskets and all other kinds of fire-arms ; charging or loading the same; fixing bayonets on fire-arms.
Richards, William Westley	4611	10th Nov. 1821	Construction of gun and pistol locks.
Richards, William Westley	6071	11th Feb. 1831	Touch-holes and primers suitable to percussion guns, pistols, and all sorts of arms fired on that principle.
Richards, William Westley	7041	22nd March 1836	Primers for discharging fire-arms by means of percussion.
Richards, William Westley	7582	8th March 1838	Primer for fire-arms.
Richards, William Westley	9177	14th Dec. 1841	Construction of gun and pistol locks, and primers for the discharge of fire-arms.
Richards, William Westley	14,027	20th March 1852	Fire-arms; and means used for discharging the same; also projectiles.
Richardson, Francis - -	4031	25th May 1816	Locks and barrels of fire-arms; bayonets.
Ritchie, William Henry -	12,648	7th June 1849	Fire-arms.
Riviere, Isaac - - -	5175	20th May 1825	Construction, arrangement, and simplification of the machinery by which guns, pistols, and other fire-arms are discharged.
Rolinson, Solomon - -	9673	20th March 1843	Manufacture of shot.
Roose, James - - -	9947	18th Nov. 1843	Manufacturing gun-barrels and ordnance.
Rose Aaron - - - -	13,299	24th Oct. 1850	Manufacturing twisted gun and pistol barrels.
Round, George - - -	7712	30th June 1838	Manufacture of certain parts of gun and pistol locks.
Rousseau, Alexander - -	9258	15th Feb. 1842	Fire-arms.

S.

Name of Patentee.	Number of Patent.	Date.	Subject-matter of Patent.
Sargant, William Lucas -	10,030	30th Jan. 1844	Manufacture of barrels for fire-arms.
Sartoris, Urbanus - -	4107	11th March 1817	Construction and use of fire-arms.
Sartoris, Urbanus - -	4336	23rd Jan. 1819	Construction and use of fire-arms.
Schlesinger, Joseph William	14,227	20th July 1852	Fire-arms, cartridges; and manufacture of powder.
Searles, James Sturman -	1837	28th Nov. 1791	Gun-triggers.
Searles, James Sturman -	2744	3rd Dec. 1803	Fire-arms, or defensive instruments.
Sears, Matthew Urlwin -	12,920	11th Jan. 1850	Construction of guns and cannons; manufacture of cartridges for the loading thereof.
Seymour, John - - -	10,109	14th March 1844	Safety-bolt, and tumbler, for the locks of certain kinds of fire-arms.
Seymour, Thomas - - -	10,109	14th March 1844	Safety-bolt, and tumbler, for the locks of certain kinds of fire-arms.
Shaw, John - - -	12,728	1st Aug. 1849	Air-guns.
Shaw, Joshua - - -	8704	17th Nov. 1840	Discharging ordnance, muskets, fowling-pieces, and other fire-arms.
Shoubridge, James - -	2763	18th May 1804	Shield or protection for the human body against sword, bayonet, or pike, also proof against musket balls.

Name of Patentee.	Number of Patent.	Date.	Subject-matter of Patent.
SHRAPNELL, HENRY - -	6675	6th Sept. 1834	Fire-arms, and ammunition for the same.
SHRAPNELL, HENRY NEED- [HAM SCROPE	14,271	23rd Aug. 1852	Ordnance and fire-arms, cartridges, and ammunition or projectiles; mode of making up or preparing the same.
SIDDON, WILLIAM - - -	2173	14th March 1797	Screwing and fastening hammer-springs and sear-springs to gun-locks and pistol-locks.
SIMMONS, JOHN LINTORN [ARABIN.	14,095	29th April 1852	Manufacture of ordnance; construction and manufacture of carriages, and traversing-apparatus for manœuvring the same.
SIMPSON, GEORGE - - -	13,377	30th Nov. 1850	Sheathing ships; protecting and confining gunpowder and compounds thereof; materials used for such purposes.
SMITH, CHARLES JAMES -	10,667	14th May 1845	Artillery, guns, pistols, and other fire-arms; apparatus to be used therewith.
SMITH, DAVID - - -	12,624	29th May 1849	Manufacture of certain articles in lead.
SMITH, GEORGE - - -	10,928	6th Nov. 1845	Manufacture of the miner's safety-fuze.
SMITH, GEORGE - - -	11,447	12th Nov. 1846	Safety-fuze.
SMITH, HENRY - - -	2658	13th Nov. 1802	Vessel or barrel for the safe carriage and conveyance of gunpowder.
SMITH, SAMUEL - - -	5978	7th Aug. 1830	Touch-hole applied to fire-arms for firing the same by percussion; cap to contain the priming.
SMITH, WILLIAM - - -	3588	28th July 1812	Gun or pistol lock.
SOMERVILLE, Rev. JOHN -	5026	4th Nov. 1824	Method of preventing the accidental discharge of fowling-pieces or other fire-arms.
SOMERVILLE, Rev. JOHN -	6825	28th April 1835	Construction of guns or muskets, and other such fire-arms.
STARKEY, THOMAS - -	9188	16th Dec. 1841	Percussion-caps for discharging fire-arms.
STOCKER, GEORGE - -	8024	9th April 1839	Guns, pistols, and other fire-arms.
SUTTON, RALPH - - -	3773	24th Dec. 1813	Security to prevent accidental discharges of fowling-pieces, which is unconnected with the lock, and applicable to all kinds of fire-arms.
SYKES, THOMAS - - -	1978	4th Feb. 1794	Lock instrument, or machine with one trigger or bolt, whereby both locks affixed to fire-arms with double barrels may be fired without a side motion or other inconvenience.
SYKES, THOMAS - -	3828	4th Aug. 1814	Construction of guns, pistols, and other fire-arms, and of implements used for loading them.

T.

Name of Patentee.	Number of Patent.	Date.	Subject-matter of Patent.
TAYLOR, JOHN - - -	11,407	8th Oct. 1846	Manufacture of explosive compounds.
TAYLOR, STEPHEN - - -	11,994	10th Dec. 1847	Construction of fire-arms, and cartridges for charging the same.
TAYLOR, STEPHEN - - -	14,289	10th Sept. 1852	Construction of fire-arms, and cartridges for charging the same.

Name of Patentee.	Number of Patent.	Date.	Subject-matter of Patent.
TEMPLEMAN, JAMES - -	1707	6th Nov. 1789	Making locks to discharge double-barrel guns and pistols, by means of one trigger, without discharging both barrels at the same instant, which may be discharged either in immediate succession or at any distant period from each other.
THOMASON, EDWARD - -	2360	28th Nov. 1799	Mechanism of the cocks of gun-locks, applicable to all kinds of fire-arms.
THOMASON, EDWARD - -	2750	7th Feb. 1804	Making pikes.
THOMSON, JAMES - -	3784	9th March 1814	Construction of fire-arms and the locks to fire-arms.
TUCKER, JOHN - - -	5839	9th Sept. 1829	Construction of cannon.
TUCKER, JOHN - - -	5864	2nd Nov. 1829	Exploding shot or projectile.
TURNER, WILLIAM - -	3400	4th March 1811	Pike or halberd with cuteaus.
TYSON, JOSEPH WASHINGTON	10,718	10th June 1845	Fire-arms and ordnance.

V.

Name of Patentee.	Number of Patent.	Date.	Subject-matter of Patent.
VAZIE, ROBERT - - -	2436	23rd Jan. 1801	Constructing, construction and application of a gun, by removing the touch-hole from the side to the centre of the butt-end of the barrel.
VON WINIWARTER JOSEPH [MAXIMILIAN RITTER	13,935	29th Jan. 1852	Locks of fire-arms and cannon; gun-matches, or mode of igniting gunpowder in guns; machinery for manufacturing the same.

W.

Name of Patentee.	Number of Patent.	Date.	Subject-matter of Patent.
WALKER, JAMES - - -	3373	7th Sept. 1810	Machine or vessel for the safe conveyance of gunpowder, and for its preservation from injury by damp.
WALKER, RICHARD- - -	6633	26th June 1834	Wadding for fire-arms.
WALKER, THOMAS - - -	14,095	29th April 1852	Manufacture of ordnance; construction and manufacture of carriages, and traversing apparatus for manœuvring the same.
WARING, THOMAS - - -	1639	5th Feb. 1788	Manufacture of long-bows; machines for the purpose.
WATERS, JOHN - - -	1284	9th March 1781	Pistol with a bayonet.
WATTS, WILLIAM - - -	1347	10th Dec. 1782	Making small shot solid throughout, perfectly globular in form, and without the imperfections usual in shot as hitherto manufactured.
WEBB, RICHARD - - -	2042	28th Feb. 1795	Construction of guns and other fire-arms.
WEBSTER, WILLIAM - -	4590	14th Sept. 1821	Mechanism of and appertaining to "Forsyth's Roller Magazine," for the discharge of fire-arms by percussion.

Name of Patentee.	Number of Patent.	Date.	Subject-matter of Patent.
WHITEHOUSE, CORNELIUS -	**10,696**	3rd June 1845	Machinery for welding and hammering; manufacture of gun-barrels and other tubes.
WHITFORD, SAMUEL - -	**7712**	30th June 1838	Making certain parts of gun and pistol locks.
WILKINSON, ISAAC - -	**723**	21st April 1758	Casting guns or cannon, fire-engines, cylinders, pipes, sugar-rolls, and such like instruments.
WILKINSON, JOHN - - -	**1063**	27th Jan. 1774	Casting and boring iron guns or cannon.
WILKINSON, JOHN - - -	**1694**	30th July 1789	Making cannon or other ordnance with metal, together with its shot, shell, or arrow.
WILKINSON, HENRY - -	**8119**	22nd June 1839	Fire-arms.
WILLSON, JAMES - - -	**1897**	5th July 1792	Construction of fire-arms for the better protection of the powder in bad weather.
WILSON, ALEXANDER - -	**2865**	3rd July 1805	Shot-belts, powder-flasks, and fire-arms.
WINSOR, FREDERICK ALBERT -	**3510**	4th Dec. 1811	Employing raw and refined sugars in the composition of sundry articles.
WOOLLETT, ALFRED - -	**12,559**	3rd April 1849	Gun-carriages.
WRIGHT, THOMAS - - -	**1003**	4th Feb. 1772	Construction of guns and all other fire-arms.
WYATT, MATTHEW COTES -	**4218**	3rd Feb. 1818	Safeguard to prevent the accidental movement forward of the cock of a gun, pistol, or other fire-arm, towards the hammer.

SUBJECT-MATTER INDEX

OF

PATENTS OF INVENTION

RELATING TO

FIRE-ARMS, PROJECTILES,

&c., &c., &c.

Subject-matter of Patent.			Number of Patent.	Date.	Name of Patentee.
			A.		
Accidental discharge of fire-arms, improvements in preventing the			1095	8th April 1775	H. NOCK, W. JOVER, and J. GREEN.
Do.	do.	do.	1707	6th Nov. 1789	J. TEMPLEMAN.
Do.	do.	do.	1837	28th Nov. 1791	J. S. SEARLES.
Do.	do.	do.	2178	12th April 1797	J. MANTON.
Do.	do.	do.	2825	28th Feb. 1805	G. DODD.
Do.	do.	do.	2920	21st March 1806	F. PLACE.
Do.	do.	do.	3155	30th July 1808	G. RICHARDS.
Do.	do.	do.	3283	11th Dec. 1809	J. MANTON.
Do.	do.	do.	3558	30th April 1812	J. MANTON.
Do.	do.	do.	3599	25th Sept. 1812	D. EGG.
Do.	do.	do.	3773	24th Dec. 1813	R. SUTTON.
Do.	do.	do.	3784	9th March 1814	J. THOMSON.
Do.	do.	do.	3828	4th Aug. 1814	T. SYKES.
Do.	do.	do.	4218	3rd Feb. 1818	M. C. WYATT.
Do.	do.	do.	4336	23rd Jan. 1819	U. SARTORIS.
Do.	do.	do.	5026	4th Nov. 1824	J. SOMERVILLE.
Do.	do.	do.	5242	15th Aug. 1825	C. DOWNING.
Do.	do.	do.	5726	8th Dec. 1828	J. DICKSON.
Do.	do.	do.	5845	15th Sept. 1829	D. LAWRENCE and J. CRUNDWELL.
Do.	do.	do.	5905	27th Feb. 1830	C. R. Baron de BERENGER.
Do.	do.	do.	6053	17th Dec. 1830	B. REDFERN.
Do.	do.	do.	6137	13th July 1831	A. DEMONDION.
Do.	do.	do.	6825	28th April 1835	J. SOMERVILLE.
Do.	do.	do.	6909	22nd Oct. 1835	S. COLT.
Do.	do.	do.	7610	10th April 1838	J. R. COOPER.
Do.	do.	do.	8024	9th April 1839	G. STOCKER and J. BENTLEY.
Do.	do.	do.	10,109	14th March 1844	T. SEYMOUR and J. SEYMOUR.
Do.	do.	do.	10,849	2nd Oct. 1845	J. HALE.
Do.	do.	do.	12,432	20th Jan. 1849	H. NEEDHAM.
Do.	do.	do.	12,613	15th May 1849	L. A. DE CHATAUVILLARD.

Subject-matter of Patent	Number of Patent	Date.	Name of Patentee.
Accidental discharge of fire-arms, improvements in preventing the—*continued*	**13,527**	24th Feb. 1851	R. Adams.
Do. do. do.	**13,994**	3rd March 1852	G. L. L. Kufahl.
Do. do. do.	**14,265**	19th Aug. 1852	T. Hunt.
Air, discharging fire-arms by means of compressed	**3833**	4th Aug. 1814	J. S. Pauly.
Do. do. do.	**4026**	14th May 1816	J. S. Pauly.
Do. do. do.	**5317**	16th Jan. 1826	B. Newmarch.
Air guns, improvements in locks for	**6046**	6th Dec. 1830	J. G. Lacy and S. Davis.
Air guns or cannons, improvements in compressing and heating air for, and in working	**11,864**	9th Sept. 1847	C. W. O'Leary.
r guns, improvements in	**12,728**	1st Aug. 1849	J. Shaw.
Armour for protecting the human body against swords, bayonets, pikes, and musket balls	**2763**	18th May 1804	J. P. Barthelemy and J. Shoubridge.
	2885	30th Oct. 1805	R. Kentish.
Arrow-shot, forming with a belt, or wings, or grooves, and rifling cannon for discharging the same	**1694**	30th July 1789	J. Wilkinson.
Arrows, improvements in making and testing long bows for discharging	**1639**	5th Feb. 1788	T. Waring.
Arrows or harpoons, for spearing fish, to be discharged from fire-arms	**2744**	3rd Dec. 1803	J. S. Searles.
Arrows charged with explosive compounds, to produce light at any required spot	**13,215**	6th Oct. 1850	A. Melville and E Callow.

B.

Subject-matter of Patent	Number of Patent	Date.	Name of Patentee.
Balls or bullets, gun for discharging square and round	**418**	15th May 1718	J. Puckle.
Balls or bullets, improvements in manufacturing	**725**	29th June 1758	H. Raminger.
Balls or bullets, belted, winged, or grooved, for rifled cannon	**1694**	30th July 1789	J. Wilkinson.
Balls or bullets, affixing a tail to	**2744**	3rd Dec. 1803	J. S. Searles.
Balls or bullets, shield or protection for the human body against	**2763**	18th May 1804	J. P. Barthelemy and J. Shoubridge.
Balls or bullets, armour for protecting the human body against	**2885**	30th Oct. 1805	R. Kentish.
Balls or bullets, machinery for rolling leaden	**2932**	1st May 1806	W. Bundy.
Balls or bullets, for rifled cannon, of cast iron coated with soft metal	**3155**	30th July 1808	G. Richards.
Balls or bullets, improved method of casting cannon	**4705**	27th Sept. 1822	B. Boothby.
Balls or bullets, peculiar construction of projectile to be substituted for	**6139**	15th July 1831	J. de Burgh, Marquis of Clanricarde.
Balls or bullets of tin and lead or iron, and of iron coated with tin, and filling hollow shot with small shot	**6675**	6th Sept. 1834	H. Shrapnel.
Balls or bullets, wrapping in muslin, by which they are to be attached to the cartridge cases	**7980**	23rd Feb. 1839	C. L. S. Baron Heurteloup.

Subject-matter of Patent.	Number of Patent.	Date.	Name of Patentee.
Balls or bullets, wrapping in muslin, by which they are to be attached to the cartridge cases	7980*	7th Jan. 1842	C. L. S. Baron HEURTELOUP. Disclaimer and Memorandum of Alteration.
Balls or bullets, applying charges of detonating powder to	8347	21st Jan. 1840	J. R. COOPER.
Balls or bullets, machinery for forming by dies and pressure	8385	12th Feb. 1840	D. NAPIER.
Balls or bullets, applying charges of fulminating powder to	9129	2nd Nov. 1841	W. GOLDEN, and J. HANSON.
Balls or bullets, manufacturing iron, by cutting, compressing, and rolling	9673	20th Mar. 1843	S. ROLINSON.
Balls or bullets, heating compressed air for working air guns or cannons by the use of red-hot	11,864	9th Sept. 1847	C. W. O'LEARY.
Balls or bullets, making hollow, and open at the rear end to receive a charge of powder	11,994	10th Dec. 1847	S. TAYLOR.
Balls or bullets, machinery for manufacturing, by punching, rolling, &c.	12,048	31st Jan. 1848	A. V. NEWTON.
Balls or bullets, combining the charge and priming with	12,613	15th May 1849	L. A. DE CHATAUVILLARD.
Balls or bullets, to be used with rifled barrels, improvements in	12,920	11th Jan. 1850	M. U. SEARS.
Balls or bullets, for rifle fire-arms, improvements in	13,161	3rd July 1850	C. W. LANCASTER.
Balls or bullets, for rifled cannon, improvements in	13,215	6th Oct. 1850	A. MELVILLE, and E. CALLOW.
Balls or bullets, for rifled fire-arms, improvements in	13,454	16th Jan. 1851	C. W. LANCASTER.
Do. do. do.	13,934	29th Jan. 1852	F. J. MANCEAUX.
Balls or bullets, improvements in - - -	14,027	20th Mar. 1852	W. W. RICHARDS.
Balls or bullets, containing the charge of powder, improvements in	14,041	24th Mar. 1852	J. MACINTOSH.
Balls or bullets, for containing the charge of powder and other projectiles	14,058	6th April 1852	J. W. DE LONGUEVILLE GIFFARD.
Balls or bullets, improvements in	14,066	17th April 1852	H. G. DELVIGNE.
Do. do. do.	14,087	24th April 1852	W. CHURCH, S. A. GODDARD, and E. MIDDLETON.
Do. do. do.	14,271	23rd Aug. 1852	H. N. S. SHRAPNEL.
Bayonets, mode of fitting on to pistols - -	1284	9th March 1781	J. WATERS.
Bayonets, improvements in applying to guns -	2744	3rd Dec. 1803	J. S. SEARLES.
Bayonets, shield or protection for the human body against	2763	18th May 1804	J. P. BARTHELEMY and J. SHOUBRIDGE.
Bayonets, armour for protecting the human body against	2885	30th Oct. 1805	R. KENTISH.
Bayonets, improvements in, and in affixing to fire-arms	3155	30th July 1808	G. RICHARDS.
Bayonets, an improvement in or addition to -	4031	25th May 1816	F. RICHARDSON.
Bayonets, machine for making sheaths for, of one piece of leather	4923	20th March 1824	J. H. PETITPIERRE.
Bayonets, improvements in, and in affixing to muskets	5905	27th Feb. 1830	C. R. Baron DE BERENGER.
Do. do. do.	6137	13th July 1831	A. DEMONDION.
Bayonets, method of carbonating and hardening the sockets and necks of	6872	6th Aug. 1835	W. MASON.
Bayonets, mode of using ramrods as substitutes for	7980	23rd Feb. 1839	C. L. S. Baron HEURTELOUP.

Subject-matter of Patent.	Number of Patent.	Date.	Name of Patentee.
Bayonets, mode of using ramrods as substitutes for	**7980***	7th Jan. 1842	C. L. S. Baron Heurteloup. Disclaimer and Memorandum of Alteration.
Bayonets, improvements in the manufacture of -	**11,173**	21st April 1846	P. Bishop.
Bayonets,improvements in,and in affixing to fire-arms	**12,613**	15th May 1849	L.A.DeChatauvillard.
Bayonets, improvements in the manufacture of -	**13,991**	27th Feb. 1852	C. Reeves
Bayonets, improvements in sheaths for -	**14,029**	22nd March 1852	J. D. Brady.
Bores of cannon, method of rifling the - -	**1694**	30th July 1789	J. Wilkinson.
Bores of cannon, machinery for rifling the - -	**2436**	2nd Aug. 1800	T. Gill.
Bores of cannon, carbonating and hardening the	**6872**	6th Aug. 1835	W. Mason.
Bores of cannon, improvements in rifling the	**13,161**	3rd July 1850	C. W. Lancaster.
Bores of cannon, improvements in machinery for rifling the	**13,454**	16th Jan. 1851	C. W. Lancaster.
Bores of cannon, machinery for rifling the	**14,087**	24th April 1852	W. Church, S. A. Goddard, and E. Middleton.
Bores of cannon, method of rifling the	**14,271**	23rd Aug. 1852	H. N. S. Shrapnel.
Bores of small arms, method of rifling the - -	**1139**	2nd Dec. 1776	P. Ferguson.
Bores of small arms, machinery for rifling the -	**2436**	2nd Aug. 1800	T. Gill.
Bores of small arms, methods of contracting and rifling the	**6675**	6th Sept. 1834	H. Shrapnel.
Bores of small arms, carbonating and hardening the	**6872**	6th Aug. 1835	W. Mason.
Bores of small arms, improvements in rifling the -	**13,161**	3rd July 1850	C. W. Lancaster.
Bores of small arms, improvements in machinery for rifling the	**13,454**	16th Jan. 1851	C. W. Lancaster.
Bores of small arms, method of rifling the - -	**13,527**	24th Feb. 1851	R. Adams.
Bores of small arms, machinery for rifling the -	**13,934**	29th Jan. 1852	F. J. Manceaux.
Bores of small arms, lining with rifled tubes, and machinery for rifling and polishing, leading, or drawboring the	**14,087**	24th April 1852	W. Church, S. A. Goddard, and E. Middleton.
Bores of small arms, improvements in, and in rifling the	**14,271**	23rd Aug. 1852	H. N. S. Shrapnel.
Boring cannon, improvements in - -	**1063**	27th Jan. 1774	J. Wilkinson.
Do. do. do.	**6826**	30th April 1835	J. Dodds.
Do. do. do.	**13,161**	3rd July 1850	C. W. Lancaster.
Do. do. do.	**13,454**	16th Jan. 1851	C. W. Lancaster.
Boring gun barrels, machinery for - -	**6826**	30th April 1835	J. Dodds.
Do. do. do.	**13,161**	3rd July 1850	C. W. Lancaster.
Do. do. do.	**13,454**	16th Jan. 1851	C. W. Lancaster.
Bows, improvements in manufacturing and ascertaining the strength of long	**1639**	5th Feb. 1788	T. Waring.
Breeches for cannon, improvements in - -	**2466**	23rd Jan. 1801	R. Vazie.
Do. do. do.	**5553**	11th Oct. 1827	J. Horton.
Do. do. do.	**9961**	25th Nov. 1843	J. Frith.
Do. do. do.	**11,864**	9th Sept. 1847	C. W. O'Leary.
Breeches for small arms, improvements in	**1003**	4th Feb. 1772	T. Wright and C. Byrne.
Do. do. do.	**1598**	25th April 1787	H. Nock.
Do. do. do.	**1865**	18th April 1792	J. Manton.
Do. do. do.	**2178**	12th April 1797	J. Manton.
Do. do. do.	**2454**	9th Dec. 1800	J. Prosser.
Do. do. do.	**2466**	23rd Jan. 1801	R. Vazie,
Do. do, do.	**2566**	16th Dec. 1801	C. Grierson.

Subject-matter of Patent.	Number of Patent.	Date.	Name of Patentee.
Breeches for small arms, improvements in—*continued*	2865	3rd July 1805	A. Wilson.
Do. do. do.	3558	30th April 1812	J. Manton.
Do. do. do.	3794	23rd March 1814	E. Heaton.
Do. do. do.	3985	29th Feb. 1816	J. Manton.
Do. do. do.	4990	27th July 1824	C.R. Baron de Berenger.
Do. do. do.	5055	18th Dec. 1824	S. Davis.
Do. do. do.	5421	7th Nov. 1826	B. Newmarch.
Do. do. do.	6166	24th Sept. 1831	W. Bingham and W. Dupe
Do. do. do.	6572	13th March 1834	J. A. Manton.
Do. do. do.	8024	9th April 1839	G. Stocker and J. Bentley.
Do. do. do.	11,864	9th Sept. 1847	C. W. O'Leary.
Do. do. do.	13,527	24th Feb. 1851	R. Adams.
Do. do. do.	14,058	6th April 1852	J. W. de Longueville Giffard.
Do. do. do.	14,265	19th Aug. 1852	T. Hunt.
Breech-loading cannons, inventions relating to -	577	4th June 1741	G. Hadley.
Do. do. do.	1139	2nd Dec. 1776	P. Ferguson.
Do. do. do.	3155	30th July 1808	G. Richards.
Do. do. do.	3755	23rd Nov. 1813	J. Bodmer.
Do. do. do.	3833	4th Aug. 1814	J. S. Pauly
Do. do. do.	4026	14th May 1816	J. S. Pauly.
Do. do. do.	4315	24th Nov. 1818	E. H. Collier.
Do. do. do.	5155	23rd April 1825	A. L. Hunout.
Do. do. do.	5839	9th Sept. 1829	J. Tucker.
Do. do. do.	7286	19th Jan. 1837	M. Poole.
Do. do. do.	12,920	11th Jan. 1850	M. U. Sears.
Do. do. do.	13,215	6th Oct. 1850	A. Melville and E. Callow.
Do. do. do.	14,087	24th April 1852	W. Church, S. A. Goddard, and E. Middleton.
Breech-loading small arms, inventions relating to	418	15th May 1718	J. Puckle.
Do. do. do.	1003	4th Feb. 1772	T. Wright and C. Byrne.
Do. do. do.	1139	2nd Dec. 1776	P. Ferguson.
Do. do. do.	2692	23rd March 1803	D. Egg.
Do. do. do.	3155	30th July 1808	G. Richards.
Do. do. do.	3755	23rd Nov. 1813	J. Bodmer.
Do. do. do.	3784	9th March 1814	J. Thomson.
Do. do. do.	3833	4th Aug. 1814	J. S. Pauley.
Do. do. do.	4026	14th May 1816	J. S. Pauley.
Do. do. do.	4107	11th March 1817	U. Sartoris.
Do. do. do.	4315	24th Nov. 1818	E. H. Collier.
Do. do. do.	4336	23rd Jan. 1819	U. Sartoris.
Do. do. do.	5155	23rd April 1825	A. L. Hunout.
Do. do. do.	6137	13th July 1831	A. Demondion.
Do. do. do.	6139	15th July 1831	J. de Burgh, Marquis of Clanricarde.
Do. do. do.	6611	22nd May 1834	C. L. S. Baron Heurteloup.

Subject-matter of Patent.	Number of Patent.	Date.	Name of Patentee.
Breech-loading small arms, inventions relating to—*con.*	6611*	7th Jan. 1842	C. L. S. Baron HEURTELOUP. Disclaimer.
Do. do. do.	6826	30th April 1835	J. DODDS.
Do. do. do.	6909	22nd Oct. 1835	S. COLT.
Do. do. do.	7286	19th Jan. 1837	M. POOLE.
Do. do. do.	8513	20th May 1840	W. BUSH.
Do. do. do.	9119	14th Oct. 1841	M. POOLE.
Do. do. do.	9129	2nd Nov. 1841	W. GOLDEN and J. HANSON.
Do. do. do.	11,994	10th Dec. 1847	S. TAYLOR.
Do. do. do.	12,613	15th May 1849	L. A. DE CHATAUVILLARD.
Do. do. do.	12,648	7th June 1849	W. H. RITCHIE.
Do. do. do.	12,668	20th June 1849	S. COLT.
Do. do. do.	12,920	11th Jan. 1850	M. U. SEARS.
Do. do. do.	13,215	6th Oct. 1850	A. MELVILLE and E. CALLOW.
Do. do. do.	13,527	24th Feb. 1851	R. ADAMS.
Do. do. do.	13,994	3rd March 1852	G. L. L. KUFAHL.
Do. do. do.	14,027	20th March 1852	W. W. RICHARDS.
Do. do. do.	14,052	31st March 1852	M. POOLE.
Do. do. do.	14,066	17th April 1852	H. G. DELVIGNE.
Do. do. do.	14,087	24th April 1852	W. CHURCH, S. A. GODDARD, and E. MIDDLETON.
Do. do. do.	14,227	20th July 1852	J. W. SCHLESINGER.
Do. do. do.	14,265	19th Aug. 1852	T. HUNT.
Do. do. do.	14,289	10th Sept. 1852	S. TAYLOR.
Bursting of fire-arms, preventing accidents from the	2042	28th Feb. 1795	R. WEBB.
Do. do. do.	4491	20th July 1820	W. DELL.
Do. do. do.	6675	6th Sept. 1834	H. SHRAPNEL.
Do. do. do.	10,849	2nd Oct. 1845	J. HALE.
Do. do. do.	11,864	9th Sept. 1847	C. W. O'LEARY.
Do do. do.	12,613	15th May 1849	L. A. DE CHATAUVILLARD.

C.

Subject-matter of Patent.	Number of Patent.	Date.	Name of Patentee.
Cannon, making shorter, lighter, more effective, and portable	503	5th Oct. 1728	H. BROWNE.
Cannon, breech-loading - - - -	577	4th June 1741	G. HADLEY.
Cannon, method of making cast iron carriages for -	683	30th May 1753	L. F. D. DE VILLERS.
Cannon, new method of casting - -	723	21st April 1758	J. WILKINSON.
Cannon, new construction of cast iron carriages for	765	5th Aug. 1761	S. REMNANT.
Cannon, method of casting and boring - -	1063	27th Jan. 1774	J. WILKINSON.
Cannon, new lock for - - - -	1095	8th April 1775	H. NOCK, W. GOWER, and J. GREEN.

Subject-matter of Patent.	Number of Patent.	Date.	Name of Patentee.
Cannon, breech-loading - - - -	1139	2nd Dec. 1776	P. FERGUSON.
Cannon, improvements in fitting locks to, and in carriages for	1218	10th April 1779	R. BLIGHT.
Cannon, spiral wedge to be applied to gun carriages for elevating, pointing, and lowering	1233	7th Sept. 1779	C. BERGER.
Cannon, method of loading to obtain repeated discharges without re-loading	1270	5th Dec. 1780	J. AITKEN.
Cannon, elastic sponge for sponging - -	1668	26th Aug. 1788	H. COOK.
Cannon, method of rifling - - -	1694	30th July 1789	J. WILKINSON.
Cannon, improvements in manufacturing welded -	1951	23rd April 1793	S. BENTHAM.
Cannon, improvements in the construction of wooden carriages for	2151	8th Dec. 1796	J. GOVER.
Cannon, improvements in the construction of wooden carriages for	2224	23rd March 1798	J. HAYCRAFT.
Cannon, improvements in making, working, and loading, and in constructing gun carriages	2428	24th July 1800	A. C. DE POGGI.
Cannon, machinery for rifling - - -	2436	2nd Aug. 1800	T. GILL.
Cannon, improvements in the manufacture of -	2466	23rd Jan. 1801	R. VAZIE.
Cannon, pistol to be used for firing off - -	2692	23rd March 1803	D. EGG.
Cannon, improvements in the construction of wooden carriages for	2803	19th Dec. 1805	J. GOVER.
Cannon, improvements in gun carriages, and in mounting, elevating, or depressing, and wo ing	2917	13th Mar. 1806	M. LOGAN.
Cannon, improvements in making and working carriages for carronades and	3005	29th Jan. 1807	R. FRIEND.
Cannon, improvements in making and working carriages for carronades and	3028	8th April 1807	R. F. HAWKINS.
Cannon, discharging by means of compositions which explode by percussion	3032	11th April 1807	A. J. FORSYTH.
Cannon, cartridge paper, an improved manufacture of	3080	12th Nov. 1807	J. DICKINSON.
Cannon, improvements in the manufacture of -	3122	28th March 1808	B. COOK.
Cannon, improvements in gun carriages, and in constructing and mounting	3134	24th May 1808	W. CONGREVE.
Cannon, breech-loading and double - -	3155	30th July 1808	G. RICHARDS.
Cannon, improved - - - - - -	3196	26th Jan. 1809	M. LOGAN.
Cannon, improvements in flint locks for - -	3233	4th May 1809	T. MOON.
Cannon, improvements in the manufacture of artillery or	3469	26th July 1811	H. JAMES and J. JONES.
Cannon, improvements in carriages for carronades and	3565	11th May 1812	W. CONGREVE.
Cannon, improvements in breech-loading, and in touch-holes and sights for	3755	23rd Nov. 1813	J. BODMER.
Cannon, loading at the breech, and discharging by compressed air	3833	4th Aug. 1814	J. S. PAULY.
Do. do. do.	4026	14th May 1816	J. S. PAULY.
Cannon, constructing so as to fire several times in succession with once loading	4315	24th Nov. 1818	E. H. COLLIER.
Cannon, improvements in priming so as to discharge by percussion	4427	15th Jan. 1820	F. FOX.
Cannon, employing self-igniting paper crackers or matches for discharging	4853	16th Oct. 1823	SIR W. CONGREVE.
Cannon, improvements in percussion locks for -	5033	6th Nov. 1824	T. CARTMELL.

Subject-matter of Patent.	Number of Patent.	Date.	Name of Patentee.
Cannon, constructing so as to obtain repeated discharges without re-loading	5155	23rd April 1825	A. L. Hunout.
Cannon, apparatus for working - - -	5379	22nd June 1826	T. Halahan.
Cannon, improvements in applying percussion locks to	5421	7th Nov. 1826	B. Newmarch
Cannon, improvements in mounting - - -	5511	26th June 1827	J. Marshall.
Cannon, improvements in the manufacture of	5553	11th Oct. 1827	J. Horton.
Cannon, breech-loading - - - - -	5839	9th Sept. 1829	J. Tucker.
Cannon, method of constructing double, and carriages for same	6491	19th Oct. 1833	T. A. G. Gillyon.
Cannon, method of constructing double and quadruple cannon, and machinery for boring		30th April 1835	J. Dodds.
Cannon, carbonating and hardening the bore of, and the elevating screws and screw-boxes for	6872	6th Aug. 1835	W. Mason.
Cannon, breech-loading - - - - -	7286	19th Jan. 1837	M. Poole.
Cannon, combining several together in the form of a star, and mounting several on one carriage	7472	14th Sept. 1837	W. Coles.
Cannon, improvements in discharging by percussion	8704	17th Nov. 1840	J. Shaw.
Cannon, welding hoisted wrought iron - -	9947	18th Nov. 1843	J. Roose.
Cannon, constructing with moveable breech -	9961	25th Nov. 1843	J. Frith.
Cannon, formed of wrought iron or steel, or wrought iron and steel, and machinery for making the same	10,013	16th Jan. 1844	T. Aspinwall.
Cannon, improvements in fuzes for firing - -	10,364	24th Oct. 1844	H. Carbines.
Cannon, improvements in percussion priming for, and in supplying to	10,667	14th May 1845	C. J. Smith.
Cannon, improvements in percussion priming for, and in supplying to	10,718	10th June 1845	J. W. Tyson.
Cannon, improvements in carriages for - -	10,728	23rd June, 1845	T. Aspinwall.
Cannon, improvements in fuzes for firing -	10,928	6th Nov. 1845	J. S. Bickford, G. Smith, and T. Davey.
Cannon, new safety fuze for firing - -	11,447	12th Nov. 1846	G. Smith.
Cannon, compressing and heating air for, and working air guns or cannons, and preventing the concussion and partly destroying the report of	11,864	9th Sept. 1847	C. W. O'Leary.
Cannon, improvements in the manufacture of fuzes for firing	12,406	11th Jan. 1849	M. Loam.
Cannon, improvements in carriages for - -	12,559	3rd April 1849	A. Woollett.
Cannons, improvements in - - -	12,613	15th May 1849	L. A. De Chatauvillard.
Cannons, applying elastic purchases to the breeching of	12,623	29th May 1849	R. E. Hodges.
Cannons, improvements in breech-loading - -	12,920	11th Jan. 1850	M. U. Sears.
Cannon, improvements in boring and rifling, and in percussion tubes	13,161	3rd July 1850	C. W. Lancaster.
Cannon, improvements in breech-loading, and in mounting	13,215	6th Oct. 1850	A. Melville and E. Callow.
Cannon, improvements in carriages for - -	13,403	12th Dec. 1850	T. H. Howels.
Cannon, improvements in machinery for boring and rifling	13,454	16th Jan. 1851	C. W. Lancaster.
Cannon, for discharging several balls without recharging, and method of firing guns under the water-line of gun-boats or other vessels.	14,041	24th March 1852	J. Macintosh.
Cannon, improvements in the manufacture of fuzes for firing.	14,065	15th April 1852	S. Davey and A. L. Chann.

Subject-matter of Patent	Number of Patent.	Date.	Name of Patentee.
Cannon, improvements in breech-loading, and in machinery for rifling	14,087	24th April 1852	W. Church, S. A. Goddard, and E. Middleton.
Cannon, improvements in manufacturing, in carriages and traversing apparatus for	14,095	29th April 1852	J. L. A. Simmons and T. Walker.
Cannon, mode of causing a three-pounder gun to impel a six-pounder shell, and method of rifling	14,271	23rd Aug. 1852	H. N. S. Shrapnel.
Cartridges, pouches and boxes for containing -	1173	28th Nov. 1777	W. Rawle.
Cartridges, cartouch boxes, or receptacles for -	2810	19th Jan. 1805	T. H. Keddie.
Cartridges, an improved manufacture of cannon cartridge paper	3080	12th Nov. 1807	J. Dickinson.
Cartridges, belts for containing - - -	3155	30th July 1808	G. Richards.
Cartridges, improvements in - - -	3833	4th Aug. 1814	J. S. Pauly.
Do. do. - - - -	4026	14th May 1816	J. S. Pauly.
Cartridges, cartouch boxes, or receptacles for -	4724	9th Nov. 1822	F. Deakin.
Cartridges, machine for making cartouch boxes, or receptacles for, of one piece of leather	4923	20th March 1824	J. H. Petitpierre.
Cartridges, improvements in - - -	5570	28th Nov. 1827	J. Jenour.
Do. do. - - - -	5708	18th Sept. 1828	E. F. Orson.
Do. do - - - -	6137	13th July 1831	A. Demondion.
Do. do. - - - - -	6139	15th July 1831	J. de Burgh, Marquis of Clanricarde.
Cartridges, improvements in ball - - -	6196	15th Dec. 1831	A. A. Moser
Cartridges, improvements in - - - -	6611	22nd May 1834	C. L. S. Baron Heurteloup.
Do. do. - - - -	6611*	7th Jan. 1842	C. L. S. Baron Heurteloup.
Cartridges, wrapping balls in muslin to attach them to the cases of	7980	23rd Feb. 1839	C. L. S. Baron Heurteloup.
Do. do. do. -	7980*	7th Jan. 1842	C. L. S. Baron Heurteloup. Disclaimer and Memorandum of Alteration.
Cartridges, improvements in - - - -	8143	6th July 1839	T. T. Berney.
Do. do. - - - -	8513	20th May 1840	W. Bush.
Do. do. - - - - -	10,364	24th Oct. 1844	H. Carbines.
Do. do. - - - - -	11,657	15th April 1847	J. Mollett.
Cartridges, making balls serve as - -	11,994	10th Dec. 1847	S. Taylor.
Cartridges, improvements in, and in making balls serve as	12,613	15th May 1849	L. A. De Chatauvillard.
Cartridges, improvements in - - -	12,648	7th June 1849	W. H. Ritchie.
Do. do. - - - - -	12,920	11th Jan. 1850	M. U. Sears.
Do. do. - - - -	13,215	6th Oct. 1850	A. Melville and E. Callow.
Do. do. - - - -	13,377	30th Nov. 1850	J. A. Elmslie and G. Simpson.
Cartridge boxes, and means of slinging the same, improvements in	14,029	22nd March 1852	J. D. Brady.
Cartridges, making balls serve as - -	14,041	24th March 1852	J. Macintosh.
Do. do. - - - -	14,058	6th April 1852	J. W. de Longueville Giffard.
Cartridges, improvements in - - -	14,227	20th July 1852	J. W. Schlesinger.
Do. do. - - -	14,271	23rd Aug. 1852	H. N. S. Shrapnel.
Do. do. - - -	14,289	10th Sept. 1852	S. Taylor.

Subject-matter of Patent.	Number of Patent.	Date.	Name of Patentee.
Casting cannon shot, improved method of -	4705	27th Sept. 1822	B. BOOTHBY.
Casting guns or cannon, improvements in -	723	21st April 1758	J. WILKINSON.
Do. do. do.	1063	27th Jan. 1774	J. WILKINSON.
Chain shot, double, cannon for discharging -	3155	30th July 1808	G. RICHARDS.
Chain shot, improvements in, and in double cannon for discharging	6491	19th Oct. 1833	T. A. G. GILLYON.
Cutting and working wood for gun stocks, machinery for	1838	26th Nov. 1791	S. BENTHAM.
Do. do. do.	1951	23rd April 1793	S. BENTHAM.
Do. do. do.	2652	30th Oct. 1802	J. BRAMAH.
Do. do. do.	6755	29th Jan. 1835	J. DODDS.
Do. do. do.	11,880	30th Sept. 1847	J. DE BARROS.

D.

Subject-matter of Patent.	Number of Patent.	Date.	Name of Patentee.
Double-barrelled fire-arms, improvements in locks for	1707	6th Nov. 1789	J. TEMPLEMAN.
Double-barrelled fire-arms, improvements in triggers for	1837	28th Nov. 179[illegible]	J. S. SEARLES.
Double-barrelled fire-arms, improved breeches for	1865	18th April 1792	J. MANTON.
Double-barrelled fire-arms, improvements in triggers for	1893	5th July 1792	J. MANTON.
Double-barrelled fire-arms, improvements in locks for	1978	4th Feb. 1794	T. SYKES.
Double-barrelled fire-arms, improvements in locks, triggers, and breeches for	2178	12th April 1797	J. MANTON.
Double-barrelled fire-arms, improvements in the manufacture of barrels for	2252	23rd July 1798	W. DUPE.
Double-barrelled fire-arms, improvements in breeches and locks for	2566	16th Dec. 1801	C. GRIERSON.
Double-barrelled fire-arms, improvements in locks for	2744	3rd Dec. 1803	J. S. SEARLES.
Double-barrelled fire-arms, improvements in breeches and locks for	2865	3rd July 1805	A. WILSON.
Double-barrelled fire-arms, elevated top piece or top rib for	2966	15th Sept. 1806	J. MANTON.
Double-barrelled fire-arms, improvements in breech-loading, and in triggers for	3155	30th July 1808	G. RICHARDS.
Double-barrelled fire-arms, improvements in locks and breeches for	3558	30th April 1812	J. MANTON.
Double-barrelled fire-arms, improvements in, and in triggers for	3828	4th Aug. 1814	T. SYKES.
Double-barrelled fire-arms, discharging by means of compressed air	3833	4th Aug. 1814	J. S. PAULY.
Double-barrelled fire-arms, improvements in, to be discharged by percussion	3985	29th Feb. 1816	J. MANTON.
Double-barrelled fire-arms, improvements in priming	4727	26th Nov. 1822	J. EGG.

Subject-matter of Patent.	Number of Patent.	Date.	Name of Patentee.
Double-barrelled fire-arms, improvements in percussion	4823	29th July 1823	J. JACKSON.
Do. do. do.	4990	27th July 1824	C. R. Baron de BERENGER.
Double-barrelled fire-arms, improvements in locks for	5026	4th Nov. 1824	J. SOMERVILLE.
Double-barrelled fire-arms, revolving primer for percussion locks of	5106	26th Feb. 1825	J. MANTON.
Double-barrelled fire-arms, constructing so as to obtain repeated discharges without re-loading	5155	23rd April 1825	A. L. HUNOUT.
Double-barrelled fire-arms, improvements in percussion locks for	5175	20th May 1825	J. RIVIERE.
Double-barrelled fire-arms, improved lock, break-off, and trigger for	6053	17th Dec. 1830	B. REDFERN.
Double-barrelled fire-arms, improvements in percussion locks for	6394	17th March 1833	C. JONES.
Do. do. do.	6436	2nd June 1833	C. JONES.
Double-barrelled fire-arms, improvements in breech-loading	6611	22nd May 1834	C. L. S. Baron HEURTELOUP.
Do. do. do.	6611*	7th Jan. 1842	C. L. S. Baron HEURTELOUP. Disclaimer.
Double-barrelled fire-arms, improvements in locks for	6825	28th April 1835	J. SOMERVILLE.
Double-barrelled fire-arms, improvements in percussion locks for	6826	30th April 1835	J. DODDS.
Double-barrelled fire-arms, improvements in priming locks of	6829	9th May 1835	J. EGG.
Double-barrelled fire-arms, improvements in percussion locks for	7610	10th April 1838	J. R. COOPER.
Double-barrelled fire-arms, improvements in priming percussion locks of	7965	11th Feb. 1839	G. H. MANTON.
Double-barrelled fire-arms, improvements in percussion locks for	7980	23rd Feb. 1839	C. L. S. Baron HEURTELOUP.
Do. do. do.	7980*	7th Jan. 1842	C. L. S. Baron HEURTELOUP. Disclaimer and Memorandum of Alteration.
Double-barrelled fire-arms, improvements in percussion locks for	8024	9th April 1839	G. STOCKER and J. BENTLEY.
Double-barrelled fire-arms, improvements in priming percussion locks of	9084	9th Sept. 1841	C. L. S. Baron HEURTELOUP.
Do. do. do.	9258	15th Feb. 1842	A. ROUSSEAU.
Double-barrelled fire-arms, improvements in percussion locks for	10,280	30th July 1844	J. BENTLEY.
Do. do. do.	10,667	14th May 1845	C. J. SMITH.
Double-barrelled fire-arms, improvements in percussion locks and stocks for	10,849	2nd Oct. 1845	J. HALE.
Double-barrelled fire-arms, improvements in priming percussion locks of	12,432	20th Jan. 1849	H. NEEDHAM.
Do. do. do.	12,543	28th March 1849	G. H. MANTON and J. HARRINGTON.
Double-barrelled fire-arms, improvements in percussion locks for	12,613	15th May 1849	L. A. DE CHATAUVILLARD.
Double-barrelled fire-arms, improvements in breech-loading	12,920	11th Jan. 1850	M. U. SEARS.
Double-barrelled fire-arms, improvements in breeches for, and in fixing and removing the same	14,265	19th Aug. 1852	T. HUNT.

Subject-matter of Patent.	Number of Patent.	Date.	Name of Patentee.
Double cannon, method of constructing - -	3155	30th July 1808	G. RICHARDS.
Do. do. do.	6491	19th Oct. 1833	T. A. GILLYON.
Do. do. do.	6826	30th April 1835	J. DODDS.

E.

Subject-matter of Patent.	Number of Patent.	Date.	Name of Patentee.
Explosive mixtures, preparing and employing as a moving power for machinery	5402	12th Aug 1826	E. HAZARD.
Explosive substances, using water and incombustible gases to prevent the accidental explosion of	11,011	20th Dec. 1845	J. BLYTH.
Explosive compounds, improvements in the manufacture of	11,407	8th Oct. 1846	J. TAYLOR.
Explosive compounds, improvements in - -	13,215	6th Oct. 1850	A. MELVILLE and E. CALLOW.
Do. do. do.	14,065	15th April 1852	S. DAVEY and A. L. CHANN.
Explosive compounds, improvements in the manufacture of	14,227	20th July 1852	J. W. SCHLESINGER.

F.

Subject-matter of Patent.	Number of Patent.	Date.	Name of Patentee.
Forging or hammering metal in the manufacture of guns, arrangement of machinery for	2466	23rd Jan. 1801	R. VAZIE.
Forging or hammering, machinery for welding barrels for fire-arms and artillery by	3469	26th July 1811	H. JAMES and J. JONES.
Forging, machinery for welding gun-barrels by -	3617	28th Nov. 1812	H. OSBORN.
Forging, guns or ordnance, improvements in - -	5553	11th Oct. 1827	J. HORTON.
Forging, method of welding twisted wrought iron cannon by	9947	18th Nov. 1843	J. ROOSE.
Forging, machinery for welding gun-barrels by -	10,696	3rd June 1845	C. WHITEHOUSE.
Furnace, to be used in the manufacture of guns -	2466	23rd Jan. 1801	R. VAZIE.
Fuzes for firing mines, &c., substituting detonating compositions for	3032	11th April 1807	A. J. FORSYTH.
Fuzes, employing self-igniting paper crackers or matches instead of	4853	16th Oct. 1823	Sir W. CONGREVE.
Fuzes, improvements in the manufacture of -	6159	6th Sept. 1831	W. BICKFORD.
Fuzes, improvements in percussion - -	7759	2nd Aug. 1838	J. DENNETT.
Fuzes, improvements in - - - - -	10,364	24th Oct. 1844	H. CARBINES.
Fuzes, improvements in manufacturing - -	10,928	6th Nov. 1845	J. S. BICKFORD, G. SMITH, and T. DAVEY.
Fuzes, new safety - - - - -	11,447	12th Nov. 1846	G. SMITH.
Fuzes, improvements in the manufacture of -	12,406	11th Jan. 1849	M. LOAM.
Fuzes, improvements in explosive compounds for, in making and in firing - - -	14,065	15th April, 1852	S. DAVEY and A. L. CHANN.

Subject-matter of Patent.	Number of Patent.	Date.	Name of Patentee.
G.			
Grinding parts of the barrels of breech-loading fire-arms, machinery for	6137	13th July 1831	A. DEMONDION.
Gun barrels, new method of rifling - -	1139	2nd Dec. 1776	P. FERGUSON.
Gun barrels, improvements in the manufacture of -	1951	23rd April 1793	S. BENTHAM.
Gun barrels, for double barrelled fire-arms, improvements in manufacturing	2252	23rd July 1798	W. DUPE.
Gun barrels, machinery for rifling - - -	2436	2nd Aug. 1800	T. GILL.
Gun barrels, machinery for hammering metal in the manufacture of	2466	23rd Jan. 1801	R. VAZIE.
Gun barrels, improvements in the manufacture of -	2911	20th Feb. 1806	J. JONES.
Do. do. do.	3122	28th March 1808	B. COOK.
Gun barrels, improvements in manufacturing skelps for making	3267	28th Sept. 1809	J. JONES.
Gun barrels, rolling skelps for making - -	3437	24th April 1811	J. BRADLEY.
Gun barrels, welding by hammering or rolling -	3469	26th July 1811	H. JAMES and J. JONES.
Gun barrels, machine for closing skelps preparatory to making into	3590	3rd Aug. 1812	H. OSBORN.
Gun barrels, improved method of testing - -	3599	25th Sept, 1812	D. EGG.
Gun barrels, machinery for welding - -	3617	28th Nov. 1812	H. OSBORN.
Gun barrels, machinery for rolling - -	3740	15th Oct. 1813	H. OSBORN.
Do. do.	3813	7th June 1814	G. HEYWOOD.
Gun barrels, improvements in - - -	4031	25th May 1816	F. RICHARDSON
Gun barrels, machinery for rolling - - -	4105	1st March 1817	H. OSBORN.
Gun barrels, improvements in - - -	4491	20th July 1820	W. DELL.
Gun barrels, of breech-loading fire-arms, machinery for grinding the	6137	13th July 1831	A. DEMONDION.
Gun barrels, methods of contracting the bore and rifling	6675	6th Sept. 1834	H. SHRAPNEL.
Gun barrels, machinery for boring - -	6826	30th April 1835	J. DODDS.
Gun barrels, carbonating and hardening the bore of	6872	6th Aug. 1835	W. MASON.
Gun barrels, improvements in welding the seams or joints of twisted	9947	18th Nov. 1843	J. ROOSE.
Gun barrels, improvements in the manufacture of -	10,030	30th Jan. 1844	W. L. SARGANT.
Gun barrels, machinery for welding - -	10,696	3rd June 1845	C. WHITEHOUSE.
Gun barrels, ornamenting iron and other metals to be used in the manufacture of	12,732	1st Aug. 1849	B. AINGWORTH.
Gun barrels, improvements in boring and rifling -	13,161	3rd July 1850	C. W. LANCASTER.
Gun barrels, improvements in welding and finishing twisted	13,299	24th Oct. 1850	A. ROOSE.
Gun barrels, machinery for boring and rifling -	13,454	16th Jan. 1851	C. W. LANCASTER.
Gun barrels, improved method of rifling - -	13,527	24th Feb. 1851	R. ADAMS.
Gun barrels, machinery for rifling - -	13,934	29th Jan. 1852	F. J. MANCEAUX.
Gun barrels, making or lining with steel the muzzles of rifled	14,027	20th March 1852	W. W. RICHARDS.
Gun barrels, lining with rifled tubes, and machinery for rifling and polishing, leading, or draw-boring	14,087	24th April 1852	W. CHURCH, S. A. GODDARD, and E. MIDDLETON
Gun barrels, enlarging breech ends and rifling bores of	14,271	23rd Aug. 1852	H. N. S. SHRAPNEL.
Gun boats, improvements in, so as to fire guns therefrom below the water line	14,041	24th March 1852	J. MACINTOSH.
Gun carriages, method of making cast iron - -	683	30th May 1753	L. F. D. DE VILLERS.

Subject-matter of Patent.	Number of Patent.	Date.	Name of Patentee.
Gun carriages, new construction of cast-iron -	765	5th Aug. 1761	S. Remnant.
Gun carriages, improvements in the construction of wooden	1218	10th April 1779	R. Blight.
Gun carriages, spiral wedge to be applied to, for elevating, pointing, and lowering cannon	1233	7th Sept. 1779	C. Berger.
Gun carriages, improvements in the construction of wooden	2151	8th Dec. 1796	J. Gover.
Do. do. do.	2224	23rd March 1798	J. Haycraft.
Gun carriages, improvements in the construction of	2428	24th July 1800	A. C. De Poggi.
Gun carriages, improvements in the construction of wooden	2803	19th Dec. 1805	J. Gover.
Gun carriages, improvements in the construction of	2917	13th March 1806	M. Logan.
Gun carriages, improvements in making and working carronade and	3005	29th Jan. 1807	R. Friend.
Gun carriages, improvements in making and working carronade and	3028	8th April 1807	R. F. Hawkins.
Gun carriages and breechings, improvements in -	3134	24th May 1808	W. Congreve.
Gun carriages, improvements in carronade and -	3565	11th May 1812	W. Congreve.
Gun carriages, improvements in working - -	5379	22nd June 1826	T. Halahan.
Gun carriages, improvements in - - -	5511	26th June 1827	J. Marshall.
Do. do. - - -	6491	19th Oct. 1833	T. A. G. Gillyon.
Gun carriages, carbonating and hardening the elevating screws and screw-boxes of	6872	6th Aug. 1835	W. Mason.
Gun carriages, improvements in - - -	7472	14th Sept. 1837	W. Coles.
Do. do. - - -	10,728	23rd June 1845	T. Aspinwall.
Do. do. - - - -	12,559	3rd April 1849	A. Woollett.
Gun carriages, applying elastic purchases to -	12,623	29th May 1849	R. E. Hodges.
Gun carriages, improvements in - - -	13,215	6th Oct. 1850	A. Melville and E. Callow.
Do. do. - - - -	13,403	12th Dec. 1850	T. H. Howels.
Gun carriages and traversing apparatus, improvements in	14,095	29th April 1852	J. L. A. Simmons and T. Walker.
Gun cotton, manufacturing the substance called -	11,407	8th Oct. 1846	J. Taylor.
Gunpowder, new method of making - -	846	22nd April 1766	T. Delaval.
Gunpowder, vessel or barrel for safe conveyance of -	2658	30th Nov. 1802	H. Smith.
Gunpowder flasks, improvements in - -	2865	3rd July 1805	A. Wilson.
Gunpowder, vessel for conveying and preserving from injury by damp	3373	7th Sept. 1810	J. Walker.
Gunpowder, employing sugar in the manufacture of	3510	4th Dec. 1811	F. A. Winsor.
Gunpowder flasks, improvements in - -	3599	25th Sept. 1812	D. Egg.
Do. do. - - -	3784	9th March 1814	J. Thomson.
Do. do. - - - -	3828	4th Aug. 1814	T. Sykes.
Gunpowder, new mode of manufacturing - -	3937	3rd July 1815	Sir W. Congreve.
Gunpowder flasks, improvements in - -	5439	20th Dec. 1826	C. R. Baron de Berenger.
Do. do. - - - -	10,667	14th May 1845	C. J. Smith.
Do. do. - - -	10,683	22nd May 1845	M. Boche.
Gunpowder, using water and incombustible gases to prevent accidental explosion in the manufacture and storing of	11,011	20th Dec. 1845	J. Blyth.
Gunpowder, manufacturing a substitute for -	11,407	8th Oct. 1846	J. Taylor.
Gunpowder, explosive compounds to be used instead of	13,215	6th Oct. 1850	A. Melville and E. Callow.

Subject-matter of Patent.	Number of Patent.	Date.	Name of Patentee.
Gunpowder, improvements in explosive compounds which may be used instead of	14,065	15th April 1852	S. DAVEY and A. L. CHANN.
Gunpowder, manufacturing a substitute for ordinary	14,227	20th July 1852	J. W. SCHLESINGER.
Gun stocks, machinery for manufacturing - -	1838	26th Nov. 1791	S. BENTHAM.
Do. do. do. -	1951	23rd April 1793	S. BENTHAM.
Do. do. do. - -	2652	30th Oct. 1802	J. BRAMAH.
Gun stocks, method of strengthening - -	3599	25th Sept. 1812	D. EGG.
Gun stocks, improvements in - - -	6554	8th Feb. 1834	W. S. GILLETT.
Do. do. - - - -	6675	6th Sept. 1834	H. SHRAPNEL.
Gun stocks, machinery for manufacturing - -	6755	29th Jan. 1835	J. DODDS.
Gun stocks, improvements in - - -	7610	10th April 1838	J. R. COOPER.
Do. do. - - - -	10,849	2nd Oct. 1845	J. HALE.
Gun stocks, machinery for manufacturing - -	11,880	30th Sept. 1847	J. DE BARROS.
Gun wads, improvements in, and in the manufacture of	1893	5th July 1792	J. MANTON.
Do. do. do.	6633	26th June 1834	R. WALKER.
Do. do. do.	9069	6th Sept. 1841	J. DREW.
Do. do. do.	9712	25th April 1843	W. BROCKEDON.
Do. do. do.	13,377	30th Nov. 1850	J. A. ELMSLIE and G. SIMPSON.
Do. do. do.	13,454	16th Jan. 1851	C. W. LANCASTER.
Do. do. do.	14,027	20th March 1852	W. W. RICHARDS.
Do. do. do,	14,271	23rd Aug. 1852	H. N. S. SHRAPNEL.

H.

Subject-matter of Patent.	Number of Patent.	Date.	Name of Patentee.
Hammering the metal in the manufacture of guns, arrangement of machinery for	2466	23rd Jan. 1801	R. VAZIE.
Hammering, machinery for welding barrels for fire-arms and artillery by	3469	26th July 1811	H. JAMES and J. JONES.
Hammering, machinery for welding gun barrels by	3617	28th Nov. 1812	H. OSBORN.
Hammering or forging metal into guns or ordnance, improvements in	5553	11th Oct. 1827	J. HORTON.
Hammering, method of welding twisted wrought iron cannon by	9947	18th Nov. 1843	J. ROOSE.
Hammering, machinery for welding gun barrels by	10,696	3rd June 1845	C. WHITEHOUSE.
Hammering, improvements in welding twisted gun and pistol barrels by	13,299	24th Oct. 1850	A. ROSE.
Helmets, improvements in - - -	14,029	22nd March 1852	J. D. BRADY.
Horseback, method of fixing and carrying fire-arms on	8205	23rd Aug. 1839	C. B. COLES.
Horse swivel guns, improvements in - -	2466	23rd Jan. 1801	R. VAZIE.
Hydrostatic press, welding cannon by means of the	10,013	16th Jan. 1844	J. ASPINWALL.

L.

Subject-matter of Patent.	Number of Patent.	Date.	Name of Patentee.
Locks for air guns, improvements in	6046	6th Dec. 1830	J. G. Lacy and S. Davis.
Do. do. do.	12,728	1st Aug. 1849	J. Shaw.
Locks for cannon, certain improvements in flint	1095	8th April 1775	H. Nock, W. Jover, and J. Green.
Locks for cannon, improvements in fitting flint	1218	10th April 1779	R. Blight.
Locks for cannon, improvements in flint	3233	4th May 1809	T. Noon.
Locks for cannon, constructing percussion	3032	11th April 1805	A. J. Forsyth.
Locks for cannon, improvements in percussion	4427	15th Jan. 1820	F. Fox.
Do. do. do.	5033	6th Nov. 1824	T. Cartmell.
Do. do. do.	5421	7th Nov. 1826	B. Newmarch.
Locks for discharging fire-arms by means of compressed air	3833	4th Aug. 1814	J. S. Pauly.
Do. do. do.	4026	14th May 1816	J. S. Pauly.
Do. do. do.	5317	16th Jan. 1826	B. Newmarch.
Locks for small arms, improvements in, and in the manufacture of flint	1003	4th Feb. 1772	T. Wright and C. Byrne.
Do. do. do.	1095	8th April 1775	H. Nock, W. Jover, and J. Green.
Do. do. do.	1707	6th Nov. 1789	J. Templeman.
Do. do. do.	1837	28th Nov. 1791	J. S. Searles.
Do. do. do.	1865	18th April 1792	J. Manton.
Do. do. do.	1893	5th July 1792	J. Manton.
Do. do. do.	1897	5th July 1792	J. Willson.
Do. do. do.	1978	4th Feb. 1794	T. Sykes.
Do. do. do.	2041	28th Feb. 1795	G. Boulton.
Do. do. do.	2042	28th Feb. 1795	R. Webb.
Do. do. do.	2072	22nd Oct. 1795	J. Grove.
Do. do. do.	2173	14th March 1797	W. Siddon.
Do. do. do.	2178	12th April 1797	J. Manton.
Do. do. do.	2360	28th Nov. 1799	E. Thomason.
Do. do. do.	2454	9th Dec. 1800	J. Prosser.
Do. do. do.	2466	23rd Jan. 1801	R. Vazie.
Do. do. do.	2566	16th Dec. 1801	C. Grierson.
Do. do. do.	2573	16th Jan. 1802	J. Hall.
Do. do. do.	2692	23rd March 1803	D. Egg.
Do. do. do.	2709	10th June 1803	J. R. Peckham.
Do. do. do.	2722	6th July 1803	J. Manton.
Do. do. do.	2744	3rd Dec. 1803	J. S. Searles.
Do. do. do.	2825	28th Feb. 1805	G. Dodd.
Do. do. do.	2865	3rd July 1805	A. Wilson.
Do. do. do.	2920	21st March 1806	F. Place.
Do. do. do.	2991	4th Dec. 1806	J. F. Matthey.
Do. do. do.	3155	30th July 1808	G. Richards.
Do. do. do.	3233	4th May 1809	T. Noon.
Do. do. do.	3286	11th Dec. 1809	J. Manton.

Subject-matter of Patent.	Number of Patent.	Date.	Name of Patentee.
Locks for small arms, improvements in, and in the manufacture of flint—*continued*	3558	30th April 1812	J. MANTON.
Do. do. do.	3588	28th July 1812	W. SMITH.
Do. do. do.	3599	25th Sept. 1812	D. EGG.
Do. do. do.	3722	19th July 1813	R. PRETYMAN.
Do. do. do.	3784	9th March 1814	J. THOMSON.
Do. do. do.	3794	23rd March 1814	E. HEATON.
Do. do. do.	3828	4th Aug. 1814	T. SYKES.
Do. do. do.	3942	21st July 1815	J. MANTON.
Do. do. do.	4031	25th May 1816	F. RICHARDSON.
Do. do. do.	4054	12th Aug. 1816	S. NOCK.
Do. do. do.	4166	26th Sept. 1817	J. MANTON.
Do. do. do.	4218	3rd Feb. 1818	M. C. WYATT.
Do. do. do.	4336	23rd Jan. 1819	U. SARTORIS.
Do. do. do.	4577	30th July 1821	J. MANTON.
Do. do. do.	4648	12th Feb. 1822	S. DAVIS.
Do. do. do.	4727	26th Nov. 1822	J. EGG.
Do. do. do.	4960	30th May 1824	J. COOK.
Do. do. do.	5026	4th Nov. 1824	J. SOMERVILLE.
Do. do. do.	5099	19th Feb. 1825	J. MOULD.
Do. do. do.	5242	15th Aug. 1825	C. DOWNING.
Do. do. do.	5905	27th Feb. 1830	C. R. Baron de BERENGER.
Do. do. do.	6053	17th Dec. 1830	B. REDFERN.
Do. do. do.	6675	6th Sept. 1834	H. SHRAPNEL.
Do. do. do.	6825	28th April 1835	J. SOMERVILLE.
Do. do. do.	6826	30th April 1835	J. DODDS.
Do. do. do.	6829	9th May 1835	J. EGG.
Do. do. do.	7712	30th June 1838	G. ROUND and S. WHITFORD.
Locks for small arms, constructing percussion -	3032	11th April 1807	A. J. FORSYTH.
Locks for small arms, improvements in, and in the manufacture of percussion	3985	29th Feb. 1816	J. MANTON.
Do. do. do.	4285	3rd Aug. 1818	J. MANTON.
Do. do. do.	4427	15th Jan. 1820	F. FOX.
Do. do. do.	4590	14th Sept. 1821	W. WEBSTER.
Do. do. do.	4611	10th Nov. 1821	W. W. RICHARDS.
Do. do. do.	4648	12th Feb. 1822	S. DAVIS.
Do. do. do.	4727	26th Nov. 1822	J. EGG.
Do. do. do.	4823	29th July 1823	J. JACKSON.
Do. do. do.	4861	13th Nov. 1823	J. DAY.
Do. do. do.	4930	30th May 1824	J. COOK.
Do. do. do.	4990	27th July 1824	C. R. Baron de BERENGER.
Do. do. do.	5026	4th Nov. 1824	J. SOMERVILLE.
Do. do. do.	5033	6th Nov. 1824	T. CARTMELL.
Do. do. do.	5055	18th Dec. 1824	S. DAVIS.
Do. do. do.	5099	19th Feb. 1825	J. MOULD.
Do. do. do.	5103	26th Feb. 1825	J. MANTON.
Do. do. do.	5155	23rd April 1825	A. L. HUNOUT.

Subject-matter of Patent.	Number of Patent.	Date.	Name of Patentee.
Locks for small arms, improvements in, and in the manufacture of percussion—*continued*	5175	20th May 1825	J. Riviere.
Do. do. do.	5242	15th Aug. 1825	C. Downing.
Do. do. do.	5421	7th Nov. 1826	B. Newmarch.
Do. do. do.	5726	8th Dec. 1828	J. Dickson.
Do. do. do.	5838	2nd Sept. 1829	G. H. Manton.
Dr. do. do.	5845	15th Sept. 1829	D. Lawrence and J. Crundwell.
Do. do. do.	5905	27th Feb. 1830	C.R.Baron de Berenger.
Do. do. do.	6046	6th Dec. 1830	J.G. Lacy and S. Davis.
Do. do. do.	6053	17th Dec. 1830	B. Redfern.
Do. do. do.	6137	13th July 1831	A. Demondion.
Do. do. do.	6166	24th Sept. 1831	W. Bingham and W. Dupe.
Do. do. do.	6196	15th Dec. 1831	A. A. Moser.
Do. do. do.	6326*	2nd Nov. 1832	L. Foucaud.
Do. do. do.	6394	7th March 1833	C. Jones.
Do. do. do.	6436	12th June 1833	C. Jones.
Do. do. do.	6572	13th March 1834	J. A. Manton.
Do. do. do.	6611	22nd May 1834	C. L. S. Baron Heurteloup.
Do. do. do.	6675	6th Sept. 1834	H. Shrapnel.
Do. do. do.	6825	28th April 1835	J. Somerville.
Do. do. do.	6826	30th April 1835	J. Dodds.
Do. do. do.	6829	9th May 1835	J. Egg.
Do. do. do.	7286	19th Jan. 1837	M. Poole.
Do. do. do.	7582	8th March 1838	W. W. Richards.
Do. do. do.	7610	10th April 1838	J. R. Cooper.
Do. do. do.	7712	30th June 1838	G. Round and S. Whitford.
Do. do. do.	7965	11th Feb. 1839	G. H. Manton.
Do. do. do.	7980	23rd Feb. 1839	C. L. S. Baron Heurteloup.
Do. do. do.	7980*	7th Jan. 1842	C. L. S. Baron Heurteloup. Disclaimer and Memorandum of Alteration.
Do. do. do.	8024	9th April 1839	G. Stocker and J. Bentley.
Do. do. do.	8119	22nd June 1839	H. Wilkinson.
Do. do. do.	8347	21st Jan. 1840	J. R. Cooper.
Do. do. do.	8513	20th May 1840	W. Bush.
Do. do. do.	8573	18th July 1840	M. Poole.
Do. do. do.	8704	17th Nov. 1840	J. Shaw.
Do. do. do.	9084	9th Sept. 1841	C. L. S. Baron Heurteloup.
Do. do. do.	9119	14th Oct. 1841	M. Poole.
Do. do. do.	9129	2nd Nov. 1841	W. Golden and J. Hanson.
Do. do. do.	9177	14th Dec. 1841	W. W. Richards.
Do. do. do.	9258	15th Feb. 1842	A. Rousseau.
Do. do. do.	9801	24th June 1843	W. Needham.

Subject-matter of Patent.	Number of Patent.	Date.	Name of Patentee.
Locks for small arms, improvements in, and in the manufacture of percussion—*continued*	10,109	14th March 1844	T. Seymour and J. Seymour.
Do. do. do.	10,280	30th July 1844	J. Bentley.
Do. do. do.	10,667	14th May 1845	C. J. Smith.
Do. do. do.	10,718	10th June 1845	J. W. Tyson.
Do. do. do.	10,849	2nd Oct. 1845	J. Hale.
Do. do. do.	11,657	15th April 1847	J. Mollett.
Do. do. do.	11,994	10th Dec. 1847	S. Taylor.
Do. do. do.	12,432	20th Jan. 1849	H. Needham.
Do. do. do.	12,543	28th March 1849	G. H. Manton and J. Harrington.
Do. do. do.	12,613	15th May 1849	L. A. De Chatauvillard.
Do. do. do.	12,648	7th June 1849	W. H. Ritchie.
Do. do. do.	12,668	20th June 1849	S. Colt.
Do. do. do.	12,781	20th Sept. 1849	J. R. Cooper.
Do. do. do.	12,920	11th Jan. 1850	M. U. Sears.
Do. do. do.	13,215	6th Oct. 1850	A. Melville and E. Callow.
Do. do. do.	13,527	24th Feb. 1851	R. Adams.
Do. do. do.	13,935	29th Jan. 1852	J. M. R. von Winiwarter.
Do. do. do.	13,994	3rd March 1852	G. L. L. Kufahl.
Do. do. do.	14,027	20th March 1852	W. W. Richards.
Do. do. do.	14,041	24th March 1852	J. Macintosh.
Do. do. do.	14,066	17th April 1852	H. G. Delvigne.
Do. do. do.	14,085	22nd April 1852	A. V. Newton.
Do. do. do.	14,227	20th July 1852	J. W. Schlesinger.
Do. do. do.	14,265	19th Aug. 1852	T. Hunt.

M.

Subject-matter of Patent.	Number of Patent.	Date.	Name of Patentee.
Mines, firing mines by means of compositions which explode by percussion	3032	11th April 1807	A. J. Forsyth.
Mines, employing self-igniting paper crackers or matches for firing the trains of	4853	16th Oct. 1823	Sir W. Congreve.
Mines, safety fuze for firing - - -	6159	6th Sept. 1831	W. Bickford.
Mines, improvements in fuzes for firing - -	10,364	24th Oct. 1844	H. Carbines.
Mines, safety fuze for firing - - -	10,928	6th Nov. 1845	J. S. Bickford, G. Smith, and T. Davey.
Mines, preparing gun cotton for exploding -	11,407	8th Oct. 1846	J. Taylor.
Mines, new safety fuze for firing - - -	11,447	12th Nov. 1846	G. Smith.
Mines, improvements in the manufacture of fuzes for firing	12,406	11th Jan. 1849	M. Loam.
Mines, improvements in explosive compounds for charging	13,215	6th Oct. 1850	A. Melville and E. Callow.
Mines, improvements in manufacturing fuzes, and in firing	14,065	15th April 1852	S. Davey and A. L. Chann.

Subject-matter of Patent.	Number. of Patent.	Date.	Name of Patentee.
Mortars, improvements in, and in beds for, and tubes for firing	**2428**	24th July 1800	A. C. De Poggi.
Mortars, instrument for, and method of accurately pointing	**7759**	2nd Aug. 1838	J. Dennett.
Mortars formed of wrought iron or steel, or wrought iron and steel, and machinery for making the same	**10,013**	16th Jan. 1844	T. Aspinwall.

N.

Needle gun, constructing fire-arms on the same principle as that known by the name of the	**3032**	11th April 1807	A. J. Forsyth.
Do. do. do.	**5421**	7th Nov. 1826	B. Newmarch.
Do. do. do.	**6166**	24th Sept. 1831	W. Bingham and W. Dupe.
Do. do. do.	**6196**	15th Dec. 1831	A. A. Moser.
Do. do. do.	**8024**	9th April 1839	G. Stocker and J. Bentley.
Do. do. do.	**8513**	20th May 1840	W. Bush.
Do. do. do.	**9129**	2nd Nov. 1841	W. Golden and J. Hanson.
Do. do. do.	**10,667**	14th May 1845	C. J. Smith.
Do. do. do.	**12,613**	15th May 1849	L. A. De Chatauvillard.
Do. do. do.	**12,648**	7th June 1849	W. H. Ritchie.
Do. do. do.	**12,920**	11th Jan. 1850	M. U. Sears.
Do. do. do.	**13,215**	6th Oct. 1850	A. Melville and E. Callow.
Do. do. do.	**13,994**	3rd March 1852	G. L. L. Kufahl.
Do. do. do.	**14,227**	20th July 1852	J. W. Schlesinger.
Do. do. do.	**14,265**	19th Aug. 1852	T. Hunt.

P.

Papa, an improved manufacture of cannon cartridge,	**3080**	12th Nov. 1807	J. Dickinson.
Percussion, firing artillery, small arms, mines, &c. by means of compositions which explode by	**3032**	11th April 1807	A. J. Forsyth.
Percussion, priming, and locks for fire-arms, improvements in	**3985**	29th Feb. 1816	J. Manton.
Do. do. do.	**4285**	3rd Aug. 1818	J. Manton.
Do. do. do.	**4427**	15th Jan. 1820	F. Fox.
Do. do. do.	**4590**	14th Sept. 1821	W. Webster.
Do. do. do.	**4611**	10th Nov. 1821	W. W. Richards.
Percussion, or flint lock, constructing locks to be used either as a	**4648**	12th Feb. 1822	S. Davis.

Subject-matter of Patent.	Number of Patent.	Date.	Name of Patentee.
Percussion locks, improvements in priming	4727	26th Nov. 1822	J. Egg.
Do. do. do.	4823	29th July 1823	J. Jackson.
Percussion locks, improvements in	4861	13th Nov. 1823	J. Day.
Percussion locks, applying springs to	4960	30th May 1824	J. Cook.
Percussion locks and priming, improvements in	4990	27th July 1824	C. R. Baron de Berenger.
Percussion locks, improvements in	5026	4th Nov. 1824	J. Somerville.
Percussion priming and locks, improvements in	5033	6th Nov. 1824	T. Cartmell.
Percussion locks, improvements in	5055	18th Dec. 1824	S. Davis.
Percussion locks for repeating fire arms, constructed so as to slide	5099	19th Feb. 1825	J. Mould.
Percussion priming and locks, improvements in	5106	26th Feb. 1825	J. Manton.
Percussion locks for repeating fire-arms	5155	23rd April 1825	A. L. Hunout.
Percussion locks, improvements in	5175	20th May 1825	J. Riviere.
Percussion priming and locks, improvements in	5242	15th Aug. 1825	C. Downing.
Percussion locks and priming chambers, improvements in	5421	7th Nov. 1826	B. Newmarch.
Percussion locks for walking-stick guns	5726	8th Dec. 1828	J. Dickson.
Percussion locks, improvements in	5838	2nd Sept. 1829	G. H. Manton.
Percussion priming and locks, improvements in	5845	15th Sept. 1829	D. Lawrence and J. Crundwell.
Percussion shot or projectile, which is caused to explode by percussion	5864	2nd Nov. 1829	J. Tucker.
Percussion priming and locks, improvements in	5905	27th Feb. 1830	C. R. Baron de Berenger.
Percussion caps and nipples for fire-arms, improvements in	5978	7th Aug. 1830	S. Smith.
Percussion locks, improvements in	6046	6th Dec. 1830	J. G. Lacy and S. Davis.
Do. do. do.	6053	17th Dec. 1830	B. Redfern.
Percussion primers and touch-holes, improvements in	6071	11th Feb. 1831	W. W. Richards.
Percussion locks and priming, improvements in	6137	13th July 1831	A. Demondion.
Percussion, improvements in fire-arms to be discharged by, and in cartridges to be used therewith	6139	15th July 1831	J. de Burgh, Marquis of Clanricarde.
Percussion locks and priming, improvements in	6166	24th Sept. 1831	W. Bingham and W. Dupe.
Do. do. do.	6196	15th Dec. 1831	A. A. Moser.
Percussion locks, applying a magazine to, containing detonating caps, for the purpose of priming -	6326*	2nd Nov. 1832	L. Foucaud.
Percussion locks, improvements in - - -	6394	7th Mar. 1832	C. Jones.
Do. do. do.	6436	12th June 1833	C. Jones.
Percussion locks and priming, improvements in -	6572	13th March 1834	J. A. Manton.
Do. do. do.	6611	22nd May 1834	C. L. S. Baron Heurteloup.
Do. do. do.	6675	6th Sept. 1834	H. Shrapnel.
Percussion locks, improvements in - - -	6825	28th April 1835	J. Somerville.
Do. do. do.	6826	30th April 1835	J. Dodds.
Percussion locks and priming, improvements in -	6829	9th May 1835	J. Egg.
Percussion locks for repeating fire-arms, improvements in -	6909	22nd Oct. 1835	S. Colt.
Percussion caps or primers, improvements in -	7041	22nd March 1836	W. W. Richards.
Percussion locks for repeating fire-arms, improvements in	7286	19th Jan. 1837	M. Poole.
Percussion locks, improved primer for - -	7582	8th March 1838	W. W. Richards.
Percussion locks and nipples, improvements in -	7610	10th April 1838	J. R. Cooper.

Subject-matter of Patent.	Number of Patent.	Date.	Name of Patentee.
Percussion fuzes for shells of rockets - -	**7759**	2nd Aug. 1838	J. Dennett.
Percussion locks and priming, improvements in -	**7965**	11th Feb. 1839	G. H. Manton.
Percussion locks and priming - - -	**7980**	23rd Feb. 1839	C. L. S. Baron Heurteloup.
Do. do. - - -	**7980***	7th Jan. 1842	C. L. S. Baron Heurteloup. Disclaimer and Memorandum of Alteration.
Percussion locks, improvements in - - -	**8024**	9th April 1839	G. Stocker and J. Bentley.
Percussion locks and priming, improvements in -	**8119**	22nd June 1839	H. Wilkinson.
Percussion locks and priming, and applying charges of detonating powder to balls	**8347**	21st Jan. 1840	J. R. Cooper.
Percussion locks, and applying percussion powder to cartridges	**8513**	20th May 1840	W. Bush.
Percussion locks, improvements in - - -	**8573**	18th July 1840	M. Poole.
Percussion locks, improvements in priming, and in caps for	**8704**	17th Nov. 1840	J. Shaw.
Percussion locks and priming, improvements in -	**9084**	9th Sept. 1841	C. L. S. Baron Heurteloup.
Percussion locks for repeating fire-arms, improvements in	**9119**	14th Oct. 1841	M. Poole.
Percussion locks, and applying charges of fulminating powder to balls	**9129**	2nd Nov. 1841	W. Golden and J. Hanson.
Percussion locks and priming, improvements in -	**9177**	14th Dec. 1841	W. W. Richards.
Percussion caps, improvements in - -	**9188**	16th Dec. 1841	T. Starkey.
Percussion locks and priming, improvements in -	**9258**	15th Feb. 1842	A. Rousseau.
Percussion locks, priming, and caps, improvements in	**9801**	24th June 1843	W. Needham.
Percussion locks, improvements in - -	**10,109**	14th March 1844	T. Seymour and J. Seymour.
Do. do.	**10,280**	30th July, 1844	J. Bentley.
Percussion locks and priming, improvements in -	**10,667**	14th May, 1845	C. J. Smith.
Do. do. do.	**10,718**	10th June 1845	J. W. Tyson.
Percussion caps, charging with gun cotton - -	**11,407**	8th Oct. 1846	J. Taylor.
Percussion locks and cartridges, improvements in -	**11,657**	15th April 1847	J. Mollett.
Percussion locks and priming for repeating fire-arms, improvements in	**11,994**	10th Dec. 1847	S. Taylor.
Percussion locks and priming, improvements in -	**12,432**	20th Jan. 1849	H. Needham.
Do. do. do.	**12,543**	28th March 1849	G. H. Manton and J. Harrington.
Do. do. do.	**12,613**	15th May 1849	L. A. De Chatauvillard.
Do. do. do.	**12,648**	7th June 1849	W. H. Ritchie.
Percussion locks for repeating fire-arms, improvements in	**12,668**	20th June 1849	S. Colt.
Percussion locks for repeating and other fire-arms, improvements in	**12,781**	20th Sept. 1849	J. R. Cooper.
Percussion locks and priming, improvements in -	**12,920**	11th Jan. 1850	M. U. Sears.
Percussion tubes for fire-arms, improvements in -	**13,161**	3rd July 1850	C. W. Lancaster.
Percussion shells and arrows, and explosive compounds, improvements in	**13,215**	6th Oct. 1850	A. Melville and E. Callow.
Percussion shells or projectiles, improvements in -	**13,454**	16th Jan. 1851	C. W. Lancaster.
Percussion locks for repeating and other fire-arms, improvements in	**13,527**	24th Feb. 1851	R. Adams.

Subject-matter of Patent.	Number of Patent.	Date.	Name of Patentee.
Percussion locks, priming, and primers for applying such priming	13,935	29th Jan. 1852	J. M. R. Von Winiwarter.
Percussion locks and caps and primers, improvements in	14,027	20th March 1852	W. W. Richards.
Percussion locks, improvements in - - -	14,041	24th March 1852	J. Macintos
Percussion, improvements in breech-loading fire-arms discharged by	14,052	31st March 1852	M. Poole.
Percussion guns, an improved breech for - -	14,058	6th April 1852	J. W. De Longueville Giffard.
Percussion locks for breech-loading, and other fire-arms	14,066	17th April 1852	H. G. Delvigne.
Percussion locks, improvements in priming - -	14,085	22nd April 1852	A. V. Newton.
Percussion shells, improvements in - - -	14,087	24th April 1852	W. Church, S. A. Goddard, and E. Middleton.
Percussion locks, improvements in - - -	14,227	20th July 1852	J. W. Schlesinger.
Do. do.	14,265	19th Aug. 1852	T. Hunt.
Percussion shells, improvements in - - -	14,271	23rd Aug. 1852	H. N. S. Shrapnel.
Percussion fire-arms, improvements in breech-loading	14,289	10th Sept. 1852	S. Taylor.
Pikes, an improved mode of making - -	2750	7th Feb. 1804	E. Thomason.
Pikes, shield or protection for the human body against	2763	18th May 1804	J. P. Barthelemy and J. Shoubridge.
Pikes, armour for protecting the human body against	2885	30th Oct. 1805	R. Kentish.
Pikes, or halberts with cuteaux, manufacturing -	3400	4th March 1811	W. Turner.
Pistols [*See* Small Arms *also*] fitting bayonets to -	1284	9th March 1781	J. Waters.
Pistols, improvements in stocks for, and a new pistol for firing off cannons	2692	23rd March 1803	D. Egg.
Pistols, improvements in locks for, and in combining swords with	2744	3rd Dec. 1803	J. S. Searles.
Pistols, improvements in stocks for - -	3599	25th Sept. 1812	D. Egg.
Pistols, inserting in walking sticks - -	3837	17th Aug. 1814	H. W. Vander Kleft.
Pistols, improvements in making holster cases for -	4724	9th Nov. 1822	F. Deakin.
Pistols, improvements in stocks for - -	6554	8th Feb. 1834	W. S. Gillett.
Pistols, applying props to, for warding off sword cuts	6675	6th Sept. 1834	H. Shrapnel.
Pistols, improvements in revolving and other -	8347	21st Jan. 1840	J. R. Cooper.
Pistols, improvements in revolving and double-barralled central fire	10,667	14th May 1845	C. J. Smith.
Pistols, of peculiar form - - - -	14,066	17th April 1852	H. G. Delvigne.
Powder flasks, improvements in - - -	2865	3rd July 1805	A. Wilson.
Do. do. do.	3599	25th Sept. 1812	D. Egg.
Do. do. do.	3784	9th March 1814	J. Thompson.
Do. do. do.	3828	4th Aug. 1814	T. Sykes.
Do. do. do.	5439	20th Dec. 1826	C. R. Baron de Berenger.
Do. do. do.	10,667	14th May 1845	C. J. Smith
Do. do. do.	10,683	22nd May 1845	M. Boche.
Priming flint locks, improvements in - - -	2692	23rd March 1803	D. Egg.
Do. do. do.	2709	10th June 1803	J. R. Peckham.
Do. do. do.	2722	6th July 1803	J. Manton.
Do. do. do.	2991	4th Dec. 1806	J. F. Matthey.
Do. do. do.	3286	11th Dec. 1809	J. Manton.
Do. do. do.	3784	9th March 1814	J. Thomson.

Subject-matter of Patent.			Number of Patent.	Date.	Name of Patentee.
Priming flint locks, improvements in—*continued*			4031	25th May 1816	F. Richardson.
Do.	do.	do.	4054	12th Aug. 1816	S. Nock.
Do.	do.	do.	4336	23rd Jan. 1819	U. Sartoris.
Do.	do.	do.	4727	26th Nov. 1822	J. Egg.
Do.	do.	do.	5099	19th Feb. 1825	J. Mould.
Priming fire-arms with compositions which explode by percussion			3032	11th April 1807	A. J. Forsyth.
Priming percussion locks, improvements in			3985	29th Feb. 1816	J. Manton.
Do.	do.	do.	4285	3rd Aug. 1818	J. Manton.
Do.	do.	do.	4427	15th Jan. 1820	F. Fox.
Do.	do.	do.	4590	14th Sept. 1821	W. Webster.
Do.	do.	do.	4611	10th Nov. 1821	W. W. Richards.
Do.	do.	do.	4727	26th Nov. 1822	J. Egg.
Do.	do.	do.	4823	29th July 1823	J. Jackson.
Do.	do.	do.	4990	27th July 1824	C. R. Baron de Berenger.
Do.	do.	do.	5033	6th Nov. 1824	T. Cartmell.
Do.	do.	do.	5099	19th Feb. 1825	J. Mould.
Do.	do.	do.	5106	26th Feb. 1825	J. Manton.
Do.	do.	do.	5242	15th Aug. 1825	C. Downing.
Do.	do.	do.	5845	15th Sept. 1829	D. Lawrence and J. Crundwell.
Do.	do.	do.	5905	27th Feb. 1830.	C. R. Baron de Berenger
Do.	do.	do.	5978	7th Aug. 1830	S. Smith.
Do.	do.	do.	6071	11th Feb. 1831	W. W. Richards.
Do.	do.	do.	6137	13th July 1831	A. Demondion.
Do.	do.	do.	6166	24th Sept. 1831	W. Bingham and W. Dupe.
Do.	do.	do.	6196	15th Dec. 1831	A. A. Moser.
Do.	do.	do.	6326*	2nd Nov. 1832	L. Foucaud.
Do.	do.	do.	6572	13th March 1834	J. A. Manton.
Do.	do.	do.	6611	22nd May 1834	C. L. S. Baron Heurteloup.
Do.	do.	do.	6675	6th Sept. 1834	H. Shrapnel.
Do.	do.	do.	6829	9th May 1835	J. Egg.
Do.	do.	do.	7041	22nd March 1836	W. W. Richards.
Do.	do.	do.	7582	8th March 1838	W. W. Richards.
Do.	do.	do.	7965	11th Feb. 1839	G. H. Manton.
Do.	do.	do.	7980	23rd Feb. 1839	C. L. S. Baron Heurteloup.
Do.	do.	do.	7980*	7th Jan. 1842	C. L. S. Baron Heurteloup. Disclaimer and Memorandum of Alteration.
Do.	do.	do.	8119	22nd June 1839	H. Wilkinson.
Do.	do.	do.	8347	21st Jan. 1840	J. R. Cooper.
Do.	do.	do.	8704	17th Nov. 1840	J. Shaw.
Do.	do.	do.	9084	9th Sept. 1841	C. L. S. Baron Heurteloup.
Do.	do.	do.	9177	14th Dec. 1841	W. W. Richards.
Do.	do.	do.	9188	16th Dec. 1841	T. Starkey.
Do.	do.	do.	9258	15th Feb. 1842	A. Rousseau.
Do.	do.	do.	9801	24th June 1843	W. Needham.

Subject-matter of Patent.	Number of Patent.	Date.	Name of Patentee.
Priming percussion locks, improvements in—*cont.*	10,667	14th May 1845	C. J. Smith.
Do. do. do.	10,718	10th June 1845	J. W. Tyson.
Do. do. do.	11,407	8th Oct. 1846	J. Taylor.
Do. do. do.	11,657	15th April 1847	J. Mollett.
Do. do. do.	11,994	10th Dec. 1847	S. Taylor.
Do. do. do.	12,432	20th Jan. 1849	H. Needham.
Do. do. do.	12,543	28th March 1849	G. H. Manton and J. Harrington.
Do. do. do.	12,613	15th May 1849	L. A. De Chatauvillard.
Do. do. do.	12,648	7th June 1849	W. H. Ritchie.
Do. do. do.	12,920	11th Jan. 1850	M. U. Sears.
Do. do. do.	13,161	3rd July 1850	C. W. Lancaster.
Do. do. do.	13,935	29th Jan. 1852	J. M. R. Von Winiwarter.
Do. do. do.	14,027	20th March 1852	W. W. Richards.
Do. do. do.	14,085	22nd April 1852	A. V. Newton.
Priming of fire-arms, improvements in protecting from wet	1095	8th April 1775	H. Nock, W. Jover, and J. Green.
Do. do. do.	1897	5th July 1792	J. Willson.
Do. do. do.	2178	12th April 1797	J. Manton.
Do. do. do.	2454	9th Dec. 1800	J. Prosser.
Do. do. do.	2466	23rd Jan. 1801	R. Vazie.
Do. do. do.	2573	16th Jan. 1802	J. Hall.
Do. do. do.	2692	23rd March 1803	D. Egg.
Do. do. do.	2991	4th Dec. 1806	J. F. Matthey.
Do. do. do.	3233	4th May 1809	T. Noon.
Do. do. do.	3286	11th Dec. 1809	J. Manton.
Do. do. do.	3558	30th April 1812	J. Manton.
Do. do. do.	3722	19th July 1813	R. Pretyman.
Do. do. do.	3784	9th March 1814	J. Thomson.
Do. do. do.	3794	23rd March 1814	F. Heaton.
Do. do. do.	4427	15th Jan. 1820	F. Fox.
Do. do. do.	4990	27th July 1824	C. R. Baron de Berenger.
Do. do. do.	5033	6th Nov. 1824	T. Cartmell.
Do, do, do.	5055	18th Dec. 1824	S. Davis.
Do. do. do.	5726	8th Dec. 1828	J. Dickson.
Do. do. do.	5905	27th Feb. 1830	C. R. Baron de Berenger.
Do. do. do.	6394	17th March 1833	C. Jones.
Do. do. do.	6436	12th June 1833	C. Jones.
Do. do. do.	6909	22nd Oct. 1835	S. Colt.
Do. do. do.	7965	11th Feb. 1839	G. H. Manton.
Do. do. do.	8704	17th Nov. 1840	J. Shaw.
Do. do. do.	9084	9th Sept. 1841	C. L. S. Baron Heurteloup.
Do. do. do.	9177	14th Dec. 1841	W. W. Richards.
Do. do. do.	9188	16th Dec. 1841	T. Starkey.
Do. do. do.	10,667	14th May 1845	C. J. Smith.
Do. do. do.	11,657	15th April 1847	J. Mollett.

Subject-matter of Patent.	Number of Patent.	Date.	Name of Patentee.
Projectiles, gun for discharging square and round -	418	15th May 1718	J. Puckle.
Projectiles: improvements in manufacturing bullets and shot	725	29th June 1758	H. Raminger.
Projectiles: improvements in manufacturing small shot	1347	10th Dec. 1782	W. Watts.
Projectiles: making and testing long bows for discharging arrows	1639	5th Feb. 1788	T. Waring.
Projectiles: making balls, shells, or arrow-shotbelted winged, or grooved, for rifled cannon	1694	30th July 1789	J. Wilkinson.
Projectiles: making tailed balls and harpoons -	2744	3rd Dec. 1803	J. S. Searles.
Projectiles: shield or protection for the human body against musket balls	2763	18th May 1804	J. P. Barthelemy and J. Shoubridge.
Projectiles: armour for protecting the human body against musket balls	2885	30th Oct. 1805	R. Kentish.
Projectiles: machinery for rolling leaden bullets and shot	2932	1st May 1806	W. Bundy.
Projectiles: making balls for rifled cannon, of cast iron coated with soft metal	3155	30th July 1808	G. Richards.
Projectiles: improved method of casting cannon shot	4705	27th Sept. 1822	R. Boothby.
Projectiles: improvements in rockets for signals, lighting an enemy's camp, &c.	4853	16th Oct. 1823	Sir W. Congreve.
Projectiles: throwing shells, rockets, and other projectiles by the force of steam	4952	15th May 1824	J. Perkins.
Projectiles: coating shot with mercury - -	5135	25th March 1825	J. Manton.
Projectiles: peculiar construction of shot or projectile to explode by percussion	5864	2nd Nov. 1829	J. Tucker.
Projectiles: peculiar construction of projectile to be substituted for shot or bullets	6139	15th July 1831	J. de Burgh, Marquis of Clanricarde.
Projectiles: making balls of tin and lead or iron, and of iron coated with tin, and fitting hollow shot with small shot	6675	6th Sept. 1834	H. Shrapnel.
Projectiles: improvements in war rockets, and in communicating with ships by means of rockets, and firing shot from mortars	7759	2nd Aug. 1838	J. Dennett.
Projectiles: wrapping balls in muslin, by which they are to be attached to cartridge cases	7980	23rd Feb. 1839	C. L. S. Baron Heurteloup.
Do. do. do.	7980*	7th Jan. 1842	C. L. S. Baron Heurteloup. Disclaimer and Memorandum of Alteration.
Projectiles: applying charges of detonating powder to balls	8347	21st Jan. 1840	J. R. Cooper.
Projectiles: machinery for forming balls and shot by dies and pressure.	8385	12th Feb. 1840	D. Napier,
Projectiles: applying charges of fulminating powder to bullets and other projectiles	9129	2nd Nov. 1841	W. Golden and J. Hanson.
Projectiles: manufacturing iron shot by cutting, compressing, and rolling	9673	20th Mar. 1843	S. Rolinson.
Projectiles: improvements in rockets - -	10,008	11th Jan. 1844	W. Hale.
Projectiles: improvements in rockets - -	10,364	24th Oct. 1844	H. Carbines.
Projectiles: using water and incombustible gases to prevent accidental explosion of the substances used in making	11,011	20th Dec. 1845	J. Blyth.
Projectiles: heating compressed air for working air guns or cannons by the use of red hot balls	11,864	9th Sept. 1847	C. W. O'Leary.
Projectiles: making ball shollow and open at the rear end to receive a charge of powder	11,994	10th Dec. 1847	S. Taylor.

Subject-matter of Patent.	Number of Patent.	Date.	Name of Patentee.
Projectiles: machinery for manufacturing balls and shot by punching, rolling, &c.	12,048	31st Jan. 1848	A. V. Newton.
Projectiles: combining the charge and priming with bullets	12,613	15th May 1849	L. A. De Chatauvillard.
Projectiles: impelling projectiles by means of elastic purchases	12,623	29th May 1849	R. E. Hodges.
Projectiles: improvements in the manufacture of shot	12,624	29th May 1849	D. Smith.
Projectiles: improvements in bullets to be used with rifled barrels	12,920	11th Jan. 1850	M. U. Sears.
Projectiles: improvements in balls for rifled fire-arms	13,161	3rd July 1850	C. W. Lancaster.
Projectiles: improvements in balls for rifled cannon, in arrows charged with explosive compounds to produce light at any required spot, and in percussion shells	13,215	6th Oct. 1850	A. Melville and E. Callow.
Projectiles: improvements in cases for rockets -	13,377	30th Nov. 1850	J. A. Elmslie and G. Simpson.
Projectiles: improvements in balls for rifled fire-arms, and in percussion shells and other projectiles	13,454	16th Jan. 1851	C. W. Lancaster.
Projectiles: improvements in balls or projectiles for rifled fire-arms	13,934	29th Jan. 1852	F. J. Manceaux.
Projectiles: improvements in balls or projectiles -	14,027	20th Mar. 1852	W. W. Richards.
Projectiles: improvements in shells and balls to contain a charge of powder	14,041	24th Mar. 1852	J. Macintosh.
Projectiles: balls for containing the charge of powder and other projectiles	14,058	6th April 1852	J. W. de Longueville Giffard.
Projectiles: improvements in bullets or projectiles -	14,066	17th April 1852	H. G. Delvigne.
Projectiles: improvements in bullets, cannon shot, shells, rockets, and other projectiles	14,087	24th April 1852	W. Church, S. A. Goddard, and E. Middleton.
Projectiles: improvements in cannon shot, shells, and balls	14,271	23rd Aug. 1852	H. N. S. Shrapnel.

R.

Ramrods, improvements in - - -	2692	23rd Mar. 1803	D. Egg.
Do. do. do.	3122	28th Mar. 1808	B. Cook.
Do. do. do.	3828	4th Aug. 1814	T. Sykes.
Do. do. do.	5726	8th Dec. 1828	J. Dickson.
Ramrods, securing bayonets on muskets by means of	5905	27th Feb. 1830	C.R.Baron de Berenger.
Ramrods, improvements in - - -	6872	6th Aug. 1835	W. Mason.
Do. do. do.	7980	23rd Feb. 1839	C. L. S. Baron Heurteloup.
Do. do. do.	7980*	7th Jan. 1842	C. L. S. Baron Heurteloup. Disclaimer and Memorandum of Alteration.
Ramrods, actuating priming apparatus by means of	12,432	20th Jan. 1849	H. Needham.
Do. do. do.	12,543	28th Mar. 1849	G. H. Manton and J. Harrington.

Subject-matter of Patent.	Number of Patent.	Date.	Name of Patentee.
Ramrods, lever apparatus to be used instead of loose	12,668	20th June 1849	S. Colt.
Ramrods, making parts of, act as spring hammers -	12,781	20th Sept. 1849	J. R. Cooper.
Ramrods, loading fire-arms by means of tubular -	13,527	24th Feb. 1851	R. Adams.
Recoil of small arms, preventing the disagreeable effects of the	1003	4th Feb. 1772	T. Wright and C. Byrne.
Recoil or shock of horse swivel guns, improvements in lessening the effects of the	2466	23rd Jan. 1801	R. Vazie.
Recoil of small arms, diminishing the effects of the	2692	23rd Mar. 1803	D. Egg.
Recoil of small arms, lessening the effects of the -	6554	8th Feb. 1834	W. S. Gillett.
Do. do. do.	6909	22nd Oct. 1835	S. Colt.
Recoil or concussion of small arms and artillery, method of preventing	11,864	9th Sept. 1847	C. W. O'Leary.
Repeating fire-arms, inventions relating to -	418	15th May 1718	J. Puckle.
Do. do. do.	1270	5th Dec. 1780	J. Aitken.
Do. do, do.	3599	25th Sept. 1812	D. Egg.
Do. do. do.	3784	9th Mar. 1814	J. Thomson.
Do. do. do.	3833	4th Aug. 1814	J. S. Pauly.
Do. do. do.	4315	24th Nov. 1818	E. H. Collier.
Do. do. do.	5099	19th Feb. 1825	J. Mould.
Do. do. do.	5155	23rd April 1825	A. L. Hunout.
Do. do. do.	6826	30th April 1835	J. Dodds.
Do. do. do.	6909	22nd Oct. 1835	S. Colt.
Do. do. do.	7286	19th Jan. 1837	M. Poole.
Do. do. do.	7472	14th Sept. 1837	W. Coles.
Do. do. do.	8024	9th April 1839	G. Stocker and J. Bentley.
Do. do. do.	8347	21st Jan. 1840	J. R. Cooper.
Do. do. do.	9119	14th Oct. 1841	M. Poole.
Do. do. do.	10,667	14th May 1845	C. J. Smith.
Do. do. do	11,994	10th Dec. 1847	S. Taylor.
Do. do. do.	12,668	20th June 1849	S. Colt.
Do. do. do.	12,781	20th Sept. 1849	J. R. Cooper.
Do. do. do.	13,527	24th Feb. 1851	R. Adams.
Do. do. do.	13,994	3rd Mar. 1852	G. L. L. Kufahl.
Do. do. do.	14,027	20th Mar. 1852	W. W. Richards.
Do. do. do.	14,041	24th Mar. 1852	J. Macintosh.
Rests for guns, improvements in - -	2865	3rd July 1805	A. Wilson.
Do. do. do.	5416	18th Oct. 1826	W. Mills.
Do. do. do.	6675	6th Sept. 1834	H. Shrapnel.
Rifled cannon, belted, winged, or grooved balls for	1694	30th July 1789	J. Wilkinson.
Rifled cannon, constructing so as to load at the breech, and making balls of cast iron coated with soft metal for	3155	30th July 1808	G. Richards.
Rifled cannon, improvements in balls for - -	13,161	3rd July 1850	C. W. Lancaster.
Do. do. do.	13,215	6th Oct. 1850	A. Melville and E. Callow.
Do. do. do.	13,454	16th Jan. 1851	C. W. Lancaster.
Rifled small arms, making bullets for, of tin and lead or iron, and of iron coated with tin	6675	6th Sept. 1834	H. Shrapnel.
Rifled small arms, improvements in the breech and ramrod to facilitate the loading of	7980	23rd Feb. 1839	C. L. S. Baron Heurteloup.

Subject-matter of Patent.	Number of Patent.	Date.	Name of Patentee.
Rifled small arms, improvements in balls to be used with	12,920	11th Jan. 1850	M. U. SEARS.
Rifled small arms, improvements in balls for -	13,161	3rd July 1850	C. W. LANCASTER.
Do. do. do.	13,454	16th Jan. 1851	C. W. LANCASTER.
Do. do. do.	13,934	29th Jan. 1852	F. J. MANCEAUX.
Rifled small arms, making or lining the muzzles with steel	14,027	20th March 1852	W. W. RICHARDS.
Rifled tubes, lining ordinary muskets or guns of large bore with	14,087	24th April 1852	W. CHURCH, S. A. GODDARD, and E. MIDLETON.
Rifling cannon, method of - - - - -	1694	30th July 1789	J. WILKINSON.
Rifling cannon, machinery for - - -	2436	2nd Aug. 1800	T. GILL.
Rifling cannon, improvements in - - -	13,161	3rd July 1850	C. W. LANCASTER.
Rifling cannon, machinery for - - -	13,454	16th Jan. 1851	C. W. LANCASTER.
Rifling cannon and rocket tubes, machinery for -	14,087	24th April 1852	W. CHURCH, S. A. GODDARD, and E. MIDDLETON.
Rifling cannon, improvements in - - -	14,271	23rd Aug. 1852	H. N. S. SHRAPNEL.
Rifling the barrels of small arms, new method of -	1139	2nd Dec. 1776	P. FERGUSON.
Rifling the barrels of small arms, machinery for -	2436	2nd Aug. 1800	T. GILL.
Rifling the barrels of small arms, improvement in -	6675	6th Sept. 1834	H. SHRAPNEL.
Do. do. do.	13,161	3rd July 1850	C. W. LANCASTER.
Rifling the barrels of small arms, machinery for -	13,454	16th Jan. 1851	C. W. LANCASTER.
Rifling the barrels of small arms, new method of -	13,527	24th Feb. 1851	R. ADAMS.
Rifling the barrels of small arms, machinery for -	13,934	29th Jan. 1852	F. J. MANCEAUX.
Do. do. do.	14,087	24th April 1852	W. CHURCH, S. A. GODDARD, and E. MIDDLETON.
Rifling the barrels of small arms, improvements in -	14,271	23rd Aug. 1852	H. N. S. SHRAPNEL.
Rockets, improvements in, for signals, lighting an enemy's camp, &c.	4853	16th Oct. 1823	Sir W. CONGREVE.
Rockets, propelling by the force of steam - -	4952	15th May 1824	J. PERKINS.
Rockets, improvements in war rockets, and in communicating with ships by	7759	2nd Aug. 1838	J. DENNETT.
Rockets, improvements in - - -	10,008	11th Jan. 1844	W. HALE.
Rockets, improvements in - - - -	10,364	24th Oct. 1844	H. CARBINES.
Rockets, using water and incombustible gases to prevent accidental explosion of the substances used in making	11,011	20th Dec. 1845	J. BLYTH.
Rockets, improvements in cases for - -	13,377	30th Nov. 1850	J. A. ELMSLIE, and G. SIMPSON.
Rockets, improvements in, and in machinery for rifling rocket tubes	14,087	24th April 1852	W. CHURCH, S. A. GODDARD, and E. MIDDLETON.
Rolling bevelled and tapered bars for making gun barrels, machinery for	2911	20th Feb. 1806	J. JONES.
Rolling leaden bullets and other shot, machinery for	2932	1st May 1806	W. BUNDY.
Rolling, tapered plated, for making gun barrels -	3267	28th Sept. 1809	J. JONES.
Rolling skelps for making gun barrels - -	3437	24th April 1811	J. BRADLEY.
Rolling, welding barrels for fire-arms and artillery by	3469	26th July 1811	H. JAMES and J. JONES.
Rolling gun barrels, machinery for - -	3740	15th Oct. 1813	H. OSBORN.
Rolling gun and pistol barrels, machinery for -	3813	7th June 1814	G. HEYWOOD.
Do. do. do.	4105	1st March 1817	H. OSBORN.

Subject-matter of Patent.	Number of Patent.	Date.	Name of Patentee.
Rolling, welding the seams or joints of twisted gun barrels by	9947	18th Nov. 1843	J. Roose.
Rolling, improvements in welding twisted gun and pistol barrels by	13,299	24th Oct. 1850	A. Rose.

S.

Subject-matter of Patent.	Number of Patent.	Date.	Name of Patentee.
Shells, belted, winged, or grooved, for rifled cannon	1694	30th July 1789	J. Wilkinson.
Shells, throwing by the force of steam - -	4952	15th May 1824	J. Perkins.
Shells, or projectiles, peculiar construction of, to explode by percussion	5864	2nd Nov. 1829	J. Tucker.
Shells, or hollow shot filled with small shot -	6675	6th Sept. 1834	H. Shrapnel.
Shells, percussion and other, to be attached to rockets	7759	2nd Aug. 1838	J. Dennett.
Shells, improvements in percussion - -	13,215	6th Oct. 1850	A. Melville and E. Callow.
Shells, improvements in percussion and other -	13,454	16th Jan. 1851	C. W. Lancaster.
Shells, improvements in - - - -	14,041	24th March 1852	J. Macintosh.
Shells, improvements in percussion and other -	14,087	24th April 1852	W. Church, S. A. Goddard, and E. Middleton.
Do. do. do.	14,271	23rd Aug. 1852	H. N. S. Shrapnel.
Shield or protection for the human body against sword, bayonet, pike, or musket ball	2763	18th May 1804	J. P. Barthelemy and J. Shoubridge.
Ships, apparatus to be applied to, for the purpose of preventing boarding by an enemy	3447	9th May 1811	G. Hawkins.
Ships, using rockets and mortars for communicating with	7759	2nd Aug. 1838	J. Dennett.
Ships, employing a portable air cannon for communicating with	11,864	9th Sept. 1847	C. W. O'Leary.
Ships, using elastic purchases to discharge projectiles for the purpose of communicating with	12,623	29th May 1849	R. E. Hodges.
Ships, mode of fitting the port-holes of, so as to admit of guns being fired below the water line	14,041	24th Mar. 1852	J. Macintosh.
Shot, improvements in manufacturing - -	725	29th June 1758	H. Raminger.
Do. do. do.	1347	10th Dec. 1782	W. Watts.
Do. do. do.	2932	1st May 1806	W. Bundy.
Shot, coating with mercury - - -	5135	25th Mar. 1825	J. Manton.
Shot, peculiar construction of projectile to be substituted for	6139	15th July 1831	J. de Burgh, Marquis of Clanricarde.
Shot, filling hollow shot with small - -	6675	6th Sept. 1834	H. Shrapnel.
Shot, machinery for forming by dies and pressure -	8385	12th Feb. 1840	D. Napier.
Shot, machinery for manufacturing by punching, rolling, &c.	12,048	31st Jan. 1848	A. V. Newton.
Shot, improvements in the manufacture of -	12,624	29th May 1849	D. Smith.
Shot belts, improvements in - - -	2865	3rd July 1805	A. Wilson.
Shot chargers, improvements in - - -	3599	25th Sept. 1812	D. Egg.
Shot belts and chargers, improvements in - -	3828	4th Aug. 1814	T. Sykes.
Shot chargers, improvements in - - -	10,683	22nd May 1845	M. Boche.

Subject-matter of Patent.			Number of Patent.	Date.	Name of Patentee.
Sights for fire-arms, improvements in - -			1139	2nd Dec. 1776	P. FERGUSON.
Do.	do.	do.	2692	23rd Mar. 1803	D. EGG.
Do.	do.	do.	2966	15th Sept. 1806	J. MANTON.
Do.	do.	do.	3755	23rd Nov. 1813	J. BODMER.
Do.	do.	do.	6675	6th Sept. 1834	H. SHRAPNEL.
Do.	do.	do.	10,849	2nd Oct. 1845	J. HALE.
Do.	do.	do.	13,934	29th Jan. 1852	F. J. MANCEAUX.
Do.	do.	do.	14,271	23rd Aug. 1852	H. N. S. SHRAPNEL.
Signals, improvements in rockets to be used for -			4853	16th Oct. 1823	Sir W. CONGREVE.
Small arms, improvements in, and in the manufacture of			418	15th May 1718	J. PUCKLE.
Do.	do.	do.	1003	4th Feb. 1772	J. WRIGHT and C. BYRNE.
Do.	do.	do.	1095	8th April 1775	H. NOCK, W. JOVER, and J. GREEN.
Do.	do.	do.	1139	2nd Dec. 1776	P. FERGUSON.
Do.	do.	do.	1270	5th Dec. 1780	J. AITKEN.
Do.	do.	do.	1284	9th Mar. 1781	J. WATERS.
Do.	do.	do.	1598	25th April 1787	H. NOCK.
Do.	do.	do.	1707	6th Nov. 1789	J. TEMPLEMAN.
Do.	do.	do.	1837	28th Nov. 1791	J. S. SEARLES.
Do.	do.	do.	1838	26th Nov. 1791	S. BENTHAM.
Do.	do	do.	1865	18th April 1792	J. MANTON.
Do.	do.	do.	1893	5th July 1792	J. MANTON.
Do.	do.	do.	1897	5th July 1792	J. WILLSON.
Do.	do.	do.	1951	23rd April 1793	S. BENTHAM.
Do.	do.	do.	1978	4th Feb. 1794	T. SYKES.
Do.	do.	do.	2041	28th Feb. 1795	G. BOLTON.
Do.	do.	do.	2042	28th Feb. 1795	R. WEBB.
Do.	do.	do.	2072	22nd Oct. 1795	J. GROVE.
Do.	do.	do.	2173	14th Mar. 1797	W. SIDDON.
Do.	do.	do.	2178	12th April 1797	J. MANTON.
Do.	do.	do.	2252	23rd July 1798	W. DUPE.
Do.	do.	do.	2360	28th Nov. 1799	E. THOMASON.
Do.	do.	do.	2436	2nd Aug. 1800	T. GILL.
Do.	do.	do.	2454	9th Dec. 1800	J. PROSSER.
Do.	do.	do.	2466	23rd Jan. 1801	R. VAZIE.
Do.	do.	do.	2566	16th Dec. 1801	C. GRIERSON.
Do.	do.	do.	2573	16th Jan. 1802	J. HALL.
Do.	do.	do.	2652	30th Oct. 1802	J. BRAMAH.
Do.	do.	do.	2692	23rd Mar. 1803	D. EGG.
Do.	do.	do.	2709	10th June 1803	J. R. PECKHAM.
Do.	do.	do.	2722	6th July 1803	J. MANTON.
Do.	do.	do.	2744	3rd Dec. 1803	J. S. SEARLES.
Do.	do.	do.	2825	28th Feb. 1805	G. DODD.
Do.	do.	do.	2865	3rd July 1805	A. WILSON.
Do.	do.	do.	2911	20th Feb. 1806	J. JONES.
Do.	do.	do.	2920	21st Mar. 1806	F. PLACE.
Do.	do.	do.	2966	15th Sept. 1806	J. MANTON.
Do.	do.	do.	2991	4th Dec. 1806	J. F. MATTHEY.

Subject-matter of Patent.	Number of Patent.	Date.	Name of Patentee.
Small arms, improvements in, and in the manufacture of—*continued*	3032	11th April 1807	A. J. Forsyth.
Do. do. do.	3122	28th March 1808	B. Cook.
Do. do. do.	3155	30th July 1808	G. Richards.
Do. do. do.	3233	4th May 1809	T. Noon.
Do. do. do.	3267	28th Sept. 1809	J. Jones.
Do. do. do.	3286	11th Dec. 1809	J. Manton.
Do. do. do.	3437	24th April 1811	J. Bradley.
Do. do. do.	3469	26th July 1811	H. James & J. Jones.
Do. do do.	3558	30th April 1812	J. Manton.
Do. do. do.	3588	28th July 1812	W. Smith.
Do. do. do.	3590	3rd Aug. 1812	H. Osborn.
Do. do. do.	3599	25th Sept. 1812	D. Egg.
Do. do. do.	3617	28th Nov. 1812	H. Osborn.
Do. do. do.	3722	19th July 1813	R. Pretyman.
Do. do. do.	3740	15th Oct. 1813	H. Osborn.
Do. do. do.	3755	23rd Nov. 1813	J. Bodmer.
Do. do. do.	3773	24th Dec. 1813	R. Sutton.
Do. do. do.	3784	9th March 1814	J. Thomson.
Do. do. do.	3794	23rd March 1814	E. Heaton.
Do. do. do.	3813	7th June 1814	G. Heywood.
Do. do. do.	3828	4th Aug. 1814	T. Sykes.
Do. do. do.	3833	4th Aug. 1814	J. S. Pauly.
Do. do. do.	3942	21st July 1815	J. Manton.
Do. do. do.	3985	29th Feb. 1816	J. Manton.
Do. do. do.	4026	14th May 1816	J. S. Pauly.
Do. do. do.	4031	25th May 1816	F. Richardson.
Do. do. do.	4054	12th Aug. 1816	S. Nock.
Do. do. do.	4105	1st March 1817	H. Osborn.
Do. do. do.	4107	11th March 1817	U. Sartoris.
Do. do. do.	4166	26th Sept. 1817	J. Manton.
Do. do. do.	4218	3rd Feb. 1818	M. C. Wyatt.
Do. do. do.	4285	3rd Aug. 1818	J. Manton.
Do. do. do.	4315	24th Nov. 1818	E. H. Collier.
Do. do. do.	4336	23rd Jan. 1819	U. Sartoris.
Do. do. do.	4427	15th Jan. 1820	F. Fox.
Do. do. do.	4491	20th July 1820	W. Dell.
Do. do. do.	4577	30th July 1821	J. Manton.
Do. do. do.	4590	14th Sept. 1821	W. Webster.
Do. do. do.	4611	10th Nov. 1821	W. W. Richards.
Do. do. do.	4648	12th Feb. 1822	S. Davis.
Do. do. do.	4727	26th Nov. 1822	J. Egg.
Do. do. do.	4823	29th July 1823	J. Jackson.
Do. do. do.	4861	13th Nov. 1823	J. Day.
Do. do. do.	4960	30th May 1824	J. Cook.
Do. do. do.	4990	27th July 1824	C. R. Baron de Berenger.
Do. do. do.	5026	4th Nov. 1824	J. Somerville.
Do. do. do.	5033	6th Nov. 1824	T. Cartmell.
Do. do. do.	5055	18th Dec. 1824	S. Davis.

Subject-matter of Patent.			Number of Patent.	Date.	Name of Patentee.
Small arms, improvements in, and in the manufacture of—*continued*			5099	19th Feb. 1825	J. Mould.
Do.	do.	do.	5106	26th Feb. 1825	J. Manton.
Do.	do.	do.	5155	23rd April 1825	A. L. Hunout.
Do.	do.	do.	5175	20th May 1825	J. Riviere.
Do.	do.	do.	5242	15th Aug. 1825	C. Downing.
Do.	do.	do.	5317	16th Jan. 1826	B. Newmarch.
Do.	do.	do.	5416	18th Oct. 1826	W. Mills.
Do.	do.	do.	5421	7th Nov. 1826	B. Newmarch.
Do.	do.	do.	5726	8th Dec. 1828	J. Dickson.
Do.	do.	do.	5838	2nd Sept. 1829	G. H. Manton.
Do.	do.	do.	5845	15th Sept. 1829	D. Lawrence and J. Crundwell.
Do.	do.	do.	5905	27th Feb. 1830	C. R. Baron de Berenger.
Do.	do.	do.	5978	7th Aug. 1830	S. Smith.
Do.	do.	do.	6046	6th Dec. 1830	J. G. Lacy, and S. Davis.
Do.	do.	do.	6053	17th Dec. 1830	B. Redfern.
Do.	do.	do.	6071	11th Feb. 1831	W. W. Richards.
Do.	do.	do.	6137	13th July 1831	A. Demondion.
Do.	do,	do.	6139	15th July 1831	J. de Burgh, Marquis of Clanricarde.
Do.	do.	do.	6166	24th Sept. 1831	W. Bingham and W. Dupe.
Do.	do.	do.	6196	15th Dec. 1831	A. A. Moser.
Do.	do.	do.	6326*	2nd Nov. 1832	L. Foucaud.
Do.	do.	do.	6394	7th March 1833	C. Jones.
Do.	do.	do.	6436	12th June 1833	C. Jones.
Do.	do.	do.	6554	8th Feb. 1834	W. S. Gillett.
Do.	do.	do.	6572	13th March 1834	J. A. Manton.
Do.	do.	do.	6611	22nd May 1834	C. L. S. Baron Heurteloup.
Do.	do.	do.	6611*	7th Jan. 1642	C. L. S. Baron Heurteloup. Disclaimer.
Do.	do,	do.	6675	6th Sept. 1834	H. Shrapnel.
Do.	do.	do.	6755	29th Jan. 1835	J. Dodds.
Do.	do.	do.	6825	28th April 1835	J. Somerville.
Do.	do.	do.	6826	30th April 1835	J. Dodds.
Do.	do.	do.	6829	9th May 1835	J. Egg.
Do.	do.	do.	6872	6th Aug. 1835	W. Mason.
Do.	do.	do.	6909	22nd Oct. 1835	S. Colt.
Do.	do.	do.	7041	22nd March 1836	W. W. Richards.
Do.	do.	do.	7286	19th Jan. 1837	M. Poole.
Do.	do.	do.	7582	8th March 1838	W. W. Richards.
Do.	do.	do.	7610	10th April 1838	J. R. Cooper.
Do.	do.	do.	7712	30th June 1838	G. Round and S. Whitford.
Do.	do.	do.	7965	11th Feb. 1839	G. H. Manton.
Do.	do.	do.	7980	23rd Feb. 1839	C. L. S. Baron Heurteloup.

Subject-matter of Patent.			Number of Patent.	Date.	Name of Patentee.
Small arms, improvements in, and in the manufacture of—*continued*			7980*	7th Jan. 1842	C. L. S. Baron HEURTELOUP. Disclaimer and Memorandum of Alteration.
Do.	do.	do.	8024	9th April 1839	G. STOCKER and J. BENTLEY.
Do.	do.	do.	8119	22nd June 1839	H. WILKINSON.
Do.	do.	do.	8347	21st Jan. 1840	J. R. COOPER.
Do.	do.	do.	8513	20th May 1840	W. BUSH.
Do.	do.	do.	8573	18th July 1840	M. POOLE.
Do.	do.	do.	8704	17th Nov. 1840	J. SHAW.
Do.	do.	do.	9084	9th Sept. 1841	C. L. S. Baron HEURTELOUP.
Do.	do.	do.	9119	14th Oct. 1841	M. POOLE.
Do.	do.	do.	9129	2nd Nov. 1841	W. GOLDEN and J. HANSON.
Do.	do.	do.	9177	14th Dec. 1841	W. W. RICHARDS.
Do.	do.	do.	9258	15th Feb. 1842	A. ROUSSEAU.
Do.	do.	do.	9801	24th June 1843	W. NEEDHAM.
Do.	do.	do.	9947	18th Nov. 1843	J. ROOSE.
Do.	do.	do.	10,109	14th March 1844	T. SEYMOUR and J. SEYMOUR.
Do.	do.	do.	10,280	30th July 1844	J. BENTLEY.
Do.	do.	do.	10,667	14th May 1845	C. J. SMITH.
Do.	do.	do.	10,718	10th June 1845	J. W. TYSON.
Do.	do.	do.	10,849	2nd Oct. 1845	J. HALE.
Do.	do.	do.	11,657	15th April 1847	J. MOLLETT.
Do.	do.	do.	11,864	9th Sept. 1847	C. W. O'LEARY.
Do.	do.	do.	11,880	30th Sept. 1847	J. DE BARROS.
Do.	do.	do.	11,994	10th Dec. 1847	S. TAYLOR.
Do.	do.	do.	12,613	15th May 1849	L. A. DE CHATAUVILLARD.
Do.	do.	do.	12,623	29th May 1849	R. E. HODGES.
Do.	do.	do.	12,648	7th June 1849	W. H. RITCHIE.
Do.	do.	do.	12,668	20th June 1849	S. COLT.
Do.	do.	do.	12,728	1st Aug. 1849	J. SHAW.
Do.	do.	do.	12,732	1st Aug. 1849	B. AINGWORTH.
Do.	do.	do.	12,781	20th Sept. 1849	J. R. COOPER.
Do.	do.	do.	12,920	11th Jan. 1850	M. U. SEARS.
Do.	do.	do.	13,161	3rd July 1850	C. W. LANCASTER.
Do.	do.	do.	13,215	6th Oct. 1850	A. MELVILLE and E. CALLOW.
Do.	do.	do.	13,299	24th Oct. 1850	A. ROSE.
Do.	do.	do.	13,454	16th Jan. 1851	C. W. LANCASTER.
Do.	do.	do.	13,527	24th Feb. 1851	R. ADAMS.
Do.	do.	do.	13,934	29th Jan. 1852	F. J. MANCEAUX.
Do,	do.	do.	13,935	29th Jan. 1852	J. M. R. von WINIWARTER.
Do.	do.	do.	13,994	3rd Mar. 1852	G. L. L. KUFAHL.
Do.	do.	do.	14,027	20th Mar. 1852	W. W. RICHARDS.
Do.	do.	do.	14,041	24th Mar. 1852	J. MACINTOSH.
Do.	do.	do.	14,052	31st Mar. 1852	M. POOLE.

Subject-matter of Patent.	Number of Patent.	Date.	Name of Patentee.
Small arms, improvements in, and in the manufacture of—*continued*	**14,058**	6th April 1852	J. W. DE LONGUEVILLE GIFFARD.
Do. do. do.	**14,066**	17th April 1852	H. G. DELVIGNE.
Do. do. do.	**14,085**	22nd April 1852	A. V. NEWTON.
Do. do. do.	**14,087**	24th April 1852	W. CHURCH, S. A. GODDARD, and E. MIDDLETON.
Do. do. do.	**14,227**	20th July, 1852	J. W. SCHLESINGER.
Do. do. do.	**14,265**	19th Aug. 1852	T. HUNT.
Do. do. do.	**14,271**	23rd Aug. 1852	H. N. S. SHRAPNEL.
Do. do. do.	**14,289**	10th Sept. 1852	S. TAYLOR.
Small arms on horseback, method of fixing and carrying	**8205**	23rd Aug. 1839	C. B. COLES.
Sponge, an elastic for sponging cannon and small arms	**1668**	26th Aug. 1788	H. COOK.
Stamping, manufacturing parts of gun and pistol locks by	**7712**	30th June 1838	G. ROUND and S. WHITFORD.
Steam, impelling rockets, shells, and other projectiles by the force of	**4952**	15th May 1824	J. PERKINS.
Swords, methods of combining with pistols and guns	**2744**	3rd Dec. 1803	J. S. SEARLES.
Swords, shield or protection for the human body against	**2763**	18th May 1804	J. P. BARTHELEMY and J. SHOUBRIDGE.
Swords, armour for protecting the human body against	**2885**	30th Oct. 1805	R. KENTISH.
Swords, machine for making scabbards for, of one piece of leather	**4923**	20th March 1824	J. H. PETITPIERRE.
Swords, method of rendering more effective -	**5905**	27th Feb. 1830	C. R. Baron de BERENGER.
Swords, improvements in the manufacture of -	**13,991**	27th Feb. 1852	C. REEVES.
Swords, improvements in scabbards for - -	**14,029**	22nd March 1852	J. D. BRADY.

T.

Subject-matter of Patent.	Number of Patent.	Date.	Name of Patentee.
Testing long bows, machine for - - -	**1639**	5th Feb. 1878	T. WARING.
Testing gun barrels and fire-arms, method of -	**3599**	25th Sept. 1812	D. EGG.
Triggers [*See* Small Arms *also*], improvements in, for double-barrelled fire-arms	**1837**	28th Nov. 1791	J. S. SEARLES.
Triggers, improvements in, for double and single barrelled fire-arms	**1893**	5th July 1792	J. MANTON.
Triggers, improvements in, for double-barrelled fire-arms	**1978**	4th Feb. 1794	T. SYKES.
Triggers, improvements in, for single and double barrelled fire-arms	**2178**	12th April 1797	J. MANTON.
Triggers, improvements in, for double-barrelled guns and pistols	**3155**	30th July 1808	G. RICHARDS.
Triggers, inprovements in, for single and double-barrelled fire-arms	**3828**	4th Aug. 1814	T. SYKES.
Triggers, arranging so that four or more may be placed close together	**5155**	23rd April 1825	A. L. HUNOUT.

Subject-matter of Patent.	Number of Patent.	Date.	Name of Patentee.
Triggers, improvements in, for single and double-barrelled fire-arms	6053	17th Dec. 1830	B. Redfern.
Triggers, improvements in - - -	6166	24th Sept. 1831	W. Bingham and W. Dupe.
Triggers, safety bolting guard for - - -	8024	9th April 1839	G. Stocker and J. Bentley.
Turning barrels for fire-arms, improved lathe for -	3469	26th July 1811	H. James and J. Jones.

W.

Subject-matter of Patent.	Number of Patent.	Date.	Name of Patentee.
Wadding for fire-arms, improvements in, and in the manufacture of	1893	5th July 1792	J. Manton.
Do. do. do.	6633	16th June 1834	R. Walker.
Do. do. do.	9069	6th Sept. 1841	J. Drew.
Do. do. do.	9712	25th April 1843	W. Brockedon.
Do. do. do.	13,377	30th Nov. 1850	J. A. Elmslie, and G. Simpson.
Do. do. do.	13,454	16th Jan. 1851	C. W. Lancaster.
Do. do. do.	14,027	20th March 1852	W. W. Richards.
Do. do. do.	14,271	23rd Aug. 1852	H. N. S. Shrapnel.
Walking stick to contain pistol, powder, and ball -	3837	17th Aug. 1814	H. W. Vander Kleft.
Walking-stick guns, improvements in - -	4861	13th Nov. 1823	J. Day.
Do. do. do.	4980	30th May 1824	J. Cook.
Do. do. do	5726	8th Dec. 1828	J. Dickson.
Welding bars together for making guns or ordnance	5553	11th Oct. 1827	J. Horton.
Welding the seams or joints of twisted wrought iron cannon	9947	18th Nov. 1843	J. Roose.
Welding cannon by means of the hydrostatic press	10,013	16th Jan. 1844	T. Aspinwall.
Welding gun barrels, improvements in - -	3469	26th July 1811	H. James, and J. Jones.
Do. do. do.	3617	28th Nov. 1812	H. Osborn.
Do. do. do.	4105	1st March 1817	H. Osborn.
Do. do. do.	9947	18th Nov. 1843	J. Roose.
Do. do. do.	10,696	3rd June 1845	C. Whitehouse.
Do. do. do.	13,299	24th Oct. 1850	A. Rose.

SPECIFICATIONS

OF

INVENTIONS

IN

FIRE-ARMS, PROJECTILES,

&c. &c. &c.

Filed under the Act 15 & 16 VICT. c. 83.

From 1st October 1852 to 31st December 1853.

TO WHICH ARE PREFIXED,

AN INDEX OF THE NAMES OF THE INVENTORS, AND AN INDEX ARRANGED ACCORDING TO THE SUBJECT-MATTER OF THE INVENTION.

LONDON:
PRINTED BY GEORGE EDWARD EYRE AND WILLIAM SPOTTISWOODE,
PRINTERS TO THE QUEEN'S MOST EXCELLENT MAJESTY.

1854.

ALPHABETICAL INDEX

OF

INVENTORS

IN

FIRE-ARMS, PROJECTILES,

&c. &c. &c.

Embracing the Series of Specifications filed from 1st *October* 1852 *to* 31st *December* 1853.

Name of Patentee.	Number of Patent.	Date.	Subject-matter of Patent.
A.			
ABRAHAM, JOHN - - -	**55**	8th Jan. 1853	Process and machinery for the manufacture of percussion caps.
ADAMS, ROBERT - - -	**1**	1st Oct. 1852	Manufacturing ball cartridges with metal powder chamber.
ADAMS, ROBERT - - -	**2712**	22nd Nov. 1853	Mechanical arrangements in revolving fire-arms to secure steadiness of aim by causing first action on trigger to overcome main spring, cock the piece, and revolve barrels; a second effort being limited to the discharge, by simply releasing the main spring.
B.			
BELLFORD, AUGUSTE EDOUARD [LORADOUX.	**329**	11th Oct. 1852	Constructing repeating fire-arms with sliding crotch and guard lever so arranged and combined as to rotate and wedge up breech, and cock gun, by one motion back and forth of trigger-guard.
BELLFORD, AUGUSTE EDOUARD [LORADOUX.	**1966**	24th Aug. 1853	Repeating primer and chamber for breech-loading fire-arms.
BELLFORD, AUGUSTE EDOUARD [LORADOUX.	**2778**	29th Nov. 1853	Interposing moveable breech between barrel and percussion apparatus of needle guns, such breech being hinged to a moveable trigger-guard acting as lever to slide the breech and expose opening therein for introduction of charge; the implements of percussion being concealed and acting on needle passing through breech, and the hammer used only for cocking the gun; mechanism for extraction of cartridge blank from chamber; also cartridges applicable to the above arrangements.

Name of Patentee.	Number of Patent.	Date.	Subject-matter of Patent.
Bentley, Joseph - - -	960	4th Dec. 1852	Mechanical arrangements in revolving fire-arms for holding hammer off nipple during act of loading, and for lifting hammer, revolving the chambers during its rising, and securing them in position in firing; also an elevating sight with scale or index, and application of gutta percha and papier mâché to manufacture of side plates for stocks of fire-arms.
Brand, Richard Farmer -	1870	11th Aug. 1853	Mechanism for loading fire-arms and ordnance at the breech and behind the point of ignition.
Brand, Richard Farmer -	2697	19th Nov. 1853	Constructing barrels of fire-arms with an enlargement in breech part thereof, to prevent expansion at point of ignition.
Brooman, Richard Archibald.	583	30th Oct. 1852	Construction and arrangement of lock and revolving chamber of fire-arms; self-loading flask to be attached to the former.
Brooman, Richard Archibald.	1332	30th May 1853	Repeating fire-arms: construction of magazine; mode of transferring loads to barrel; primer or "cap magazine."
Burr, Arthur Richard -	219	5th Oct. 1852	Machinery for coiling rods of iron or other metal to be used in manufacture of gun and pistol barrels, &c., and for forging or welding the same when so coiled; forming lap joint in twisted barrels, and welding twisted coil upon plain tube to form barrel.

C.

Name of Patentee.	Number of Patent.	Date.	Subject-matter of Patent.
Church, William - -	1277	25th May 1853	Sight for cannon or other ordnance.
Clerville, François Aristide.	3044	31st Dec. 1853	Constructing guns and pistols with a moveable or fly breech; and mechanical arrangements for adapting and operating the same.
Coles, Richard George -	2987	24th Dec. 1853	Substituting vulcanized india-rubber or other elastic material for steel in manufacture of main and other springs used in locks of fire-arms.
Colt, Samuel - - -	535	3rd March 1853	Mechanism for rotating the breech cylinder of repeating fire-arms.
Colt, Samuel - - -	538	3rd March 1853	Breech cylinders of repeating fire-arms; use of creeping fulcrum for transmitting motion to plunger employed in loading such fire-arms.
Colt, Samuel - - -	1703	16th July 1853	Machinery for boring the barrels, chambers, nipple holes, &c., of repeating fire-arms.
Cooper, Joseph Rock - -	131	19th Jan. 1853	Constructing fire-arms with breech moving on horizontal axis, to obtain angular position for loading.
Crickmay, Charles - -	1499	20th June 1853	Constructing breeches of fire-arms to facilitate loading; arrangement of cock, trigger, and connected parts to obtain central discharge; mode of packing cartridges in stocks of fire-arms.
Crofts, John - - - -	577	7th March 1853	Stop-catch for ensuring action of revolving or repeating fire-arms.

Name of Patentee.	Number of Patent.	Date.	Subject-matter of Patent.
D.			
DAVIS, MARCUS - - -	59	1st Oct. 1852	Application of hollow iron or other metal tubing to manufacture of wheels, axles, and other parts of carriages; attaching elastic and other substances to the same to prevent noise, vibration, and wear; method of lubricating axle-arms, &c.
F.			
FANSHAWE, HENRY RICHARDSON.	653	16th Mar. 1853	Employment of electric or electro-galvanic fluid as means of igniting charges of fire-arms; and arrangements subservient thereto.
FEATHER, ROBERT BARNARD -	884	26th Nov. 1852	Construction of iron vessels, or vessels partly of iron and partly of wood, and their decks; padding or lining the sides of ships or boats, to render them impervious to shot.
FLETCHER, ROBERT - -	2176	20th Sept. 1853	Sliding breech for breech-loading fire-arms; and cartridges and primer applicable thereto.
G.			
GEDGE, JOHN - - -	555	5th March 1853	Plate for closing in and protecting lock-works of breech-loading fire-arms; also grooved slide or plunger for such fire-arms.
GILBY, JOHN - - -	620	12th Mar. 1853	Method of securing and operating moveable breeches of fire-arms; self-acting priming apparatus, &c.
GODDARD, SAMUEL ASPINWALL	988	7th Dec. 1852	Repeating pistol.
GRAHAM, EDMUND HUGH -	2590	8th Nov. 1853	Arrangement of series of ball and powder chambers in concentric circles on side of gun barrel; and so combining percussion cock, rotary charge receiver and magazine with trigger guard as to obtain simultaneous motion, and bring hammer to full cock; also substitute in certain cases for rotary charge receiver.
H.			
HADDAN, JOHN COOPE -	1000	26th April 1853	Cartridges and wadding for fire-arms.
HADDAN, JOHN COOPE -	2829	6th Dec. 1853	Various constructions of cartridge cases to secure ready fracture and escape of powder from rear-end on compression by ramrod, &c.; apparatus for manufacture of wadding from fluid pulp or water-leaf by moulding.
HALE, WILLIAM - -	940	19th April 1853	Musket for firing rockets from the shoulder with safety to the person; pistol for firing rockets from the hand; also fire-arm for the discharge of rockets, giving direction and rotation to rockets in their flight by means of spiral wires or rods.
HALL, JOHN - - -	577	7th Mar. 1853	Stop-catch for ensuring action of revolving or repeating fire-arms.

Name of Patentee.	Number of Patent.	Date.	Subject-matter of Patent.
Harris, William - -	69	1st Oct. 1852	Constructing repeating pistols or rifles with hinge-joint frame to facilitate removal of barrel from revolving chamber for purpose of loading.
Harris, Philip - -	964	21st April 1853	Applying principle of excentric or cam to construction of triggers, to give increased certainty and command in discharging.
Harvey, William James -	1298	26th May 1853	Constructing fire-arms with revolving series of short barrels put in motion by the trigger, and one discharge barrel and guard made removeable for purpose of loading the former.
Haslewood, Edward -	86	12th Jan. 1853	Constructing breech-loading fire-arms with charging chambers of varying inner diameter, having one barrel common to all; also projectiles with projecting ring adapted to form of such chambers, which ring, on discharge, is severed and left in the breech.
Horrocks, James Dunlop -	1404	8th June 1853	Manufacturing percussion caps with lining of gutta percha, &c.
Horrocks, John, junior -	1404	8th June 1853	Manufacturing percussion caps with lining of gutta percha, &c.
Hunt, Thomas - -	125	1st Oct. 1852	Constructing lock and barrels of revolving fire-arms to produce succession of discharges by action of same needle.
Hunt, Thomas - -	628	12th Mar. 1853	Mounting sight of fire-arms on universal joint and with lateral motion by means of a screw for adjustment; engraved scale for the same on moveable and adjustable surface.

J.

Name of Patentee.	Number of Patent.	Date.	Subject-matter of Patent.
Jones, Charles John -	2351	12th Oct. 1853	Constructing fire-arms with series of barrels one above the other in same plane with the lock, the hammer being so mounted as to strike the caps in succession by repeatedly pulling the trigger.
Jones, Richard - -	2351	12th Oct. 1853	Constructing fire-arms with series of barrels one above the other in same plane with the lock, the hammer being so mounted as to strike the caps in succession by repeatedly pulling the trigger.

K.

Name of Patentee.	Number of Patent.	Date.	Subject-matter of Patent.
Kerr, John James - -	1376	4th June 1853	Manufacturing powder, shot, and ball cartridges with cylindrical case of metal tin coated with fibre or flock.
Krupp, Alfred - -	1094	17th Dec. 1852	Manufacturing cannon of solid cast steel, shaped by hammer, and bored and finished in lathe; also making barrel of cast steel, with breech and casing of cast-iron or gun-metal.

L.

Name of Patentee.	Number of Patent.	Date.	Subject-matter of Patent.
LANCASTER, CHARLES WILLIAM.	47	7th Jan. 1853	Appendage to bullet moulds for cutting off refuse metal, and thus perfecting the bullet on its leaving the mould.
LANCASTER, CHARLES WILLIAM.	1988	27th Aug. 1853	Machinery for forming a tube into a gun or pistol barrel the interior of which shall be spirally elliptical.
LARBAUD, CHARLES - -	1102	5th May 1853	Mechanism, termed "a system of trigger," applicable to play-arms, for discharging projectiles by force of compressed air.
LAWRENCE, JAMES - - -	1000	8th Dec. 1852	Projectiles cast of two or more metals; the one by its softness yielding to the bore of the fire-arm, and the other by its hardness acting with superior effect on percussion.
LEETCH, JAMES - - -	2197	23rd Sept. 1853	Constructing breech-loading fire-arms with moveable breech or charge chamber in one entire piece separate from and hinged to the barrel, such breech being held in position by a locking bolt, to be withdrawn after discharge by placing hammer on half or full cock.
LEMAIRE-DAIMÉ, JACQUES LOUIS.	1344	1st June 1853	Arrangement of piston and cylinder to form play-arms, making discharge by force of compressed air.
LEWIS, WILLIAM JAMES -	171	2nd Oct. 1852	Sight for rifles, &c. hinged to barrel, and with distance lines marked thereon, each line having a hole in its centre whose vertical height varies according to appearance of given object at distances so denoted.
LOTT, FREDERICK - - -	1390	6th June 1853	Constructing cartridges weaker at one end than the other, to allow of powder escaping and entering breech in act of ramming.

M.

Name of Patentee.	Number of Patent.	Date.	Subject-matter of Patent.
MACCARTHY, JOHN JAMES ALEXANDER.	591	8th March 1853	Constructing muskets or rifles with charge chamber open at top, and secured by sliding cylindrical shield, drawn by spring, and having touchhole perforated therein, the end of cock being pointed and of sufficient length to penetrate touchhole, and thereby dispense with percussion cap; also permanently affixing bayonet to such fire-arm. Constructing cannon with charge chamber and barrel in separate parts, united by hinge-tooth, lock, or screw, &c.; also ammunition pouch, composed of sheet steel, and with compartments to receive the charges.

Name of Patentee.	Number of Patent.	Date.	Subject-matter of Patent.
MACINTOSH, JOHN - - -	633	3rd Nov. 1852	Impregnating the end covering of projectiles or cartridges with meal powder to cause ignition of charge without previous fracture or puncture of said covering; also constructing hollow and cylindrical projectiles with screw-formed interior to give rotary motion thereto.
MACINTOSH, JOHN - -	3008	28th Dec. 1853	Constructing fire-arms and instruments used for projecting rockets with rotating barrels for the purpose of communicating rotary motion to projectiles at moment of discharge.
MARSTON, WILLIAM WALKER-	1320	28th May 1853	Constructing breech-loading fire-arms with sliding breech-pin grooved into sides of a chamber formed in and opening vertically through upper part of stock, such breech-pin having a perforated cone on which to explode percussion cap by usual means, and being moved from or retained against rear end of barrel by lever, which in latter case forms blocking-piece to resist recoil of explosion; also cartridge adapted to such fire-arm, with wad or swab at rear end, which, being left behind, serves to clean out barrel when expelled by succeeding discharge.
MOLE, ROBERT - -	436	19th Oct. 1852	Manufacture of matchets and swords.
MOORE, WILLIAM - -	69	1st Oct. 1852	Constructing repeating pistols or rifles with hinge-joint frame to facilitate removal of barrel from revolving chambers for purpose of loading.

N.

Name of Patentee.	Number of Patent.	Date.	Subject-matter of Patent.
NEEDHAM, HENRY - -	2184	20th Sept. 1853	Forming triggers of revolving fire-arms with two projections to act upon notches in tumbler and lifter, to secure steadiness of aim and enable barrels to be revolved by pulling up cock; also placing a tube on under side of barrel, &c., to prevent breech from sticking.
NEEDHAM, JOSEPH - -	184	2nd Oct. 1852	Constructing breech-loading fire-arms with trough formed by longitudinal section of barrel for reception of tubular cases containing lock mechanism, such mechanism comprising a revolving spiral cam for actuating the needle, a cocking lever, a shifting cylindrical cap to form joint with breech-end of barrel, a safety bolt, &c.; also various constructions of cartridges applicable to such arrangements.
NEWTON, WILLIAM EDWARD -	540	3rd March 1853	Self-acting primer for fire-arms, consisting of a rotating disc or wheel provided with recesses for receiving detonating composition, which recesses are brought in turn over nipple.
NEWTON, WILLIAM EDWARD -	2046	5th Sept. 1853	Method of bringing charge chamber of breech-loading fire-arms into vertical position for loading, returning breech to barrel, and locking the same, by means of cams, catches, and springs operated by a single lever.

Name of Patentee.	Number of Patent.	Date.	Subject-matter of Patent.
NEWTON, WILLIAM EDWARD -	2313	8th Oct. 1853	Constructing breech-loading fire-arms with moveable or vibrating breech, operated by trigger-plate, provided at bottom with hollow punch communicating with touch-hole for perforating cartridge, and so arranged that by act of turning to receive charge the nipple shall become coincident with a priming tube from which a cap may be forced on by tension of spring; also forming cartridges with an envelope of cloth, slit at ball end, to prevent leading of barrel.
NEWTON, ALFRED VINCENT -	2839	6th Dec. 1853	Constructing barrels of cannon and small-arms with hinged breech-plate opening for admission of charge, and held in place during discharge, in cannon, by hinged strap or bridle and packed cylindrical projection entering bore of barrel, and in small arms by moveable wedge operated by a lever; also combination of carrier and priming tube with lipped hammer, which by act of cocking receives disc primers in succession, and explodes the same on nipple.
NORTON, JOHN - - -	936	2nd Dec. 1852	Construction of hollow expanding shot or projectiles, each containing its own cartridge or explosive charge.
NORTON, JOHN - -	2264	4th Oct. 1853	Firing cannon and explosive charges by tensional separation of two cords whose ends are cemented together by explosive composition.
NORTON, JOHN - - -	2409	19th Oct. 1853	Casting cannon and other fire-arms with rifle grooves therein complete.

O.

Name of Patentee.	Number of Patent.	Date.	Subject-matter of Patent.
OMMANNEY, HENRY MORTLOCK.	131	1st Oct. 1852	Manufacturing guns, cannon, and other ordnance of iron by casting and annealing.

P.

Name of Patentee.	Number of Patent.	Date.	Subject-matter of Patent.
PENNELL, THOMAS - -	1038	29th April 1853	Enclosing lock mechanism of repeating fire-arms in case; arrangements for connecting stationary barrel with frame, and for loading; connecting case and barrel of rifles by means of a folding sight.
PHILLIPS, MORETON HASSALL	1211	16th May 1853	Gun for projecting missiles by the retractile force of a tube of vulcanized india-rubber, or other elastic material, secured at one end to a button in the muzzle of a second barrel, and drawn back by a rod passing through a catch attached to other end of tube; with trigger arrangements applicable thereto.
PIGOT, RICHARD GREVILLE -	879	12th April 1853	Composing caltraps of six spikes or blades, four of which placed and moveable on cylindrical axis, the other two connected to cylinder at ends by pin joints; the whole folding together and portable.

Name of Patentee.	Number of Patent.	Date.	Subject-matter of Patent.
POOLE, MOSES - - -	6	1st Oct. 1852	Application of india-rubber combined with sulphur, &c., to manufacture of stocks and furniture and to coating the metal parts of fire-arms.
POOLE, MOSES - - -	2690	19th Nov. 1853	Constructing breech-loading fire-arms with sliding breech-pin grooved into sides of a chamber in the metal stock, and opening vertically through its upper part, such breech-pin having a perforated cone on which to explode percussion cap by usual means, and being moved from or retained against rear end of barrel by lever, which in latter case forms blocking-piece to resist recoil of explosion; also cartridge adapted to such fire-arm, with fibrous packing at rear-end, which, being left behind, forms wad for next cartridge, and forces out dirt or residue of previous explosion.
PRYSE, CHARLES - - -	635	4th Nov. 1852	Applying bolt to cock or hammer of saloon fire-arms to retain discharged cap in its place.

R.

Name of Patentee.	Number of Patent.	Date.	Subject-matter of Patent.
REDMAN, RICHARD - -	635	4th Nov. 1852	Applying bolt to cock or hammer of saloon fire-arms to retain discharged cap in its place.
REEVES, CHARLES EVANS -	1447	15th June 1853	Constructing fire arms to be loaded at breech by means of two tubes, an external one revolving by means of a handle, and an internal one immoveable and fixed to barrel and plug, each tube having an opening to receive charge, which is fired by needle attached to tumbler, &c.; also making cone or other shaped ball with hollow for reception of explosive compound, &c.
REEVES, CHARLES, junior -	960	21st April 1853	Manufacturing swords with tang of same width and as continuation of blade, and forming handle of said sword by attaching a piece of wood to each side of tang.
REEVES, CHARLES, junior -	2550	3rd Nov. 1853	Manufacture of swords and bayonets by cutting out in one piece the whole of the metal of which a bayonet, or socket and neck of a bayonet, or a sword, with its hilt, tang, and fastening, is formed.
RIERA, MARIANO - - -	1123	6th May 1853	Constructing guns and pistols with sliding barrel and interior of breech in form of a cone, on base of which barrel rests, the top being furnished with hole the diameter of and immediately beneath capsule; appropriate arrangements of hammer, trigger, &c.; partial application of the above to cannon and ordinary fire-arms; field-piece called a "batidor," with moveable breech and apparatus for operating the same; acorn-shaped cartridges or projectiles applicable to the above inventions.

Name of Patentee.	Number of Patent.	Date.	Subject-matter of Patent.
		S.	
SHARPS, CHRISTIAN - -	712	11th Nov. 1852	Constructing breech-loading guns with adjustable tube or charge chamber tightly fitting into recess in barrel, and forming joint with a sliding or moveable breech; also facing joint-surfaces with platina, to prevent oxidisation.
SHEATH, ISAAC BRENTNALL -	996	26th April 1853	Constructing revolving chamber of fire-arms in two parts, one solid with barrel, the other with stock, and united by screws, the axle rod of such chamber being fixed in that part solid with stock; also revolving or stopping such chamber by means of a spring lever fixed to trigger.
SITWELL, WILLIAM HURT -	1033	28th April 1853	Giving rotary motion to projectiles by means of plain or screw-formed air-passage through centre thereof, or by placing iron wad or button in rear of the same.
SMITH, JOHN - - - - -	2176	20th Sept. 1853	Sliding breech for breech-loading fire-arms, and cartridges and primer applicable thereto.
STIRLING, JOHN DAVIES MORRIES.	66	10th Jan. 1853	Employing zinc coated with tin in manufacture of percussion caps.
SWINBURN, JOHN FIELD -	1881	12th Aug. 1853	Constructing a sight for rifles by the combination of a solid block or foot with a leaf or upright piece formed with a spring, which is made to press vertically upon upper surface of said block and retain leaf in position, while permitting of backward and forward motion.
		T.	
TRANTER, WILLIAM - -	212	28th Jan. 1853	Adapting and applying to fire-arms a double trigger or two triggers, one portion or one of the same being used for raising the cock, and the other for discharge, combining therewith a safety stop spring; also combining such trigger or triggers with chamber of breech-loading guns and with use of needle: elongating tube or socket of revolving pistols to secure position of chamber. Manufacture of gun barrels out of spirals of iron, &c., crossed round a mandril.
TRANTER, WILLIAM - -	2921	16th Dec. 1853	Adaptation to breech-loading fire-arms of a plug or stopper inserted in frame, to hold nipple or tube, and to be operated by spring and set screw; or, in place thereof, a revolving chamber. inserted and operated as above, in combination with a lock in which spring, trigger spring, and tumbler act in middle of frame instead of on a separate plate. Constructing bullets or waddings with deposit of grease to be discharged against inner surface of barrel on explosion; also formation of oval or conical bullets with longitudinal perforation of varying diameter.

Name of Patentee.	Number of Patent.	Date.	Subject-matter of Patent.
TURNER, THOMAS - -	1881	12th Aug. 1853	Constructing a sight for rifles by the combination of a solid block or foot with a leaf or upright piece formed with a spring, which is made to press vertically upon upper surface of said block and retain leaf in position, while permitting of backward and forward motion.

W.

Name of Patentee.	Number of Patent.	Date.	Subject-matter of Patent.
WEBLEY, PHILIP - -	305	4th Feb. 1853	Arrangements for turning barrel of repeating fire-arms to facilitate change of charge chambers, and for strengthening parts of such fire-arms to resist concussion; also fixing sear by joint to hammer instead of trigger, and other changes in lock mechanicism.
WEBLEY, JAMES - -	743	29th Mar. 1853	Mechanism for engaging and disengaging trigger from cock or hammer; applying screw to prevent total withdrawal of bolt which secures the barrel in its place; connecting barrel with body of pistol by jointed pin sliding in cylindrical hole in frame; making middle part of axle of revolving chambers of square or prismatic form to receive accumulations of dirt and oil from ends of axle.
WEBLEY, PHILIP -	2127	14th Sept. 1853	Forming driver by which rotation of charge chamber of repeating fire-arms is effected in one piece with stop, to govern motion of such chamber; also mechanism for forcing charge cylinder into close contact with barrel during discharge by action of hammer, and releasing the same by spring pressure on withdrawal of hammer from nipple.
WILLIAMS, THOMAS - -	2924	16th Dec. 1853	Constructing revolving pistols with lock enclosed by moveable side piece, and with cock or hammer having ratchet lever or pall, which revolves chambers by acting directly on ratchet wheel attached thereto; also mechanism for locking revolving chambers both when pistol is at full cock and at instant of discharge.
WINIWARTER, GEORGE -	306	4th Feb. 1853	Explosive compounds for fire-arms, as substitutes for gunpowder, and application of same to cylindro-conical bullets to be used as shells, and to manufacture of gun matches or primers.
WINIWARTER, GEORGE -	371	12th Feb. 1853	Application to breech-loading needle fire-arms of rings or washers of vulcanized India-rubber, combined with mechanical arrangements for keeping barrel firmly in position and preventing escape of gases on discharge; also method of applying grease to needle to prevent corrosion.
WINIWARTER, GEORGE -	452	22nd Feb. 1853	Construction of stocks of fire-arms of metal alone or in combination with gutta percha, straw pulp, &c.
WOLLOWICZ, ANTOINE -	319	5th Feb. 1853	Cap or primer holder formed by strap of India-rubber or other flexible material, and provided with apertures for reception of caps so lodged as to be readily transferred therefrom on to nipple of fire-arm.

SUBJECT-MATTER INDEX

OF

INVENTIONS

IN

FIRE-ARMS, PROJECTILES,

&c. &c. &c.

***Embracing the Series of Specifications filed from* 1*st October* 1852 *to* 31*st December* 1853.**

Subject-matter of Patent.	Number of Patent.	Date.	Name of Patentee.
A.			
Accidental discharge of fire-arms, improvements in preventing the	184	2nd Oct. 1852	J. Needham.
Do. do. do.	583	30th Oct. 1852	R. A. Brooman.
Do. do. do.	960	4th Dec. 1852	J. Bentley.
Do. do. do.	212	28th Jan. 1853	W. Tranter.
Do. do. do.	535	3rd March 1853	S. Colt.
Do. do. do.	964	21st April 1853	P. Harris.
Do. do. do.	1298	26th May 1853	W. J. Harvey.
Do. do. do.	1447	15th June 1853	C. E. Reeves.
Do. do. do.	2197	23rd Sept. 1853	J. Leetch.
Do. do. do.	2313	8th Oct. 1853	W. E. Newton.
Do. do. do.	2590	8th Nov. 1853	E. H. Graham.
Air-guns, improvements in toy - - - - -	1102	5th May 1853	C. Larbaud.
Do. do. - - - - - -	1344	1st June 1853	J. L. Lemaire Daimé.
B.			
Balls or bullets, improvements in, for the purpose of attaching cartridges thereto	1	1st Oct. 1852	R. Adams.
Do. do. do.	184	2nd Oct. 1852	J. Needham.
Balls or bullets for containing a charge of powder, and other projectiles, improvements in	633	3rd Nov. 1852	J. Macintosh.

Subject-matter of Patent.	Number of Patent.	Date.	Name of Patentee.
Balls or bullets, improvements in rendering ships and boats more impervious to	884	26th Nov. 1852	R. B. FEATHER.
Balls or bullets, making hollow expanding, containing charges of powder	—	2nd Dec. 1852	J. NORTON.
Balls or bullets, or projectiles of two or more metals	1000	8th Dec. 1852	J. LAWRENCE.
Balls or bullets, an improvement in moulds for making	47	7th Jan. 1853	C. W. LANCASTER.
Balls or bullets, an improvement in - - -	86	12th Jan. 1853	E. HASLEWOOD.
Balls or bullets, an improvement in, to be combined with cartridges	212	28th Jan. 1853	W. TRANTER.
Balls or bullets, method of using as shells - -	306	4th Feb. 1853	G. WINIWARTER.
Balls or bullets, method of combining with cartridges	555	5th March 1853	J. GEDGE.
Balls or bullets, method of combining with cartridges	591	8th March 1853	J. J. A. MACCARTHY.
Balls or bullets for cannon and small arms, improvements in	1033	28th April 1853	W. H. SITWELL.
Balls or bullets for cannon and small arms, explosive and other projectiles, or	1123	6th May 1853	M. RIERA.
Balls or bullets, gun for impelling by the action of an elastic tube	1211	16th May 1853	M. H. PHILLIPS.
Balls or bullets, combining explosive compounds with hollow	1447	15th June 1853	C. E. REEVES.
Balls or bullets, improved, to be combined with cartridges for rifled barrels	2829	6th Dec. 1853	J. C. HADDAN.
Balls or bullets, improvements in - - -	2921	16th Dec. 1853	W. TRANTER.
Balls or bullets, giving rotary motion to barrels or tubes used for discharging	3008	28th Dec. 1853	J. MACINTOSH.
Bayonets, attaching permanently to muskets, so as to dispense with sheaths	591	8th March 1853	J. J. A. MACCARTHY.
Bayonets, improvements in the manufacture of sword-bayonets and	2550	3rd Nov. 1853	C. REEVES.
Bayonets, improvements in securing to gun barrels	2921	16th Dec. 1853	W. TRANTER.
Boats, improvements in rendering more impervious to shot	884	26th Nov. 1852	R. B. FEATHER.
Bores of gun barrels, method of, and machinery for, giving an elliptical form to	1988	27th Aug. 1853	C. W. LANCASTER.
Boring barrels, chambers, &c., of repeating firearms, machinery for	1703	16th July 1853	S. COLT.
Breeches of small arms, improvements in the -	2351	12th Oct. 1853	R. JONES and C. J. JONES.
Breech-loading cannon, improvements in - -	371	12th Feb. 1853	G. WINIWARTER.
Do. do. do.	591	8th March 1853	J. J. A. MACCARTHY.
Do. do. do.	1123	6th May 1823	M. RIERA.
Do. do. do.	1870	11th Aug. 1853	R. F. BRAND.
Do. do. do.	2046	5th Sept. 1853	W. E. NEWTON.
Do. do. do.	2697	19th Nov. 1853	R. F. BRAND.
Do. do. do.	2839	6th Dec. 1853	A. V. NEWTON.
Breech-loading small arms, improvements in, and in the manufacture of	69	1st Oct. 1852	W. MOORE and W. HARRIS.
Do. do. do.	184	2nd Oct. 1852	J. NEEDHAM.
Do. do. do.	329	11th Oct. 1852	A. E. L. BELLFORD.
Do. do. do.	583	30th Oct. 1852	R. A. BROOMAN.
Do. do. do.	635	4th Nov. 1852	C. PRYSE and R. REDMAN.

Subject-matter of Patent.			Number of Patent.	Date.	Name of Patentee.
Breech-loading small arms, improvements in, and in the manufacture of			**712**	11th Nov. 1852	C. Sharps.
Do.	do.	do.	**960**	4th Dec. 1852	J. Bentley.
Do.	do.	do.	**988**	7th Dec. 1852	S. A. Goddard.
Do.	do.	do.	**86**	12th Jan. 1853	E. Haslewood.
Do.	do.	do.	**131**	19th Jan. 1853	J. R. Cooper.
Do.	do.	do.	**212**	28th Jan. 1853	W. Tranter.
Do.	do.	do.	**305**	4th Feb. 1853	P. Webley.
Do.	do.	do.	**371**	12th Feb. 1853	G. Winiwarter.
Do.	do.	do.	**535**	3rd March 1853	S. Colt.
Do.	do.	do.	**538**	3rd March 1853	S. Colt.
Do.	do.	do.	**540**	3rd March 1853	W. E. Newton.
Do.	do.	do.	**555**	5th March 1853	J. Gedge.
Do.	do.	do.	**577**	7th March 1853	J. Hall and J. Crofts.
Do.	do.	do.	**591**	8th March 1853	J. J. A. Maccarthy.
Do.	do.	do.	**620**	12th March 1853	J. Gilby.
Do.	do.	do.	**653**	16th March 1853	H. R. Fanshawe.
Do.	do.	do.	**743**	29th March 1853	J. Webley.
Do.	do.	do.	**996**	26th April 1853	J. B. Sheath.
Do.	do.	do.	**1038**	29th April 1853	T. Pennell.
Do.	do.	do.	**1123**	6th May 1853	M. Riera.
Do.	do.	do.	**1298**	26th May 1853	W. J. Harvey.
Do.	do.	do.	**1320**	28th May 1853	W. W. Marston.
Do.	do.	do.	**1332**	30th May 1853	R. A. Brooman.
Do.	do.	do.	**1447**	15th June 1853	C. E. Reeves.
Do.	do.	do.	**1499**	20th June 1853	C. Crickmay.
Do.	do.	do.	**1703**	16th July 1853	S. Colt.
Do.	do.	do.	**1870**	11th Aug. 1853	R. F. Brand.
Do.	do.	do.	**1966**	24th Aug. 1853	A. E. L. Bellford.
Do.	do.	do.	**2127**	14th Sept. 1853	P. Webley.
Do.	do.	do.	**2176**	20th Sept. 1853	R. Fletcher and J. Smith.
Do.	do.	do.	**2184**	20th Sept. 1853	H. Needham.
Do.	do.	do.	**2197**	23rd Sept. 1853	J. Leetch.
Do.	do.	do.	**2313**	8th Oct. 1853	W. E. Newton.
Do.	do.	do.	**2590**	8th Nov. 1853	E. H. Graham.
Do.	do.	do.	**2690**	19th Nov. 1853	M. Poole.
Do.	do.	do.	**2697**	19th Nov. 1853	R. F. Brand.
Do.	do.	do.	**2712**	22nd Nov. 1853	R. Adams.
Do.	do.	do.	**2778**	29th Nov. 1853	A. E. L. Bellford.
Do.	do.	do.	**2839**	6th Dec. 1853	A. V. Newton.
Do.	do.	do.	**2921**	16th Dec. 1853	W. Tranter.
Do.	do.	do.	**2924**	16th Dec. 1853	T. Williams.
Do.	do.	do.	**3044**	31st Dec. 1853	F. A. Clerville.

C.

Subject-matter of Patent.	Number of Patent.	Date.	Name of Patentee.
Caltraps, improvements in - - - -	879	12th April 1853	R. G. Pigot.
Cannon, method of annealing cast-iron - -	131	1st Oct. 1852	H. M. Ommanney.
Cannon, manufacturing of cast-steel, or partly of cast-steel with a casing of cast-iron, gun-metal, &c.	1094	17th Dec. 1852	A. Krupp.
Cannon, improvements in breech-loading - -	371	12th Feb. 1853	G. Winiwarter.
Cannon, constructing breech-loading, preventing recoil of, and mode of firing	591	8th March 1853	J. J. A. Maccarthy.
Cannon, discharging by electricity - -	653	16th March 1853	H. R. Fanshawe.
Cannon, improvements in play-arms, such as -	1102	5th May 1853	C. Larbaud.
Cannon, improvements in breech-loading and other	1123	6th May 1853	M. Riera.
Cannon, improved sight for - - -	1277	25th May 1853	W. Church.
Cannon, improvements in play-arms, such as -	1344	1st June 1853	J. L. Lemaire Daimé.
Cannon, improvements in breech-loading - -	1870	11th Aug. 1853	R. F. Brand.
Do. do. do.	2046	5th Sept. 1853	W. E. Newton.
Cannon, improvements in firing - - -	2264	4th Oct. 1853	J. Norton.
Cannon, manufacturing rifled, by moulding and casting	2409	19th Oct. 1853	J. Norton.
Cannon, improvements in breech-loading, ordnance, or	2697	19th Nov. 1853	R. F. Brand.
Do. do. do.	2839	6th Dec. 1853	A. V. Newton.
Cannon, giving rotary motion to - - -	3008	28th Dec. 1853	J. Macintosh.
Cartridges, improvements in ball - - -	1	1st Oct. 1852	R. Adams.
Do. do. do.	184	2nd Oct. 1852	J. Needham.
Do. do. do.	633	3rd Nov. 1852	J. Macintosh.
Cartridges, balls, or projectiles, containing charges of powder to be used as	936	2nd Dec. 1852	J. Norton.
Cartridges, combining percussion caps and balls with	212	28th Jan. 1853	W. Tranter.
Cartridges, improvements in - - -	555	5th March 1853	J. Gedge.
Cartridges, combining percussion caps and balls with, and improvements in pouches for	591	8th March 1853	J. J. A. Maccarthy.
Cartridges, using waterproof, and firing by electricity	653	16th Mar. 1853	H. R. Fanshawe.
Cartridges, improvements in - - -	1000	26th April 1853	J. C. Haddan.
Cartridges, new forms of, and appending charges of powder and percussion priming to balls	1123	6th May 1853	M. Riera.
Cartridges, improvements in - - -	1320	28th May 1853	W. W. Marston.
Do. do. do.	1376	4th June 1853	J. J. Kerr.
Do. do. do.	1390	6th June 1853	F. Lott.
Cartridges, improvements in, and in charging balls with explosive compounds	1447	15th June 1853	C. E. Reeves.
Cartridges, forming a chamber in gun-stocks to receive	1499	20th June 1853	C. Crickmay.
Cartridges, combining percussion primers with -	2176	20th Sept. 1853	R. Fletcher and J. Smith.
Cartridges, improvements in - - -	2313	8th Oct. 1853	W. E. Newton.
Do. do. do.	2690	19th Nov. 1853	M. Poole.

Subject-matter of Patent.	Number of Patent.	Date.	Name of Patentee.
Cartridges, improvements in self-priming - -	2778	29th Nov. 1853	A. E. L. Bellford.
Cartridges, improvements in - - - - -	2829	6th Dec. 1853	J. C. Haddan.
Do. do. do.	2921	16th Dec. 1853	W. Tranter.
Cartridges, combining percussion priming and balls with - - - - - -	3044	31st Dec. 1853	F. A. Clerville.
Casting rifled cannon and other fire-arms - -	2409	19th Oct. 1853	J. Norton.

D.

Subject-matter of Patent.	Number of Patent.	Date.	Name of Patentee.
Double barrelled fire-arms, improvements in breech-loading	184	2nd Oct. 1852	J. Needham.
Double barrelled fire-arms, apparatus for discharging by electricity	653	16th Mar. 1853	H. R. Fanshawe.
Double barrelled fire-arms, improvements in breech-loading	2690	19th Nov. 1853	M. Poole.
Do. do. do.	2778	29th Nov. 1853	A. E. L. Bellford.

E.

Subject-matter of Patent.	Number of Patent.	Date.	Name of Patentee.
Elastic action of a spring, play arms for impelling projectiles by the	1102	5th May 1853	C. Larbaud.
Elastic action of a tube of vulcanized india-rubber or other suitable material, gun for impelling projectiles by the	1211	16th May 1853	M. H. Phillips.
Electricity, discharging small arms and cannon by -	653	16th Mar. 1853	H. R. Fanshawe.
Explosive compounds to be substituted for gunpowder	306	4th Feb. 1853	G. Winiwater.
Explosive compounds, improvements in the manufacture of	1447	15th June 1853	C. E. Reeves.
Explosive compounds, improvements in firing -	2264	4th Oct. 1853	J. Norton.

F.

Subject-matter of Patent.	Number of Patent.	Date.	Name of Patentee.
Forging gun barrels, improved machinery for -	219	5th Oct. 1852	A. R. Burr.

Subject-matter of Patent.	Number of Patent.	Date.	Name of Patentee.
G.			
Gun barrels, coating with India-rubber compositions	6	1st Oct. 1852	M. Poole.
Gun barrels, improvements in the manufacture of -	219	5th Oct. 1852	A. R. Burr.
Gun barrels, manufacturing of spirals of iron or iron and steel wire	212	28th Jan. 1853	W. Tranter.
Gun barrels, machinery for boring - -	1703	16th July 1853	S. Colt.
Gun barrels, method of, and machinery for giving an eliptical form to the bores of	1988	27th Aug. 1853	C. W. Lancaster.
Gun barrels, giving rotary motion to, instead of rifling	3008	28th Dec. 1853	J. Macintosh.
Gunpowder flasks, fixing apparatus to fire-arms as substitutes for the ordinary detached	583	30th Oct. 1852	R. A. Brooman.
Gunpowder, explosive compounds to be substituted for	306	4th Feb. 1853	G. Winiwarter.
Do. do. do.	1447	15th June 1853	C. E. Reeves.
Gun stocks, manufacturing of India-rubber compositions	6	1st Oct. 1852	M. Poole.
Gun stocks, forming stock frames of one piece of metal, and using gutta percha and papier mâché for the side plates of	960	4th Dec. 1852	J. Bentley.
Gun stocks, forming of metal alone or combined with gutta percha, board made from straw pulp, &c.	452	22nd Feb. 1853	G. Winiwarter.
Gun stocks, forming a chamber in to receive cartridges	1499	20th June 1853	C. Crickmay.
Gun-wads, improvements in the manufacture of -	1000	26th April 1853	J. C. Haddan.
Do. do. do.	2829	6th Dec. 1853	J. C. Haddan.
Do. do. do.	2921	16th Dec. 1853	W. Tranter.
H.			
Hammering, improvements in welding and finishing gun barrels by	219	5th Oct. 1852	A. R. Burr.
L.			
Locks for air-guns, improvements in - -	1102	5th May 1853	C. Larbaud.
Locks for cannon, improvements in - - -	591	8th March 1853	J. J. A. Maccarthy.
Locks for discharging fire-arms by electricity -	653	16th March 1853	H. R. Fanshawe.
Locks for small arms, improvements in, and in the manufacture of percussion	6	1st Oct. 1852	M. Poole.

Subject-matter of Patent.	Number of Patent.	Date.	Name of Patentee.
Locks for small arms, improvements in, and in the manufacture of percussion	69	1st Oct. 1852	W. Moore and W. Harris.
Do. do. do.	125	1st Oct. 1852	T. Hunt.
Do. do. do.	184	2nd Oct. 1852	J. Needham.
Do. do. do.	329	11th Oct. 1852	A. E. L. Bellford.
Do. do. do.	583	30th Oct. 1852	R. A. Brooman.
Do. do. do.	635	4th Nov. 1852	C. Pryse and R. Redman.
Do. do. do.	980	4th Dec. 1852	J. Bentley.
Do. do. do.	86	12th Jan. 1853	E. Haslewood.
Do. do. do.	212	28th Jan. 1853	W. Tranter.
Do. do. do.	305	4th Feb. 1853	P. Webley.
Do. do. do.	540	3rd March 1853	W. E. Newton.
Do. do. do.	743	29th March 1853	J. Webley.
Do. do. do.	964	21st April 1853	P. Harris.
Do. do. do.	1038	29th April 1853	T. Pennell.
Do. do. do.	1123	6th May 1853	M. Riera.
Do. do. do.	1298	26th May 1853	W. J. Harvey.
Do. do. do.	1332	30th May 1853	R. A. Brooman.
Do. do. do.	1447	15th June 1853	C. E. Reeves.
Do. do. do.	1499	20th June 1853	C. Crickmay.
Do. do. do.	1966	24th Aug. 1853	A. E. L. Bellford.
Do. do. do.	2127	14th Sept. 1853	P. Webley.
Do. do. do.	2184	20th Sept. 1853	H. Needham.
Do. do. do.	2197	23rd Sept. 1853	J. Leetch.
Do. do. do.	2351	12th Oct. 1853	R. Jones and C. J. Jones.
Do. do. do.	2590	8th Nov. 1853	E. H. Graham.
Do. do. do.	2712	22nd Nov. 1853	R. Adams.
Do. do. do.	2839	6th Dec. 1853	A. V. Newton.
Do. do. do.	2921	16th Dec. 1853	W. Tranter.
Do. do. do.	2924	16th Dec. 1853	T. Williams.
Do. do. do.	2987	24th Dec. 1853	R. G. Coles.

M.

Mines, improvements in firing - - -	2264	4th Oct. 1853	J. Norton.

N.

Needle gun, improvements in fire-arms constructed on the principle of the	125	1st Oct. 1852	T. Hunt.
Do. do. do.	184	2nd Oct. 1852	J. Needham.

Subject-matter of Patent.	Number of Patent.	Date.	Name of Patentee.
Needle gun, improvements in fire-arms constructed on the principle of the	212	28th Jan. 1853	W. Tranter.
Do. do. do.	371	12th Feb. 1853	G. Winiwarter.
Do. do. do.	591	8th March 1853	J. J. A. Maccarthy.
Do. do. do.	1447	15th June 1853	C. E. Reeves.
Do. do. do.	2778	29th Nov. 1853	A. E. L. Bellford.

P.

Subject-matter of Patent.	Number of Patent.	Date.	Name of Patentee.
Percussion locks, improvements in - - -	69	1st Oct. 1852	W. Moore and W. Harris.
Do. do. do.	125	1st Oct. 1852	T. Hunt.
Percussion locks and priming, improvements in -	184	2nd Oct. 1852	J. Needham.
Do. do. do.	329	11th Oct. 1852	A. E. L. Bellford.
Do. do. do.	583	30th Oct. 1852	R. A. Brooman.
Percussion locks, improvements in - - -	635	4th Nov. 1852	C. Pryse and R. Redman.
Percussion or fulminating powder, charging expanding shot or projectiles with	936	2nd Dec. 1852	J. Norton.
Percussion locks for repeating fire-arms, improvements in	960	4th Dec. 1852	J. Bentley.
Percussion caps, method of manufacturing - -	55	8th Jan. 1853	J. Abraham.
Percussion caps, improvements in the manufacture of	66	10th Jan. 1853	J. D. M. Stirling.
Percussion locks and priming, improvements in -	86	12th Jan. 1853	E. Haslewood.
Percussion caps, attaching to cartridges - -	212	28th Jan. 1853	W. Tranter.
Percussion locks, improvements in - - -	305	4th Feb. 1853	P. Webley.
Percussion compositions, to be used for loading and priming fire-arms	306	4th Feb. 1853	G. Winiwarter.
Percussion caps or primers, using elastic or flexible straps for holding	319	5th Feb. 1853	A. Wollowicz.
Percussion priming, improvements in - - -	540	3rd March 1853	W. E. Newton.
Percussion priming, combining with cartridges -	591	8th March 1853	J. J. A. Maccarthy.
Percussion locks, improvements in priming - -	620	12th March 1853	J. Gilby.
Percussion locks, improvements in - - -	743	29th March 1853	J. Webley.
Percussion locks of repeating fire-arms, method of encasing	1038	29th April 1853	T. Pennell.
Percussion locks, and combining charges of powder and percussion priming with balls	1123	6th May 1853	M. Riera.
Percussion locks of repeating fire-arms, improvements in	1298	26th May 1853	W. J. Harvey.
Percussion locks and priming, improvements in -	1332	30th May 1853	R. A. Brooman.
Percussion caps, improvements in the manufacture of	1404	8th June 1853	J. Horrocks and J. D Horrocks.
Percussion locks and combining explosive compounds with balls and cartridges, improvements in	1447	15th June 1853	C. E. Reeves.
Percussion locks, improvements in - - -	1499	20th June 1853	C. Crickmay.
Percussion locks, improvements in priming - -	1966	24th Aug. 1853	A. E. L. Bellford.
Percussion locks of repeating fire-arms, improvements in	2127	14th Sept. 1853	P. Webley.

Subject-matter of Patent.	Number of Patent.	Date.	Name of Patentee.
Percussion priming, applying to cartridges - -	2176	20th Sept. 1853	R. FLETCHER and J. SMITH.
Percussion locks of repeating fire-arms, improvements in	2184	20th Sept. 1853	H. NEEDHAM.
Do. do. do.	2197	23rd Sept. 1853	J. LEETCH.
Percussion locks, improvements in priming	2313	8th Oct. 1853	W. E. NEWTON.
Percussion locks, to discharge one or more series of barrels, one above the other	2351	12th Oct. 1853	R. JONES and C. J. JONES.
Percussion locks of repeating fire-arms, improvements in	2590	8th Nov. 1853	E. H. GRAHAM.
Do. do. do.	2712	22nd Nov. 1853	R. ADAMS.
Percussion priming, combining with cartridges -	2778	29th Nov. 1853	A. E. L. BELLFORD.
Percussion locks, improvements in priming -	2839	6th Dec. 1853	A. V. NEWTON.
Percussion locks of repeating fire-arms, improvements in	2921	16th Dec. 1853	W. TRANTER.
Percussion locks of revolving pistols - - -	2924	16th Dec. 1853	T. WILLIAMS.
Percussion locks, improvements in - - -	2987	24th Dec. 1853	R. G. COLES.
Percussion priming, combining with cartridges -	3044	31st Dec. 1853	F. A. CLERVILLE.
Pistols [*see* Small Arms *also*], improvements in repeating	988	7th Dec. 1852	S. A. GODDARD.
Pistols, new kind of, termed Haleon pistols - -	940	19th April 1853	W. HALE.
Pistols, improvements in revolving - - -	2924	16th Dec. 1853	T. WILLIAMS.
Powder flasks, fixing apparatus to fire-arms as substitutes for the ordinary detached	583	30th Oct. 1852	R. A. BROOMAN.
Priming percussion locks, improvements in - -	184	2nd Oct. 1852	J. NEEDHAM.
Do. do. do.	329	11th Oct. 1852	A. E. L. BELLFORD.
Do. do. do.	583	30th Oct. 1852	R. A. BROOMAN.
Do. do. do.	55	8th Jan. 1853	J. ABRAHAM.
Do. do. do.	66	10th Jan. 1853	J. D. M. STIRLING.
Do. do. do.	86	12th Jan. 1853	E. HASLEWOOD.
Do. do. do.	212	28th Jan. 1853	W. TRANTER.
Do. do. do.	306	4th Feb. 1853	G. WINIWARTER.
Do. do. do.	319	5th Feb. 1853	A. WOLLOWICZ.
Do. do. do.	540	3rd March 1853	W. E. NEWTON.
Do. do. do.	591	8th March 1853	J. J. A. MACCARTHY.
Do. do. do.	620	12th March 1853	J. GILBY.
Do. do. do.	1123	6th May 1853	M. RIERA.
Do. do. do.	1332	30th May 1853	R. A. BROOMAN.
Do. do. do.	1404	8th June 1853	J. HORROCKS and J. D. HORROCKS.
Do. do. do.	1447	15th June 1853	C. E. REEVES.
Do. do. do.	1966	24th Aug. 1853	A. E. L. BELLFORD.
Do. do. do.	2176	20th Sept. 1853	R. FLETCHER and J. SMITH.
Do. do. do.	2313	8th Oct. 1853	W. E. NEWTON.
Do. do. do.	2778	29th Nov. 1853	A. E. L. BELLFORD.
Do. do. do.	2839	6th Dec. 1853	A. V. NEWTON.
Do. do. do.	3044	31st Dec. 1853	F. A. CLERVILLE.
Projectiles, improvements in balls for the purpose of attaching cartridges thereto	1	1st Oct. 1852	R. ADAMS.
Do. do. do.	184	2nd Oct. 1852	J. NEEDHAM.

Subject-matter of Patent.	Number of Patent.	Date.	Name of Patentee.
Projectiles, improvements in balls containing a charge of powder, and other projectiles	633	3rd Nov. 1852	J. MACINTOSH.
Projectiles, improvements in rendering ships and boats more impervious to cannon shots	884	26th Nov. 1852	R. B. FEATHER.
Projectiles, hollow expanding shot or projectiles, containing charges of powder	936	2nd Dec. 1852	J. NORTON.
Projectiles, making projectiles of two or more metals	1000	8th Dec. 1852	J. LAWRENCE.
Projectiles, an improvement in bullet moulds -	47	7th Jan. 1853	C. W. LANCASTER.
Projectiles, an improvement in projectiles or balls -	86	12th Jan. 1853	E. HASLEWOOD.
Projectiles, an improvement in balls to be combined with cartridges	212	28th Jan. 1853	W. TRANTER.
Projectiles, combining new explosive compounds with hollow bullets for use as shells	306	4th Feb. 1853	G. WINIWARTER.
Projectiles, method of combining balls with cartridges	555	5th March 1853	J. GEDGE.
Do. do. do.	591	8th March 1853	J. J. A. MACCARTHY.
Projectiles, apparatus for discharging rockets and other self-impelled projectiles	940	19th April 1853	W. HALE.
Projectiles, improvements in musket balls, rockets, and other projectiles	1033	28th April 1853	W. H. SITWELL.
Projectiles, improvements in explosive and other projectiles	1123	6th May 1853	M. RIERA.
Projectiles, gun for impelling projectiles by the action of an elastic tube	1211	16th May 1853	M. H. PHILLIPS.
Projectiles, combining explosive compounds with hollow balls or projectiles	1447	15th June 1853	C. E. REEVES.
Projectiles, method of firing or exploding submarine and other shells	2264	4th Oct. 1853	J. NORTON.
Projectiles, improved balls to be combined with cartridges for rifled barrels	2829	6th Dec. 1853	J. C. HADDAN.
Projectiles, improvements in balls or projectiles -	2921	16th Dec. 1853	—
Projectiles, giving rotary motion to tubes or barrels for discharging projectiles	3008	28th Dec. 1853	J. MACINTOSH.

R.

Subject-matter of Patent.	Number of Patent.	Date.	Name of Patentee.
Ramrods, coating with india-rubber compositions -	6	1st Oct. 1852	M. POOLE.
Ramrods, locking the breeches and barrels of repeating pistols together by means of	988	7th Dec. 1852	S. A. GODDARD.
Ramrods, apparatus to be applied to rotating breech fire-arms instead of the ordinary	538	3rd March 1853	S. COLT.
Ramrods, apparatus to be applied to repeating fire-arms, instead of the ordinary	1038	29th April 1853	T. PENNELL.
Recoil of cannon, method of counteracting the -	591	8th March 1853	J. J. A. MACCARTHY.
Repeating fire-arms, improvements in, and in the manufacture of	69	1st Oct. 1852	W. MOORE and W. HARRIS.
Do. do. do.	125	1st Oct. 1852	T. HUNT.
Do. do. do.	184	2nd Oct. 1852	J. NEEDHAM.
Do. do. do.	329	11th Oct. 1852	A. E. L. BELLFORD.

Subject-matter of Patent.	Number of Patent.	Date.	Name of Patentee.
Repeating fire-arms, improvements in, and in the manufacture of	583	30th Oct. 1852	R. A. BROOMAN.
Do. do. do.	960	4th Dec. 1852	J. BENTLEY.
Do. do. do.	988	7th Dec. 1852	S. A. GODDARD.
Do. do. do.	86	12th Jan. 1853	E. HASLEWOOD.
Do. do. do.	212	28th Jan. 1853	W. TRANTER.
Do. do. do.	305	4th Feb. 1853	P. WEBLEY.
Do. do. do,	535	3rd March 1853	S. COLT.
Do. do. do.	538	3rd March 1853	S. COLT.
Do. do. do.	540	3rd March 1853	W. E. NEWTON.
Do. do. do.	577	7th March 1853	J. HALL and J. CROFTS.
Do. do. do.	743	29th March 1853	J. WEBLEY.
Do. do. do.	996	26th April 1853	J. B. SHEATH.
Do. do. do.	1038	29th April 1853	T. PENNELL.
Do. do. do.	1332	30th May 1853	R. A. BROOMAN.
Do. do. do.	1703	16th July 1853	S. COLT.
Do. do. do.	2127	14th Sept. 1853	P. WEBLEY.
Do. do. do.	2184	20th Sept. 1853	H. NEEDHAM.
Do. do. do.	2351	12th Oct. 1853	R. JONES and C. J. JONES.
Do. do. do.	2590	8th Nov. 1853	E. H. GRAHAM.
Do. do. do.	2712	22nd Nov 1853	R. ADAMS.
Do. do. do.	2921	16th Dec. 1853	W. TRANTER.
Do. do. do.	2924	16th Dec. 1853	T. WILLIAMS.
Rifled cannon, expanding shot or projectiles for, containing charges of powder	936	2nd Dec. 1852	J. NORTON.
Rifled cannon, manufacturing by moulding and casting	2409	19th Oct. 1853	J. NORTON.
Rifled small arms, expanding shot or projectiles for, containing charges of powder	936	2nd Dec. 1852	J. NORTON.
Rifled small arms, manufacturing barrels of, by moulding and casting	2409	19th Oct. 1853	J. NORTON.
Rifled small arms, improved balls to be combined with cartridges for	2829	6th Dec. 1853	J. C. HADDAN.
Rifling cannon, giving rotary motion to, instead of	3008	28th Dec. 1853	J. MACINTOSH.
Rifling the barrels of small arms, method of producing an eliptical bore, to be substituted for	1988	27th Aug. 1853	C. W. LANCASTER.
Rifling the barrels of small arms, giving rotary motion to, instead of	3008	28th Dec. 1853	J. MACINTOSH.
Rockets, improved apparatus for firing - -	940	19th April 1853	W. HALE.
Rockets, improvements in - - -	1033	28th April 1853	W. H. SITWELL.
Rockets, giving rotary motion to tubes for discharging	3008	28th Dec. 1853	J. MACINTOSH.
Rolling gun barrels, improved machinery for -	219	5th Oct. 1852	A. R. BURR.
Rolling gunbarrels, method of and machinery for, so as to produce an elliptical bore	1988	27th Aug. 1853	C. W. LANCASTER.

Subject-matter of Patent.	Number of Patent.	Date.	Name of Patentee.

S.

Subject-matter of Patent.	Number of Patent.	Date.	Name of Patentee.
Shells, method of combining new explosive compounds with hollow bullets to be used as	**306**	4th Feb. 1853	G. Winiwarter.
Shells, improvements in firing submarine and other	**2264**	4th Oct. 1853	J. Norton.
Ships, improvements in, rendering more impervious to shot	**884**	26th Nov. 1852	R. B. Feather.
Sights for fire-arms, improvements in - -	**171**	2nd Oct. 1852	W. J. Lewis.
Do. do. do.	**960**	4th Dec. 1852	J. Bentley.
Do. do. do.	**86**	12th Jan. 1853	E. Haslewood.
Do. do. do.	**628**	12th March 1853	T. Hunt.
Do. do. do.	**940**	19th April 1853	W. Hale.
Do. do. do.	**1277**	25th May 1853	W. Church.
Do. do. do.	**1881**	12th Aug. 1853	T. Turner and J. F. Swinburne.
Slings of gutta percha or volcanized india-rubber for muskets and other fire-arms	**591**	8th March 1853	J. J. A. Maccarthy.
Small arms, improvements in, and in the manufacture of	**6**	1st Oct. 1852	M. Poole.
Do. do. do.	**69**	1st Oct. 1852	W. Moore and W. Harris.
Do. do. do.	**125**	1st Oct. 1852	T. Hunt.
Do. do. do.	**171**	2nd Oct. 1852	W. J. Lewis.
Do. do. do.	**184**	2nd Oct. 1852	J. Needham.
Do. do. do.	**219**	5th Oct. 1852	A. R. Burr.
Do. do. do.	**329**	11th Oct. 1852	A. E. L. Bellford.
Do. do. do.	**583**	30th Oct. 1852	R. A. Brooman.
Do. do. do.	**635**	4th Nov. 1852	C. Pryse and R. Redman.
Do. do. do.	**712**	11th Nov. 1852	C. Sharps.
Do. do. do.	**960**	4th Dec. 1852	J. Bentley.
Do. do. do.	**988**	7th Dec. 1852	S. A. Goddard.
Do. do. do.	**86**	12th Jan. 1853	E. Haslewood.
Do. do. do.	**131**	19th Jan. 1853	J. R. Cooper.
Do. do. do.	**212**	28th Jan. 1853	W. Tranter.
Do, do. do.	**305**	4th Feb. 1853	P. Webley.
Do. do. do.	**371**	12th Feb. 1853	G. Winiwarter.
Do. do. do.	**452**	22nd Feb. 1853	G. Winiwarter.
Do. do. do.	**535**	3rd March 1853	S. Colt.
Do. do. do.	**538**	3rd March 1853	S. Colt.
Do. do. do.	**540**	3rd March 1853	W. E. Newton.
Do. do. do.	**555**	5th March 1853	S. Gedge.
Do. do. do.	**577**	7th March 1853	J. Hall and J. Crofts.
Do. do. do.	**591**	8th March 1853	J. J. A. Maccarthy.
Do. do. do.	**620**	12th March 1853	J. Gilby.
Do. do. do.	**628**	12th March 1853	T. Hunt.
Do. do. do.	**653**	16th March 1853	H. R. Fanshawe.
Do. do. do.	**743**	29th March 1853	J. Webley.

Subject-matter of Patent.	Number of Patent.	Date.	Name of Patentee.
Small arms, improvements in, and in the manufacture of	940	19th April 1853	W. HALE.
Do. do. do.	964	21st April 1853	P. HARRIS.
Do. do. do.	996	26th April 1853	J. B. SHEATH.
Do. do. do.	1038	29th April 1853	T. PENNELL.
Do. do. do.	1102	5th May 1853	C. LARBAUD.
Do. do. do.	1123	6th May 1853	M. RIERA.
Do. do. do.	1211	16th May 1853	M. H. PHILLIPS.
Do. do. do.	1298	26th May 1853	W. J. HARVEY.
Do. do. do.	1320	28th May 1853	W. W. MARSTON.
Do. do. do.	1332	30th May 1853	R. A. BROOMAN.
Do. do. do.	1344	1st June 1853	J. L. LEMAIRE DAIMÉ.
Do. do. do.	1447	15th June 1853	C. E. REEVES.
Do. do. do.	1499	20th June 1853	C. CRICKMAY.
Do. do. do.	1703	16th July 1853	S. COLT.
Do. do. do.	1870	11th Aug. 1853	R. F. BRAND.
Do. do. do.	1881	12th Aug. 1853	T. TURNER and J. F. SWINBURN.
Do. do. do.	1966	24th Aug. 1853	A. E. L. BELLFORD.
Do. do. do.	1988	27th Aug. 1853	C. W. LANCASTER.
Do. do. do.	2127	14th Sept. 1853	P. WEBLEY.
Do. do. do.	2176	20th Sept. 1853	R. FLETCHER and J. SMITH.
Do. do. do.	2184	20th Sept. 1853	H. NEEDHAM.
Do. do. do.	2197	23rd Sept. 1853	J. LEETCH.
Do. do. do.	2313	8th Oct. 1853	W. E. NEWTON.
Do. do. do.	2351	12th Oct. 1853	R. JONES and C. J. JONES.
Do. do. do.	2409	19th Oct. 1853	J. NORTON.
Do. do. do.	2590	8th Nov. 1853	E. H. GRAHAM.
Do. do. do.	2690	19th Nov. 1853	M. POOLE.
Do. do. do.	2697	19th Nov. 1853	R. F. BRAND.
Do. do. do.	2712	22nd Nov. 1853	R. ADAMS.
Do. do. do.	2778	29th Nov. 1853	A. E. L. BELLFORD.
Do. do. do.	2839	6th Dec. 1853	A. V. NEWTON.
Do. do. do.	2921	16th Dec. 1853	W. TRANTER.
Do. do. do.	2924	16th Dec. 1853	T. WILLIAMS.
Do. do. do.	2987	24th Dec. 1853	R. G. COLES.
Do. do. do.	3008	28th Dec. 1853	J. MACINTOSH.
Do. do. do.	3044	31st Dec. 1853	F. A. CLERVILLE.
Sword-bayonets, improvements in the manufacture of	2550	3rd Nov. 1853	C. REEVES.
Swords, improvements in securing to gun barrels -	2921	16th Dec. 1853	W. TRANTER.
Swords, improvements in the manufacture of -	436	19th Oct. 1852	R. MOLE and R. MOLE.
Do. do. do.	960	21st April 1853	C. REEVES.
Do. do. do.	2550	3rd Nov. 1853	C. REEVES.

Subject-matter of Patent.	Number of Patent.	Date.	Name of Patentee.

T.

Subject-matter of Patent.	Number of Patent.	Date.	Name of Patentee.
Toy arms, improvements in - - - - - -	635	4th Nov. 1852	C. PRYSE and R. REDMAN.
Do. do.	1102	5th May 1853	C. LARBAUD.
Triggers [*see* Small Arms *also*], peculiar arrangement of double	212	28th Jan. 1853	W. TRANTER.
Triggers, improvements in - - - - - -	964	21st April 1853	P. HARRIS.
Triggers, new system of for play arms - - -	1102	5th May 1853	C. LARBAUD.
Triggers for repeating fire-arms, improvements in -	2184	20th Sept. 1853	H. NEEDHAM.
Triggers of repeating fire-arms, improvements in mounting the	2313	8th Oct. 1853	W. E. NEWTON.

W.

Subject-matter of Patent.	Number of Patent.	Date.	Name of Patentee.
Wadding for fire-arms, improvements in the manufacture of	1000	26th April 1853	J. C. HADDAN.
Do. do. do.	2829	6th Dec. 1853	J. C. HADDAN.
Do. do. do.	2921	16th Dec. 1853	W. TRANTER.
Waggons, improvements in military - - -	59	1st Oct. 1852	M. DAVIS.
Welding gun barrels, improvements in - -	219	5th Oct. 1852	A. R. BURR.

LONDON:
Printed by GEORGE EDWARD EYRE and WILLIAM SPOTTISWOODE,
Printers to the Queen's most Excellent Majesty.

A.D. 1718 . . . . N° 418.

Portable Gun.

PUCKLE'S SPECIFICATION.

GEORGE, by the grace of God, of Great Brittaine, France, and Ireland King, Defender of the Faith, &c.

TO ALL TO WHOM THESE PRESENTS SHALL COME, greeting.

WHEREAS James Puckle, of our City of London, Gent., hath by his petic̃on, humble rep̃sented unto us that he hath at great expences invent "**A Portable Gun or Machine called a Defence, thatt Discharges soe often and soe many Bulletts, and can be soe Quickly Loaden as renders it next to Impossible to to Cary any Ship by Boarding**;" and haveing humbly prayed us to grant him our Royall Letters Patents for the sole makeing and vending the said gun in England & Ireland, dureing the terme of fourteene yeares; but that he thinks it not safe to specify wherein the new Invention consists; but proposes that soe soon as the Patent shall be passed, the same shall be by him ascertained, under his hand & seale, to be enrolled in our High Court of Chancery, to which the Patent may referr, to make the grant therein certain.

KNOW YE, that wee, being willing to give encouragement to all arts and inventions which may be of publick use and benifitt, of our especiall grace, certaine knowledge, & meer moc̃on have given and granted, and by these Presents, for us, our heires and successors, do give and grant unto the said James Puckle, his executors, administrators, or assignes, especiall lycence, full power, sole priviledge & authority, that he, the said James Puckle, his executors, administrators, and assignes, and every of them, by himselfe and themselves, or by his and their deputy and deputies, servants & agents, or such others as he, the said James Puckle, his executors, administrators, or assignes, shall at any time agree with, & noe others, from time to time and at all times dureing the terme of yeares herein expressed, shall and lawfully may make, exercise, work, vend, use, and enjoy his said new invented portable gun or machine called a defence, as aforesaid, with that part of our Kingdom of Great Brittaine called England, our Dñion, Town of Berwick-upon-Tweed, and our Kingdom of Ireland, in such manner and with such materialls as shall be ascertained to be the said new Invention, by writeing under the hand and seale of the said James Puckle, and inrolled in our High Court of Chancery within three calender months from the date of these Presents, and that he, the s^{d} James Puckle, his executors, administrators, and assignes, shall and may have and enjoy the whole profitt, benifitt, com̄odity, and advantage from time to time comeing, growing, accreuing, and ariseing by reason of the said Invention, for and dureing the full terme of yeares herein menc̃oned; to have, hold, exercise, and enjoy the said lsyence, powers, priviledges, & advantages herein-before granted or menc̃oned to be granted unto the said James Puckle, his executors, administrators, and assignes, for and dureing and untill the full end terme of fourteen yeares from the date of these Presents next and imediately ensueing, and fully to be compleat and ended, according to the Statute in such case made and provided. And to the end that the said James Puckle, his executors, administrators, and assignes, and every of them, may have and enjoy the full benifitt and the sole use and exercise of the said Invention, according to our gracious intention herein-before declared, wee doe by these p̃resents, for us, our heires and successors, require and strictly com̄and all and every person and persons, bodys politick and corporate, and all other our subiects whatsoever, of what estate, quality, or degree, name or condition, soever they be, within that said part of our Kingdom of Great Brittaine called England, our Dem̃on of Wales, Towne of Berwick-upon-Tweed, and our Kingdom of Ireland aforesaid, that neither they or any of them at any time dureing the continuance of the said terme of fourteen yeares hereby granted, either directly or indirectly, doe make, use, or put in practice the said Invention, or any part of the same, so attained unto by the said James Puckle as aforesaid, nor shall in any wise counterfeit, imitate, or resemble the same, nor shall make or cause to be made any adition thereunto, or substraction from the same, whereby to pretend himselfe or themselves the inventor or inventors, devisor or devisors thereof, without the lycence, consent, or agreement of the said James Puckle, his executors, administrators, or assignes, in writeing under his or their hands and seales first had and obtained in that behalfe, upon such paines and penaltyes as can or may be justly inflicted in such offenders for their contempt of this our Royall com̃and, and further to be answerable to the said James Puckle, his executors, administrators, and assignes, according to law, for his and their damages thereby occasioned. And moreover wee doe by these Presents, for us, our heires and successors, will and comand all and singular our justices of the peace, mayors, sheriffs, bayliffs, constables, headborroughs, and all other officers and ministers whatsoever, of us, our heires and successors, for the time being, that they or any of them do not, nor shall at any time hereafter dureing the said terme hereby granted, in any wise molest, trouble, or hinder the said James Puckle, his executors, adm̃ors, and assignes, or any of them, or his or their deputies, servants, or agents, in or about the due & lawfull makeing, use, or exercise of the aforesaid Invention, or any thing relateing thereto; provided alwayes, and these our Letters Patents are and shall be upon this condition, that if at any time dureing the said terme hereby granted it shall be made appear to us, our heires or successors, or any six or more of our or their Privy Councell, that this our grant is contrary to law, & prejudicial or inconvenient to our subiects in generall, or that the said Invention is not a new Invention as to the publick use and exercise thereof within that said part of our Kingdom of Great Brittaine called England, our Dominion of Wales, Town of Berwick-upon-Tweed, and our Kingdom of Ireland aforesaid, or not invented and found out by the said James Puckle as aforesaid, then upon signification and declaration thereof, to be made by us, our heires or successors, under our or their signett or privy seale, or by the Lords and others of our or their Privy Councill, or any six or more of them, under their hands, these our Letter Patente shall forthwith cease, determine, and be utterly void to all intents and purposes, any thing herein-before contained to the contrary in any wise notwithstanding: Provided also, that these our Letters Patents, or any thing herein contained, shall not extend or be construed to extend to give priviledges unto the said James Puckle, his executors, administrators, or assignes, or any of them, to use or imitate any invention or work whatsoever which hath heretofore been found out or invented by any other of our subjects whatsoever, and publickly used or exercised within that part of our Kingdom of Great Brittaine called England, our Dominion of Wales, Towne of Berwick-upon-Tweed, and our Kingdom of Ireland aforesaid, unto whom like Letters Patents or priviledges have been already granted for the sole use, exercise, and benifitt thereof: It being our will and pleasure that the said James Puckle, his executors, administrators, and assignes, and all & every other person or persons to whom like Letters Patents or priviledges have been already granted as aforesaid, shall distincttly use and practice their severall Inventions by them invented and found out according to the true intent and meaninge of the same respective Letters Patents and of these presents. And lastly, wee doe by these presents, for us, our heires and successors, grant unto the said James Puckle, his executors, administrators, and assignes, that these our Letters Patents, or the inrollment or exemplification thereof, shall be in and by all things good, firme, valid, sufficient, and effectuall in the law, according to the true intent and meaneing hereof, and shall be taken, construed, and adjudged in the most favourable and benificiall sence for the best advantage of the said James Puckle, his executors, administrators, or assignes, as well in all Courts of Record as elsewhere, and by all and singular the officers & ministers whatsoever of us, our heires and successors, within that said part of our Kingdom of Great Brittaine called England, our Dominion of Wales, Town of Berwick-upon-

Tweed, and our Kingdom of Ireland aforesaid, and amongst all & every the subjects of us, our heires & successors, whatsoever or wheresoever, notwithstanding the not full and certaine describeing the nature or quality of the s^d^ Invention, or of the materialls thereto conducing and belonging. In witness whereof we have caused these our Letters to be made Patents.

Witness ourselfes at Westminster, ye Fifteenth day of May, in the fourth year of our reigne.

By writt of Privy Seale.
COCKS.

Et memorandum quod per mandat̃ honõbil̃ viro Josepho Jekyll, mil̃, ℘ WA
Mag̃ro Rotuloȝ, in hec verba: (28th of July 1718: Lett this Patent be inrolled pro salva custodia tantum)—J. JEKYLL. Indentur̃ predic̃t de verbo in verbum prout prescribetur irr̃ die et anno predic̃t.

LONDON:
Printed by GEORGE EDWARD EYRE and WILLIAM SPOTTISWOODE,
Printers to the Queen's most Excellent Majesty. 1854.

A.D. 1718. MAY 15. Nº 418.

PUCKLE'S SPECIFICATION.

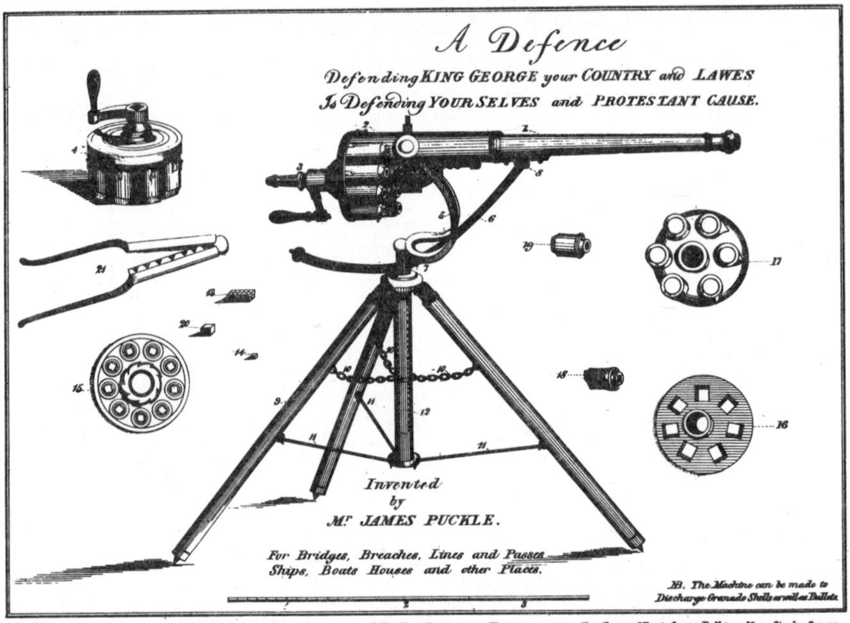

Nº 1. The Barrel of the Gun.
2. The Sett of Chambers Charg'd put on ready for Fireing.
3. The Screw upon which every Sett of Chambers play off and in.
4. a Sett of Chambers ready Charg'd to be Slip'd on when the first Sett are pull'd off to be recharg'd.
5. The Crane to rise fall and Turn the Gun round.
6. The Curb to Level and fix the Guns.
7. The Screw to rise and fall it.
8. The Screw to take out the Crane when the Gun with the Trepied is to be folded up.
9. The Trepied whereon it Plays.
10. The Chain to prevent the Trepieds extending too far out.
11. The hooks to fix the Trepied and Unhook when the same is folded up in order to be carried with the Gun upon a Man's Shoulder.
12. The Tube wherein the Pivot of the Crane turns.
13. a Charge of Twenty Square Bullets.
14. a single Bullet.
15. The front of the Chambers of a Gun for a Boat.
16. The plate of the Chambers of the Gun for a Ship shooting Square Bullet against Turks.
17. For Round Bullets against Christians.
18. a Single Square Chamber.
19. a Single round Chamber.
20. a single Bullet for a Boat.
21. The Mould for Casting Single Bullets.

Whereas our Soveraign Lord King George by his Letters Pattents bearing date the Fifteenth day of May in the Fourth Year of his Majesties Reign was Graciously pleas'd to Give & Grant unto me James Puckle of London Gent my Exors Admors & Assignes the sole priviledge & Authority to Make Exercise Work & use a Portable Gun or Machine (by me lately Invented) call'd a Defence in that part of his Majesties Kingdom of Great Brittain call'd England his Dominion of Wales, Town of Berwick upon Tweed and his Majesties Kingdom of Ireland in such manner & with such Materials as Should be ascertain'd to be the sd. New Invention by writing under my Hand & Seal and Inrolled in the High court of chancery within Three Calendar Months from the date of the sd pattent as in & by his Majties Letters Pattents Relacon being thereunto had Doth & may amongst Other things more fully & at large appear NOW I the said James Puckle Do hereby Declare that the Materials whereof the sd Machine is Made are Steel Iron & Brass and that the Trepied whereon it Stands is Wood & Iron And that in the above print (to which I hereby Refer) the said Gun or Machine by me Invented is Delienated & Described. July the 25th 1718./.

4B. 180

Jas Puckle

The enrolled drawing is not colored.

Malby & Sons, Lith.

LONDON. Printed by GEORGE EDWARD EYRE and WILLIAM SPOTTISWOODE.
Printers to the Queen's most Excellent Majesty. 1854

A.D. 1772 . . . . N° 1003.

Fire-arms.

WRIGHT AND BYRNE'S SPECIFICATION.

TO ALL TO WHOM THESE PRESENTS SHALL COME, we, THOMAS WRIGHT, of the Parish of Saint John, Clerkenwell, in the County of Middlesex, Watchmaker, and CHARLES BYRNE, of the Parish of Saint Andrew by the Wardrobe, in the City of London, Gunmaker, send greeting.

WHEREAS His most Excellent Majesty King George the Third, did, by His Letters Patent under the Great Seal of Great Britain, bearing date the Fourth day of February, in the twelfth year of His reign, give and grant unto us, the said Thomas Wright and Charles Byrne, His special licence that we, the said Thomas Wright and Charles Byrne, during the term of years therein expressed, should and lawfully might use, exercise, and vend, within England, Wales, and Town of Berwick upon Tweed, and also in all His Majesty's Colonies and Plantations abroad, our Invention of "A NEW CONSTRUCTION OF GUNS AND ALL OTHER FIRE-ARMS, MUCH PREFERABLE TO ANY HITHERTO MADE, BEING CAPABLE OF CARRYING THEIR SHOT MUCH FURTHER, OF LOADING AND PRIMING MUCH QUICKER, AND OF PREVENTING THE DISAGREEABLE EFFECTS OF RECOILING;" in which said Letters Patent there is contained a provisoe, obliging the said Thomas Wright and Charles Byrne, or one of us, under our hands and seals, or under the hand and seal of one of us, to cause a particular description of the nature of our said Invention, and how the same is to be performed, to be inrolled in His Majesty's High Court of Chancery within four calendar months after the date of the said recited Letters Patent, as in and by the same (relation being thereunto had) may more fully and at large appear.

NOW KNOW YE, that in compliance with the said provisoe, we, the said Thomas Wright and Charles Byrne, do hereby declare that our said Invention of a new construction of guns and all other firearms, much preferable to any hitherto made, being capable of carrying their shot much further, of loading and priming much quicker, and of preventing the disagreeable effects of recoiling, is described in the plan thereof hereunto annexed.

In witness whereof, we, the said Thomas Wright and Charles Byrne, have hereunto set our hands and seals, the Twenty-third day of May, One thousand seven hundred and seventy-two.

THO. WRIGHT. (L.S.)
CHAS BYRNE. (L.S.)

T. HARRIS.

AND BE IT REMEMBERED, that on the same Twenty-third day of May, in the year above written, the aforesaid Thomas Wright and Charles Byrne came before our said Lord the King, in His Chancery, and acknowledged the Specification of their Invention aforesaid, and all and every thing therein contained, in form above written. And also the Specification aforesd was stampt according to the tenor of the Statute made in the sixth year of the reign of the late King and Queen William and Mary of England, and so forth.

Inrolled the Twenty-fifth day of May, in the year above written.

LONDON:
Printed by GEORGE EDWARD EYRE and WILLIAM SPOTTISWOODE,
Printers to the Queen's most Excellent Majesty. 1854.

A.D. 1772. FEBRUARY 4. Nº 1003.

WRIGHT & BYRNE'S SPECIFICATION.

(1 SHEET)

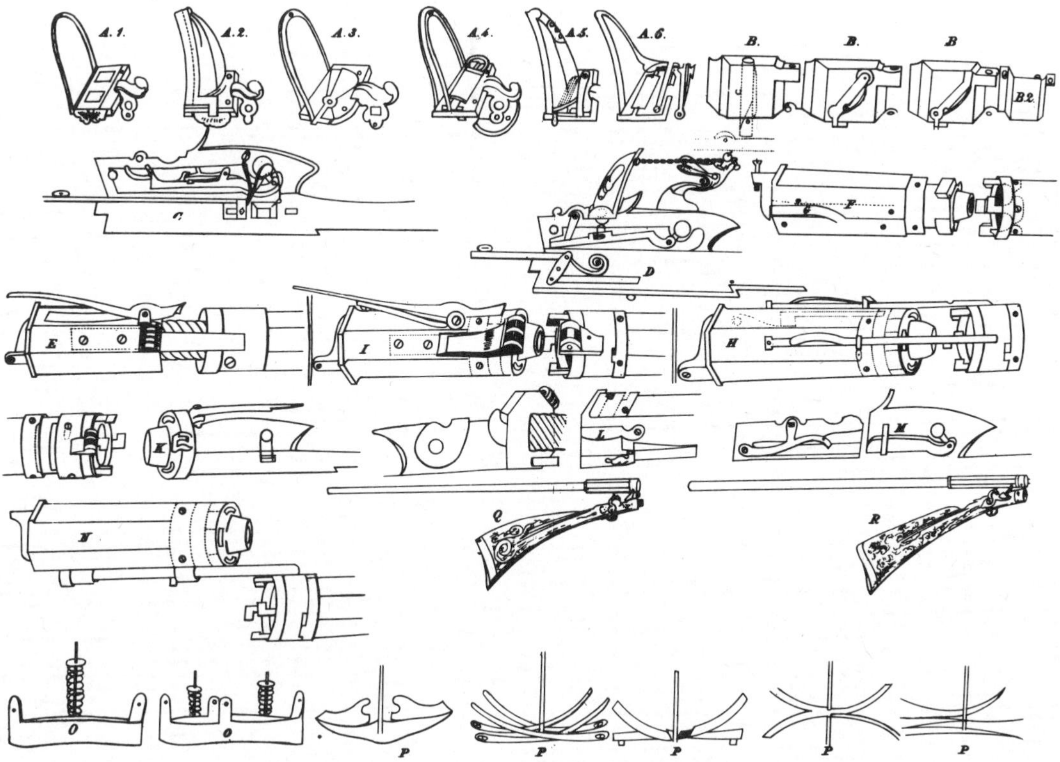

A.1. Hammer with slide with Holes thro for letting down the powder contained in A. 2. The Hammer with its back on A. 3. Ditto with Lever slide A. 4. Ditto with Cylinder and Spiral spring to it, A. 5. Ditto with Cylinder, Lever, Spring, and an aperture in the back for admitting the powder A. 6. Ditto with Cylinder Lever and different Springs BBB Pans with their different Levers, Valves, Springs, and friction Wheels B. 2. for covering ditto C. The Hind end or inside of Lock part which is fastened to the Britch end of the Barrel Z. L. by Screws, with the Lever that acts from the Tumbler to close the Hammer on the pan and prize of drawing up the Lock with a Bolt to secure the fore-part & Hindpart of the Barrel when either screwed or locked with a friction Wheel to make it slide easy D Ditto with a different Lever bolt, and Bolt Spring also a Chain from the Cock to the Hammer, to bring down the Hammer by cocking E Screw Barrel with Joint Catch & Director F, Lock Barrel with G, the Spring for the Sliding bolt, H, represents F, complete, with its Slides, Springs and Bolts, I, Lock Barrel with joint to slide, with a moveable Slope to secure the Cramps or Hooks in fastening and catch to lock the whole K Ditto to load at the Breach L, Screw Barrel to load at the Breach with its Catch & Hooks M Inside of Lock to Ditto with its Slide & Levers for bringing down the Hammer by cocking N, Lock barrel with Cramps and Sliding Pin to direct Ditto, OO, Heel plates with Spiral Springs and directing pins, PPPPP different Springs to Ditto, Q, Screw Barrel piece complete at R Lock barrel piece complete.

Tho. Wright.
Chas. Byrne

The enrolled drawing is partly colored.

Malby & Sons lith.

LONDON: Printed by GEORGE EDWARD EYRE and WILLIAM SPOTTISWOODE Printers to the Queen's most Excellent Majesty. 1854.

A.D. 1776 . . . . N° 1139.

Breech-loading Fire-arms.

FERGUSON'S SPECIFICATION.

TO ALL TO WHOM THESE PRESENTS SHALL COME, I, Patrick Ferguson, Esquire, Captain of His Majesty's Seventieth Regiment of Foot, send greeting.

WHEREAS His most Excellent Majesty King George the Third, by His Letters Patent under the Great Seal of Great Britain, bearing date at Westminster, the Second day of December, in the seventeenth year of His reign, did give and grant unto me, the said Patrick Ferguson, my executors, adm̃ors, and assigns, his special license, that I, the said Patrick Ferguson, my executors, adm̃ors, and assigns, during the term of years therein expressed, should and lawfully might make, use, exercise, and vend, within England, Wales, and Town of Berwick upon Tweed, and also within all His Colonies and Plantations abroad, my Invention of "Various Improvements upon Fire-arms, by means of which they Load with more Ease, Safety, and Expedition, Fire with greater Certainty, and possess various other Advantages;" in which said Letters Patent there is contained a proviso, obliging me, the said Patrick Ferguson, under my hand and seal to cause a particular description of the nature of my said Invention, and in what manner the same are to be performed, to be inrolled in His Majesty's High Court of Chancery within four calendar months next and immediately after the date of the said recited Letters Patent, as in and by the same, relation being thereunto had, may more fully and at large appear.

NOW KNOW YE, that in compliance with the said proviso, I, the said Patrick Ferguson, do hereby declare that my said Invention of improvements upon fire-arms is executed in the following manner, and agreeable to the map or plan thereof laid down in the schedule hereunto annexed, and the reference to the said plan hereunder mentioned, that is to say:

No. 1, is a section along the barrel of a gun, cannon, or pistol, which loads at the breech by means of a screw plug that descends by one turn as low as the bottom of the bore to admit the ball and then the powder after, which by turning the plug the contrary way the breech is closed, and the piece loaded. Something similar to this has been attempted, but has always failed for want of the following improvements, and indeed has never been seriously studied as an object of public utility, or rendered in any degree fit for service.

The improvements are, First, a projection of the barrel towards the guard (X), to receive the screw plug long enough to allow the plug to descend sufficiently low occasionally to be cleaned after much firing without coming entirely out of the grooves, by which means the difficulty, loss of time, and embarrassment of hitting the proper groove (in occasions of hurry and danger), to re-enter the plug is avoided.

Secondly, that part of the screw plug which forms the breech of the barrel is made smooth and hollow by which means it neither retains so much of the smoak or foulness, nor in going round does it lodge it into the female screw so as to clog it and impede the motion of the plug.

Thirdly, there are various channels cut across the outside of the screws of the plug in such directions as not to communicate and occasion any part of the charge to blow out, at the same time that they are contrived to go along the whole surface almost of the female screw in mounting and descending, thereby loosening and receiving any foulness that may have lodged.

Fourthly, there is a hollow or reservoir behind the screw plug at the breech of the barrel into which the smoak which had forced its way through the grooves is thrown. when the plug turns round so as to clear the grooves or female screw from being clogged by the foulness that might otherwise lodge in them.

Fifthly, there is a rising formed on the breech of the barrel, which answers as an elevating sight, which is never in the way, nor to be moved out of its place (like the sights coñonly used which are let into the barrel), and into this elevated sight a sliding sight is let in, which, by being gradually pulled out in proportion to the distance of the object, may be adjusted to all manner of distances, whereas those sights usually in use only answer for two or three given distances.

By means of these improvements the piece is to be loaded and fired with much more safety, facility and certainty, and can be discharged as often as ever can be necessary without any occasion for the cleaning, so that the publick is furnished with an arm which unites expedition, safety, and facility in using with the greatest certainty in execution, the two great desiderata in gunnery never before united.

No. 2, represents another screw plug made to ascend and descend enough in half a turn to be loaded with channels in a different direction that answer equally well, and in fact the situation of these channels may be varied fifty different ways upon the same principle.

No. 3, is a section along the barrel of a cannon, gun, or pistol, which loads at the breech by means either of a perpendicular or horizontal turn cock, in which there is a hole which may be brought in a line with the bore of the barrel so as to admit the charge at the breech by turning the cock round. These cocks will of course require as well as the screw plugs to have grooves or channels to clear them from clogging in going round from the foulness which would otherwise lodge.

This construction is particularly adapted to artillery. No. 4, is a plan of the outside of the breech of a cannon, gun, or pistol, that is loaded either above, or below, or at the side by means of a turn cock, by turning which round, the orifice behind is brought forwards to afford a free passage for the charge into the barrel, and by bringing the cock round again the breech is closed.

No. 5, is the outside plan of the breech of a cannon, gun, or pistol, which loads with a cross slider, either at the upper part of the barrel, at the side, or at the bottom. A, is a section of the same along the barrel: B. is a section across the barrel.

No. 6, is the outside plan of the breech of a cannon, gun, or pistol, which loads with a slider along the barrel either above, below, or at the side; C, is a section of the same along the barrel; and D, a section across the barrel.

No. 7, is a cross section of the barrel of a gun, cannon, or pistol, which loads at the breech by means of a traverse slider which goes across the bore.

The above manners of loading will apply to fire-arms of every kind, whether cannon, swivels, wall pieces, guns, or pistols, but some of them are better adapted to one sort of fire-arms, and some to another.

No. 8, is a section across a rifle barrel, representing a new method of rifling, where the female rifle or groove, in place of being made less than the male or projecting rifle, as has been hitherto the practice, is made much larger. By means of this narrowness of the projecting rifle the ball is less cut, and although kept equally steady in its spiral motion meets with less resistance in the barrel, consequently occasions less recoil and goes with more force; and by means of the wideness of the female rifle or groove the barrel is less apt to become foul so as to recoil or fire without certainty, and is easier cleaned, and moreover the projections formed in the ball by the cut of the rifle being twice as large as they commonly are, will not be liable to strip; whereas in the common method of rifling from the smallness of the projections or knobs formed on the ball, it is exposed to be stripped perfectly smooth so as to destroy entirely the effect of the rifle in point of certainty when the barrel is in the least foul or the ball not extreamly tight.

In witness whereof, the said Patrick Ferguson hath hereunto set his hand and seal, this Seventh day of March, One thousand seven hundred and seventy-seven.

PAT. (L.S.) FERGUSON.

Sealed and delivered (being first duly stampt) in the presence of

SARAH VALE,
WM HARDY.

A.D. 1776. December 2. No. 1139.

FFERGUSON'S Specification.

(1 SHEET)

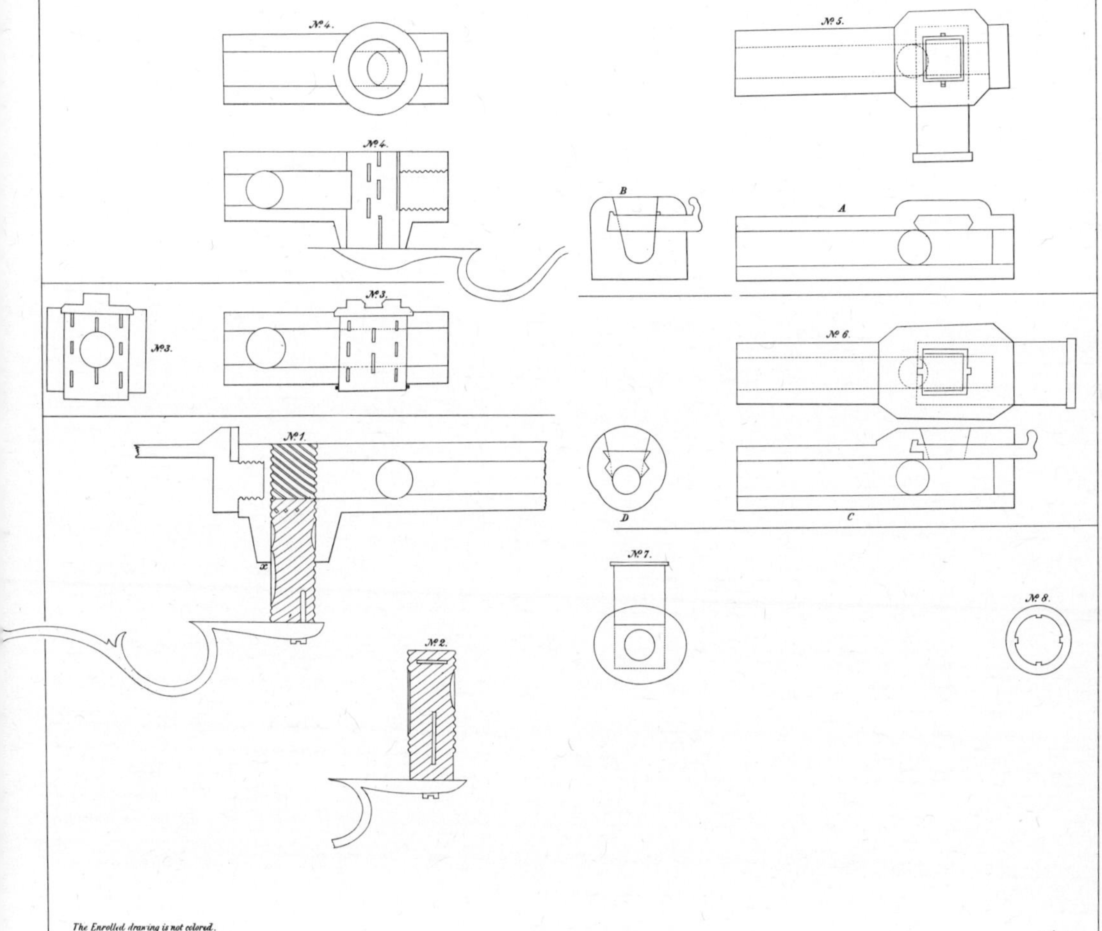

The Enrolled drawing is not colored.

Malby & Sons lith.

London: Printed by George Edward Eyre and William Spottiswoode, Printers to the Queen's most Excellent Majesty. 1854.

A.D. 1787 . . . . N° 1598.

Breech for Fire-arms.

NOCK'S SPECIFICATION.

TO ALL TO WHOM THESE PRESENTS SHALL COME, HENRY NOCK, of Ludgate Street, in the Parish of Saint Martin, Ludgate, in the City of London, Gun Maker, sendeth greeting.

WHEREAS His present Majesty King George the Third, by His Letters Patent under the Great Seal of Great Britain, bearing date at Westminster the Twenty-fifth day of April, in the twenty-seventh year of His reign, hath, for Himself, His heirs and successors, given and granted unto the said Henry Nock, his ex̃ors, adm̃ors, and assigns, His especial licence, full power, sole privilege, and authority to make, use, exercise, and vend, within that part of Great Britain called England, Dominion of Wales, and Town of Berwick-upon-Tweed, his new invented "BREECHING, APPLICABLE TO ALL KINDS OF GUNS AND OTHER FIRE-ARMS, WHEREBY THE POWDER IS FIRED OFF IN THE CENTRE AND MUCH QUICKER, AND THE CHARGE THROWN CONSIDERABLY STRONGER, AND THE GUNS AND OTHER FIRE-ARMS WILL CONTINUE TO FIRE MUCH LONGER WITHOUT CLEANING THAN BY ANY OTHER METHOD HERETOFORE CONSTRUCTED OR USED," together with the whole profit, benefit, commodity, and advantage from time to time coming, growing, accruing, and arising by reason of the said Invention, for and during the term of fourteen years, according to the form of the Statute in such case made and provided; in which said Letters Patent, amongst other things, it is provided, that if the said Henry Nock shall not particularly describe and ascertain the nature of his said Invention, and in what manner the same is to be performed, by an instrument in writing under his hand and seal, and cause the same to be inrolled in the High Court of Chancery within one calendar month next and immediately after the date of the said Letters Patent, that then the said Letters Patent, and all liberties and advantages whatsoever thereby granted, sho'd utterly cease, determine, and become void, anything therein-before contained to the contrary thereof in anywise notwithstanding.

NOW KNOW YE, that he, the said Henry Nock, in pursuance of the intention of the said Letters Patent, and in compliance with the proviso therein contained, doth hereby declare that the nature of his said Invention of a new breeching as above set forth, and in what manner the same is to be performed, is as follows (that is to say):

Figure 1, is a section of the breeching, applicable to all kinds of guns and other fire-arms for the purpose heretofore specified; this section is of itself sufficiently plain to render the principle obvious to every workman, though it will be rendered more so by the following description:—Take a piece of solid iron, or other strong metal proper for the purpose, and of such dimensions that when fitted in its place it will admit being reduced to the shape and size of the barrel for which it is intended. A hole is to be drilled through the centre of the aforesaid piece, and the outside to be turned down, and a proper screw cut upon it to fit to the barrel; this screw is represented at A, A. The centre hole above mentioned must be so enlarged as to form a proper chamber (vide letter C) for the piece to which it is to be applied. Letter D, is the anti-chamber, which is a hole forming an angle with said chamber C; it is here represented as at right angles with the chamber C, and communicating with it. At letter F, is a screw adapted to reach to the centre hole. This screw may either be rivetted fast to the breeching, or be made to take in or out at pleasure to examine the anti-chamber and inside of the touch-hole. Letter G, is a screw adapted to meet the said cross screw of the anti-chamber, which screw may be taken out at pleasure to examine the inside of the barrel. The breeching being then put in or fixed to the barrel the outside circumference B, B, B, B, may be made into any convenient shape, and be case-hardened or left soft at pleasure. Figure 2, is an elevation of the breeching. Figure 3, is a perspective view of the breeching. Figure 4, is a breeching with the anti-chamber D, made obliquely, which also fires the powder in the centre. Or the breeching may be made in the following manner to fire the powder in the centre:—Figure 5, is an elevation of section 6, and 7. The part A, A, is a representation of section 6, where a part of the breeching A, A, is left plain without a screw, the whole going into the barrel. The part B, B, represents section 7, where the breeching is screwed the whole length, the whole going into the barrel. Figure 6, is a section of a breeching by which the powder may be fired in the centre, the part A, A, being left plain without a screw, and the other part B, B, being screwed, and the whole length from C, to C, going into the barrel. Figure 7, is a section of a breeching on the same construction as Figure 6, only screwed the whole length from C, to C, which goes into the barrel.

In witness whereof, I, the said Henry Nock, have hereunto set my hand and seal, the Twenty-third day of May, in the twenty-seventh year of the reign of our Sovereign Lord George the Third, by the grace of God of Great Britain, France, and Ireland King, Defender of the Faith, and so forth, and in the year of our Lord One thousand seven hundred and eighty-seven.

HENRY NOCK. (L.S.)

Signed, sealed, and delivered (being first duly stampt) in the presence of

BENJ[N] EVERINGHAM.
THO. JOHNSON, Ely Place.

A.D. 1787. April 25. No. 1598.

NOCK'S SPECIFICATION.

(1 SHEET)

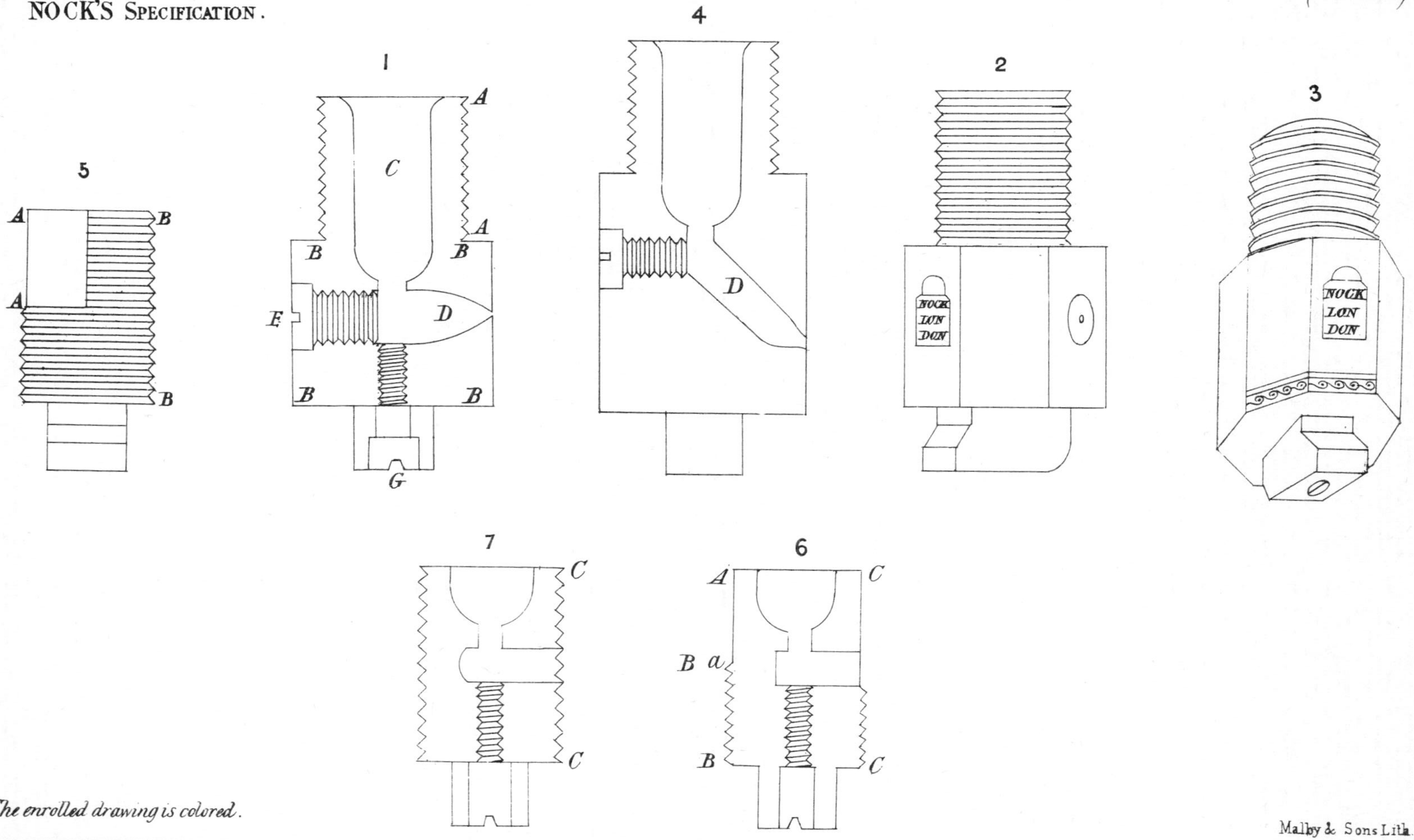

The enrolled drawing is colored.

Malby & Sons Lith.

LONDON: Printed by GEORGE EDWARD EYRE and WILLIAM SPOTTISWOODE,
Printers to the Queen's most Excellent Majesty. 1854.

A.D. 1807 . . . . N° 3032.

Apparatus for Discharging Artillery, &c. by Means of Detonating Compounds.

FORSYTH'S SPECIFICATION.

TO ALL TO WHOM THESE PRESENTS SHALL COME, I, ALEXANDER JOHN FORSYTH, Clerk, of Belhelvie, Aberdeenshire, in Scotland, send greeting.

WHEREAS His present Majesty King George the Third, in and by His Letters Patent under the Great Seal of Great Britain, bearing date at Westminster, the Eleventh day of April now last past the date hereof, for Himself, His heirs and successors, did give and grant unto me, the said Alexander John Forsyth, my ex̃ors, adm̃ors, and assigns, His special licence, full power, sole privilege and authority, that I, the said Alexander John Forsyth, my ex̃ors, adm̃ors, and assigns, and every of them, by myself and themselves, or by my and their deputy or deputies, servants, or agents, or such others as I, the said Alexander John Forsyth, my ex̃ors, adm̃ors, or assigns, should at any time agree with, and no other, from time to time and at all times thereafter during the term of years therein expressed, should and lawfully might make, use, exercise, and vend my Invention of "**AN ADVANTAGEOUS METHOD OF DISCHARGING OR GIVING FIRE TO ARTILLERY AND ALL OTHER FIRE-ARMS, MINES, CHAMBERS, CAVITIES, AND PLACES IN WHICH GUNPOWDER OR OTHER COMBUSTIBLE MATTER IS OR MAY BE PUT FOR THE**

Purpose of Explosion," within that part of the United Kingdom of Great Britain and Ireland called England, His Majesty's Dominion of Wales, and Town of Berwick-upon-Tweed, and also in all His Majesty's Colonies and Plantations abroad; in which said Letters Patent there is contained a proviso, obliging me, the said Alexander John Forsyth, particularly to describe and ascertain the nature by my said Invention, and in what manner the same is to be performed, of an instrument in writing under my hand and seal, and to cause the same to be inrolled in His Majesty's High Court of Chancery within six calendar months next and immediately after the date of the said recited Letters Patent, as in and by the same, reference being thereunto had, will more fully appear.

NOW KNOW YE, that in compliance with the said proviso, I, the said Alexander John Forsyth, do hereby declare that the nature of my said Invention of an advantageous method of discharging or giving fire to artillery and all other fire-arms, mines, chambers, cavities, and places in which gunpowder or other combustible matter is or may be put for the purpose of explosion, is expressed in the foregoing general description thereof, and the manner in which the same is to be performed is as follows, that is to say:

First, as to the chemical plan and principles thereof, instead of permitting the touch-hole or vent of the pieces of artillery, fire-arms, mines, chambers, cavities, or places to communicate with the open air, and instead of giving fire to the charge by a lighted match, or by flint and steel, or by any other matter in a state of actual combustion applied to a priming in an open pan, I do close the touch-hole or vent by means of a plug or sliding piece, or other fit piece of metal or suitable material or materials, so as to exclude the open air, and to prevent any sensible escape of the blast or explosive gas or vapor outwards, or from the priming or charge, and as much as as possible to force the said priming to go in the direction of the charge, and to set fire to the same, and not to be wasted in the open air; and as a priming I do make use of some or one of those chemical compounds which are so easily inflammable as to be capable of taking fire and exploding without any actual fire being applied thereto, and merely by a blow, or by any sudden or strong pressure or friction given or applied thereto without extraordinary violence; that is to say, I do make use of some one of the compounds of combustible matter, such as sulphur or sulphur and charcoal, with an oxymuriatic salt, for example, the salt formed of dephlogisticated marine acid and potash (or potasse), which salt is otherwise called oxymuriate of potash; or I do make use of such of the fulminating metallic compounds as may be used with safefy, for example, fulminating mercury, or of common gunpowder mixed in due quantity with any of the before-mentioned substances, or with an oxymuriatic salt as aforesaid, or of suitable mixtures of any of the before-mentioned compounds; and these compounds or mixtures of compounds I find to be much better for priming than gunpowder used alone, which cannot be made to explode without some sparks or actual fire applied thereto, or else without such a degree of extraordinary and violent percussion as cannot conveniently be made use of in gunnery, or with any of the fire-arms or artillery that are in most general use. But it is to be observed that I do not lay claim to the Invention of any of the said compounds or matters to be used for priming, my Invention in regard thereto being confined to the use and application thereof to the purposes of artillery and fire-arms as aforesaid; and the manner of priming and exploding which I use is to introduce into the touch-hole or vent, or into a small and strong chamber or place between the said touch-hole and vent, and the plug or sliding piece, or other piece by which the communication with the external air is cut off, a small portion of some or one of the chemical compounds herein-before mentioned (for example, as for priming to a musket, about the eighth part of a grain), and when the required discharge is to be made I do cause the said chemical compound or priming to take fire and explode by giving a stroke or sudden and strong pressure to the same, communicated by and through the said plug or sliding piece, or other piece before mentioned or described, in consequence of which the fire of the priming is immediately communicated to the contents or charge placed within the said piece of artillery, fire-arm, mine, chamber, cavity, or place, and the discharge accordingly follows.

And, secondly, I do hereby further declare, for the better illustration of my said Invention, and as auxiliary to the use thereof in relation to the mechanical parts thereof, that I have hereunto annexed Drawings or Sketches exhibiting several constructions which may be made and adopted in conformity to the foregoing plan and principles out of an endless variety which the subject admits of. Figure 1st, represents the section of a piece of artillery where the charge is inflamed upon the upper side. Figure 2d, represents the section of a piece of artillery where the charge is inflamed at one of its ends through a touch-hole in the line of direction of the barrel or bore of the piece. Figure 3d, is a representation of the same thing as Figure 2d, except that the narrow part of the touch-hole is lengthened more than in Figure 2d. The same letters answer for the description of Figures 1, 2, & 3d.

In Figure 1st, A, A, represents the section of a piece of artillery; B, the touch-hole, of a cylindrical bore, excepting at the bottom, where it becomes a small perforation leading to the chamber. C, is a cylindrical punch or plug fitting the bore of B, and (if need be) clothed, packed, or faced near its upper part with leather or binding, or any other material proper to render its fitting more correct, and its motion easy and smooth. The lower part of C, is made to fit the lower part of the touch-hole B, and there is a a small internal cavity at the lower part C, in which a portion of the said chemical compound may be lodged by dipping the said punch or plug therein; or a sufficient quantity of the said chemical compound is let fall into B, part of which lodges upon the shoulder at B, Figure 1st, or in the bottom of the bore at B, Figures 2d, and 3d; and in this state the said punch or plug C, is to be inserted in the touch-hole; and a smart blow being given on its upper end, the said chemical compound being suddenly compressed between the two faces nearly in contact will explode and give fire to the remaining portion, and also to the charge. The apparatus, Figures 1, 2, & 3, is chiefly to be recommended for its simplicity, but it is attended with several obvious inconveniences.

Figure 4th, represents the section of an apparatus which may be used to prime and discharge a musket, or any other fire-arm, a great number of times, even altho' the breech of the same is under water. A, A, represents a section of the barrel of the piece; D, the chamber of the barrel, which is contracted to a narrow touch-hole at I and open up to about double the diameter of the narrowmost part of the touch-hole at B, *b*. Into B, *b*, is introduced the rod or or plug C, C, C, at *b*; where the touch-hole is opened wider there is another hole F, of the same diameter as B, *b*, and at right angles to B, *b*. A cylindrical piece of any proper metal with a hole nearly through it in the line of direction of the axis passes through the solid breech of the barrel as near to I, and *b*, as can be done with safety and convenience. There is also another hole in one side of the cylinder at right angles to its axis, and corresponding with the hole F, in the breech of the gun when the cylinder is fixed in its proper place; or, as the cylinder must be accurately fitted into the breech, and move round freely this last-mentioned hole may stand either at F, or K. This first-mentioned hole in the cylinder E, is filled with any of the chemical compounds before mentioned; when the piece is to be primed the hole in the side of the cylinder E, is brought directly over F, and the powder falls out of the cylinder through F, into the small cavity at *b*, between the point of the rod or plug C, C, C, and the contracted part of the touch-hole at I, where the piece is to be discharged; the side hole in the cylinder E, is turned round to K, and the rod or plug C, C, C, impelled forward by the stroke of a hammer or spiral spring as at G, G, G, or by any other mechanical contrivance, so that by its pressure or concussion against the shoulder at I, it fires the priming contained in the cavity at *b*. H, H, is a screw, by which the power of the action of the spiral spring G, G, G, may be increased or diminished as necessary.

Figures 5th, and 6th, represent another apparatus, which may be used in the same manner as the preceding. The inner circle B, represents a section of a flat cylindrical piece, having a central stem on the opposite side or face to be screwed into the barrel where the touch-hole is commonly placed (another section of the cylinder B, is given Figure 7th). The axis or central line of the said stem is perforated or drilled up to half the hickness of the cylindrical piece so as to meet another hole D, B, drilled or made in the edge of the said piece. The space A, A, between the inner and outer circle denotes another flat piece, of which the inner part is hollowed out so as very accurately to fit the outside of the first-mentioned cylindrical piece, and to be capable of revolving upon it when turned by the handle C. At E, is a plug or punch inserted in the hole in the edge of the moveable piece, and kept from immediately touching the same by means of a spring of any suitable form. On the opposite part of the circumference of the moveable piece, and through the handle at C, there is bored a hole or cavity K, quite through to its inner surface, in which a considerable quantity of the chemical compound before mentioned is put and confined. G, is a cock or arm for giving a stroke upon E; it may be set, discharged, and impelled by the same machinery and means as are used with respect to the cock of a common gun lock, or the stroke may be given by any other means. The several parts of the mechanism are secured in their places by caps and screws, or by any other well-known means commonly used in works of this kind. In the use of the apparatus last described, the hammer G, is to be raised and cocked, and the handle C, brought round to the position, as in Figure 6th, where it is stopped by a pin or projecting piece H. At this instant the hole or cavity K, is brought immediately over the channel or hole D, B, into which a priming falls out of K, in consequence of the slight stroke or jar produced by the sudden stoppage of C. Immediately after this operation the handle is to be returned to its first position, as at Figure 5th, which brings the plug or punch E, directly over the channel or hole D, B, in which position only it is possible to give the stroke so as to inflame the priming and discharge the piece. When this last effect is required to be produced the trigger is to be drawn, and then G strikes E, and the contents of D, B, taking fire explode through the touch-hole, and set fire to the charge without allowing the escape of the gas or vapor in any direction but into the chamber and towards the muzzle of the piece.

In witness whereof, I, the said Alexander John Forsyth, have hereunto set my hand and seal, this Fourth day of July, in the year of our Lord One thousand eight hundred and seven.

ALEXR J. FORSYTH. (L.S.)

A.D. 1807. April 11. No. 3032.

FORSYTH'S SPECIFICATION

(1 SHEET.)

FIG. 1ST.

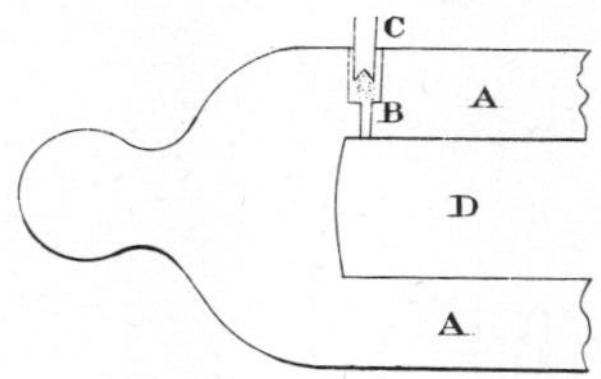

FIG. 3RD.

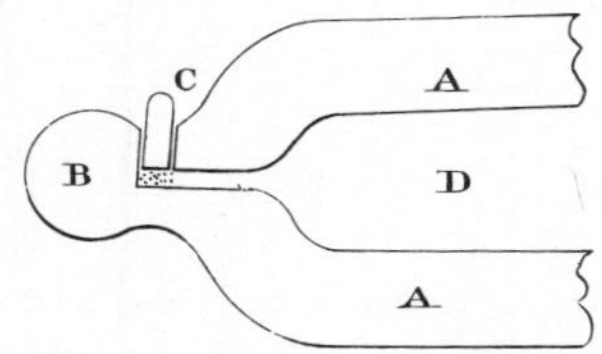

FIG. 5TH.

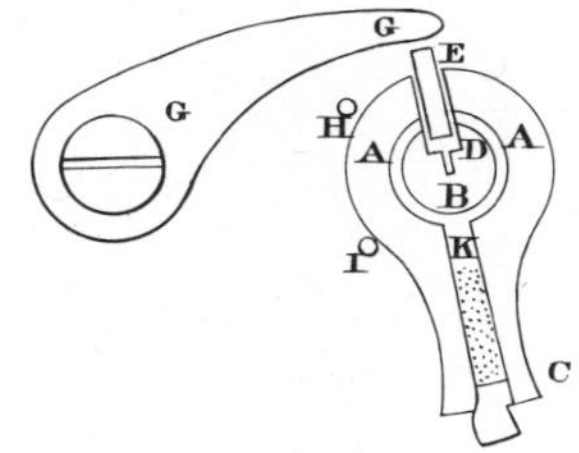

FIG. 2ND.

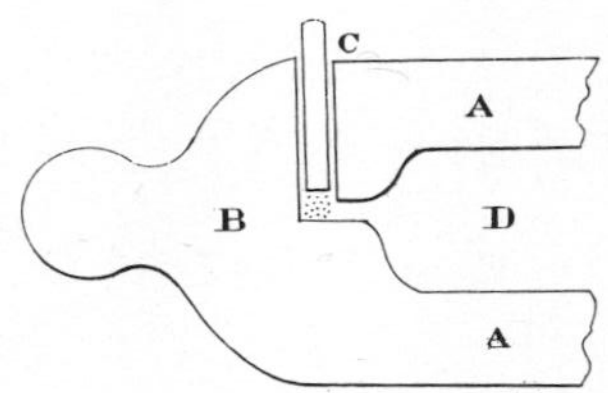

FIG. 4TH.

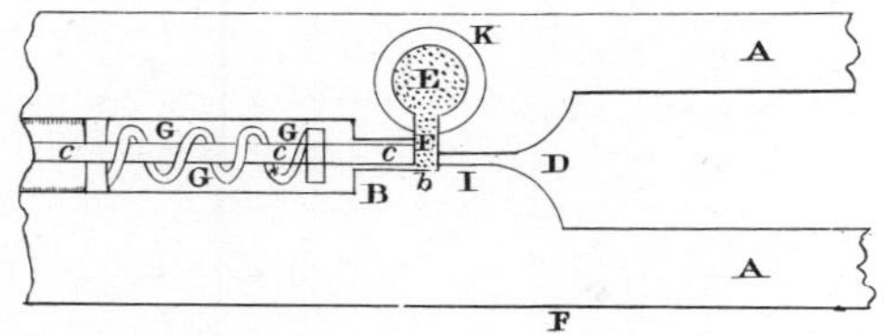

FIG. 6TH.

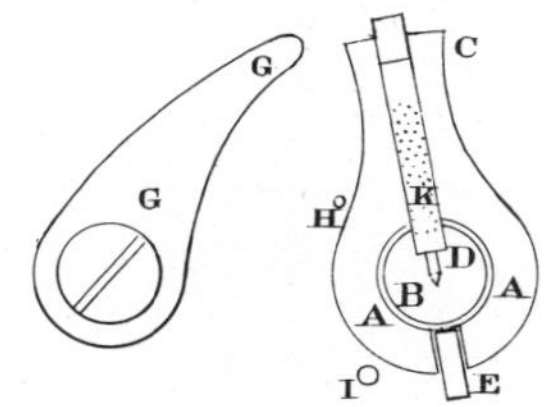

FIG. 7TH.

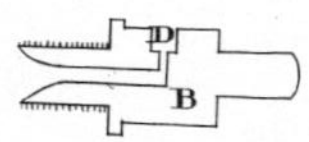

The enrolled drawing, is colored.

Malby & Sons, lith.

London: Printed by George Edward Eyre and William Spottiswoode.
Printers to the Queen's most Excellent Majesty. 1854.

A.D. 1809 . . . . N° 3286.

Lock for Fire-arms.

MANTON'S SPECIFICATION.

TO ALL TO WHOM THESE PRESENTS SHALL COME, I, John Manton, of Dover Street, in the Parish of Saint George, Hanover Square, and County of Middlesex, Gunmaker, do send greeting.

WHEREAS His most Excellent Majesty King George the Third, by His Letters Patent under the Great Seal of the United Kingdom of Great Britain and Ireland, bearing date of Westminster, the Eleventh day of December, in the fiftieth year of His reign, did give and grant unto me, the said John Manton, my executors, administrators, and assigns, His special licence, full power, sole privilege and authority, that I, the said John Manton, my executors, administrators, and assigns, during the term of years therein expressed, should and lawfully might make, use, exercise, and vend, within England, Wales, and the Town of Berwick upon Tweed, my "**Improved Lock for Guns and Pistols**;" in which Letters Patent there is contained a proviso, obliging me, the said John Manton, under my hand and seal, to cause a particular description of the nature of my said Invention, and in what manner the same is to be performed, to be inrolled in His Majesty's

High Court of Chancery within two calendar months next and immediately after the date of the said in part recited Letters Patent, as in and by the same (relation being thereunto had) may more fully and at large appear.

NOW KNOW YE, that in compliance with the said proviso, I, the said John Manton, do hereby declare that my said Invention is described in the plan or Drawings in the margin of these presents, and the description thereof hereunder written, that is to say:

The Drawing, Fig. 1, shews the outside of the lock when at half cock; Fig. 2, the outside when struck down; and Fig 3, the plan of the inside. Letters A, represent the outside of the pan; B, the hammer; C, the cock; D, the pin or screw which confines the hammer and hammer spring, and on which the hammer acts; E, the inside end of the pan, which comes opposite the touch hole, the hammer B, acting downwards opens that side of the pan nearest to the cock C, to admit the sparks to the prime; the manner of its opening is described in the Drawing, Fig. 2; the hammer returning to its jointing fills up the opening in the pan. The hammer is furnished with a strong steel face, fastened by a stud in the back and a small screw thro' the hammer. At the end of the hammer face, nearest the pan, is a small grove or notch sunk in the hammer, to carry off any wet that may come down upon it. The hammer is fixed to the plate by the same screw D, that fastens the hammer spring on the inside; the hole in the shank of the hammer being screwed, it turns on the hammer spring pin D, which comes thro' the plate about three eighths of an inch. There is a projection on the inside of the hammer spring one fourth of an inch long, which comes through a square hole in the plate into a hole in the shank of the hammer and forces it to return to its jointing with the pan; when the lock is brought to half cock, the cock C, is flat on the inside, and is barely one eighth of an inch thick, it passes between the plate and the hammer when it comes down. The jaws project outwards to answer the hammer; a bulge is left on the breast of the cock to render the fitting of the squares of the tumbler more strong and perfect. When the lock is struck down, the flint comes in contact with the hammer face, near the end, and forces it down sufficiently to admit the sparks to the pan. The pan is three fourths or seven eighths of an inch long from the outside of the plate. The Drawing, Fig. 3, shews the inside of the lock so very plain and clear that it cannot be misunderstood by any lockmaker. The inside of the pan E, is round, and the same size from end to end; about one third is cut out to receive part of the hammer, as described in the Drawing, Fig. 1. The main spring has a stud, like others. The end of the stud side is leveled, to fit under the end of the nib, by which it is prevented from rising. The crane of the tumbler has a roller in the end, on which the main spring acts. The bridle has a strong leg on the inside with a round stud, which fits into the plate near the sear nose, to prevent it from twisting when the tumbler comes in contact with the eye to stop the cock. The sear acts on the tumbler in the usual way, but the shank is nearly vertical, instead of horizontal, as they are in general. The sear spring acts on a shoulder left on the outside of the sear for that purpose, and forces the sear nose to the tumbler. The pan of this lock is primed from the touch hole, from the compression of the air in loading.

The following are the principal advantages derived from this improved lock: First, the pan being solid with the plate at top, protects the prime from wet; secondly, the hammer opening downwards and the flint acting in a direct line with the pan, the sparks communicate quicker to the prime; thirdly, the hammer returns to its jointing with the pan when the lock is brought to half cock, without any additional trouble to the user; fourthly, the lowness and compactness of the lock altogether renders it much less difficult to protect from wet, and much less liable to accidents by catching, in cover shooting, than locks of the present construction.

In witness whereof, I, the said John Manton, hath hereunto set my hand and seal, this Tenth day of February, in the fiftieth year of the reign of our Sovereign Lord King George the Third, and in the year of our Lord One thousand eight hundred and ten.

JOHN (L.S.) MANTON.

A.D. 1809. DEC 11. No. 3286. (1 SHEET.)

MANTON'S SPECIFICATION

FIG. 1.

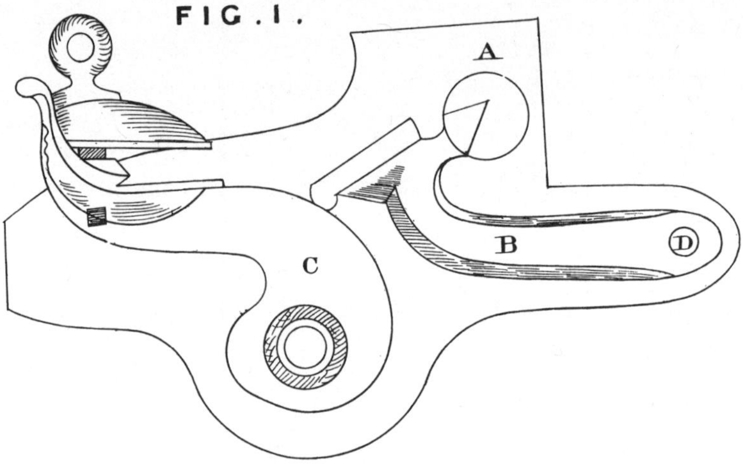

FIG. 2.

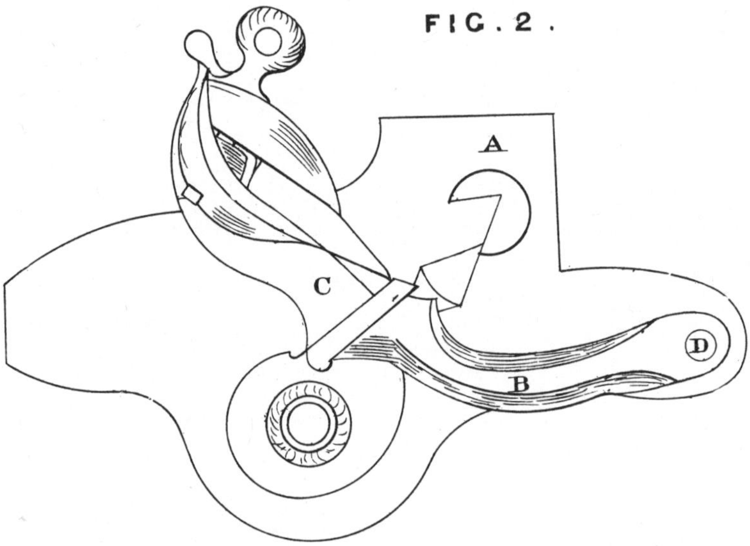

FIG. 3.

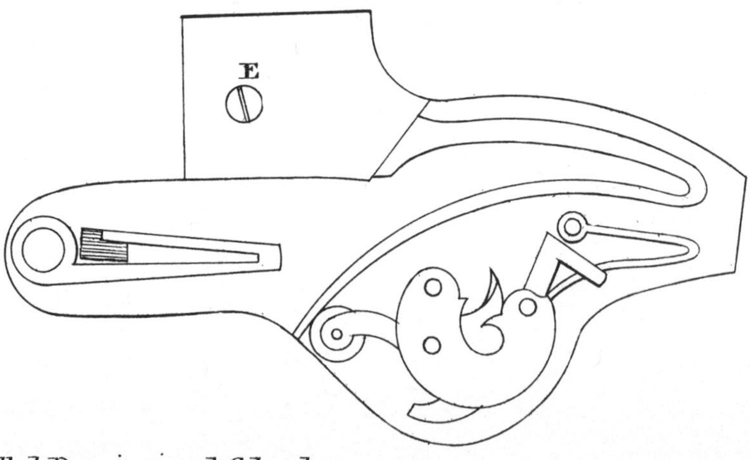

The Enrolled Drawing, is not Colored.

Malby & Sons, lith.

LONDON: Printed by GEORGE EDWARD EYRE and WILLIAM SPOTTISWOODE.
Printers to the Queen's most Excellent Majesty. 1854.

A.D. 1812 . . . . Nº 3558.

Locks and Breeches of Fire-arms; also Musical Spring for Trigger Plate.

MANTON'S SPECIFICATION.

TO ALL TO WHOM THESE PRESENTS SHALL COME, I, JOSEPH MANTON, of Davies Street, in the Parish of Saint George, Hanover Square, in the County of Middlesex, Gun Maker, send greeting.

WHEREAS His present Majesty King George the Third did, by His Letters Patent under the Great Seal of the United Kingdom of Great Britain and Ireland, bearing date at Westminster, the Thirtieth day of April, in the year of Our Lord One thousand eight hundred and twelve, and in the fifty-second of His reign, for Himself and His successors, give and grant unto me, the said Joseph Manton, my ex̃ors, adm̃ors, and assigns, His special license, full power, sole priviledge and authority, that I, the said Joseph Manton, my ex̃ors, adm̃ors, and assigns, and every of them, by myself and themselves, or by my and their deputy or deputies, servants or agents, or such others as I, my ex̃ors, adm̃ors, and assigns, should, at any time agree with, and no others, from time to time and at all times thereafter during the term of years therein expressed, should and lawfully might make, use, exercise, and vend, within England, Wales, and the Town of Berwick upon

Tweed, my new invented "IMPROVEMENTS IN GUNS AND PISTOLS;" in which Letters Patent there is contained a proviso, obliging me, the said Joseph Manton, by an instrument in writing under my hand and seal, particularly to describe and ascertain the nature of my said Invention, and in what manner the same is to be performed, and to cause the same to be inrolled in His Majesty's High Court of Chancery within two calender months next and immediately after the date of the said recited Letters Patent, as in and by the same, relation being thereunto had, may more fully and at large appear.

NOW KNOW YE, that in compliance with the said proviso, I, the the said Joseph Manton, do hereby declare that my said improvements upon guns and pistols consist in the following particulars:

First, what I call a gravitating stop, which being applied to the lock of a gun or pistol will render the same less liable to be accidentally discharged while loading if it should have been inadvertently set at the full cock.

Secondly, in an inverted breeching, which contains proper cavities to permit the escape of any water which may fall upon the breeching of a gun or pistol when used in rainy weather. And also,—thirdly, a lip added to the hammer of guns and pistols, which will turn off the water into the said cavities, and render it less liable to enter into the pan.

Fourth, what I call a musical spring or bar, which being applied within the trigger plate will cause the speaking or click (that is, the sound produced by drawing back the cock,) to be a pleasant and musical sound. The particulars of these Inventions are explained by the Drawing hereunto annexed. Figure 1, is a side view of a double barreled gun to which the improvements are applied; the gravitating stock is marked A, B, C, it is a bent lever moveable on the centre pin C, which is supported by a bridge D, screwed against the plate of the lock; the circular part A, is a weight or counterpoise which governs the action of the gravitating stop; the arm B, is so situated that it may be moved towards the centre of the cock, and passing under a hook B, formed out of the cock, it becomes a stop to the cock and prevents its striking the hammer. Although the trigger is discharged whenever the gun is used for shooting, the weight A, disposes the gravitating stop as in Figure 1, in which position it has no action; the arm which carries the weight A, fall against a part of the bridge D, and removes the arm B, so far from the centre of the cock that it does not interfere with the hook E, the gun may therefore be fired off without any interruption from the stop B. Now suppose the gun placed in such a position that its barrel is vertical, or nearly so, as it will be when the but end or heel plate of the stock is placed flat on level ground (or if the barrel is ever so much inclined so that the trigger is on the upper side), then the gravitating stop will prevent the discharge of the gun, because it assumes the position shewn in Figure 6; the weight A, now inclines the stop B, to rest under the hook E, of the cock, and thus prevents the gun from being discharged by holding up the cock, even if the trigger should be accidentally drawn. The gun maker should in general dispose the gravitating stop in such a position that when the gun barrel is inclined about an angle of eighty degrees to the horizon, the weight A, will be just upon the point of falling over towards the cock; it is evident that in this case the gun cannot be used to shoot an object at a greater distance than eighty degrees, but as very few sportsmen will require so great an elevation as this, the gravitating stop may be made so as to fall over at a less angle, in which case the gun will be more safe from accidents, though it cannot be used to shoot at so great an elevation. Other sportsmen may wish to have a gun which will shoot perpendicular; in this case the gravitating stop must be made accordingly, so as not to fall over when the barrel of the gun is vertical; the adjustment of the stop must be made by trial, bending one or other of its arms until it will fall over when the barrel is in the intended position. It is necessary that the gravitating stop should be perfectly free in its movements, for which purpose the pivots must be well fitted, and it should be raised so as not to rub against the lock plate in its motion; this is shewn in Figure 7, which is an edge view, shewing how it clears both the lock plate and the bridge D, in consequence of the part which surrounds the centre pin being so much thicker; the gravitating stop is also shewn detached in the same Figure, the same letters denoting the same parts.

Respecting the gravitating stop, I have farther to observe that it may, if required, be adapted to catch the cock at half cock, which will render the gun less liable to go off at half cock while loading, though I do not attach much value to this idea, because no well made lock will be in any danger of discharging at the half cock; it will be effected by filing away the hook of the cock or the end of the arm B, until the stop will fall under it at the half cock, in the same manner as is represented in Figure 6, at full cock.

Secondly, the inverted breeching is shewn in Figures 2, 3, 4, and 5. The water which falls upon the barrels F, G, Figure 2, drains down by the side of top piece H, and in the common double barrelled gun is liable to enter the pan, but in my improved inverted breeching this water drains down two inclined planes I, K, which pass beneath the top piece H, and are brought together into one mortice formed between the breeches of both barrels, and the mortice being continued down through the stock permits the water to pass away from the gun, the mortice between the breechings L, M, is shewn in Figures 3, and 5, at N. The continuation of the mortice down through the stock is shewn at O, Figure 3; it is a square tube of metal fitted into the wood, and round the upper end of it a quantity of leather is placed, upon which the breeches of the barrels L, M, are pressed down when in their places, and this forms a water-tight joining of the tube O, P, with the mortice N. The manner in which I apply the inverted breeching to a single barrelled gun or pistol is shewn in Figures 8, 9, and 10, which represent a breech of this kind detached from the barrel into which it is fastened by the screw A, in the common manner, and B, is the hook by which it is attached to the false breech; C, C, is the flat side filed to the breeching in which the touch hole is formed by a piece of platina D, Figure 9, screwed and rivetted in the usual manner. Now my improvement consists in a passage or mortice F, the lower surface of which forms an inclined plane E, F, G, down which the water that may drop upon it runs and escapes from the gun on the side opposite to the lock; at G, a small lip is formed to turn the water fairly over the wood of the stock and prevent it getting in the joint between the wood and the breech. In guns and pistols for military purposes it will not be necessary to leave the square or flat H, on the top of the breech.

Thirdly, the lip formed on the hammer is explained in all the Figures where the hammer is shewn, but particularly in Figure 3, which is a transverse section taken through the touch hole of the double barrelled gun. *a, b, c,* is a section of the hammer, shewing the manner in which its lip *a,* overhangs the upper edge of the inclined plane K, and turns off any water which may drop upon it, rendering it less liable to get into the pan; the lip of the hammer is applied to the single breeching, Figure 9, in exactly the same manner, overhanging the sharp edge E, of the inclined plane E, F, G. The form of the hammer, and the manner in which the lip is made out of it, is explained by Figures 11, 12, 13, which represent it in different views. On the left hand side of Figure 3, one of the pans is shewn open; the same letters here refer to its lip as in the opposite side. Figure 14, represents the fourth particular of my improvements, namely, the musical spring or bar; in this Figure *d, d,* represents the trigger plate, which is screwed into the stock of the gun by a screw which is tapped into the socket; *g, q, r,* are the triggers which pass through the trigger plate, and are moveable on the centre; *e, f,* is the small spring, which keeps them always in contact with the seer; all these are the parts of the common trigger; but my improvement consists in a spring or bar *k,* screwed to the plate, and each of the triggers has a small pin *i,* fixed into it, when the cock is drawn back, and the seer gives the trigger a sudden jirk by falling into the bents of the tumbler; those pins strike upon the spring or musical bar *k,* and as the triggers return instantly, they suffer the spring to vibrate and produce a vibrating sound which will be more pleasing to the ear than the dead click produced by the trigger in the common manner striking upon the trigger plate *d, d.* The spring *k,* may be made curved, as in the Figure, or it may be made straight, at the fancy of the maker, because as it is only intended to produce a vibration by the triggers striking upon it, its form and size will only affect the sound it will produce, and this is merely a matter of fancy, and may be applied to single guns and pistols as well as double. It is scarcely necessary to observe that the spring must be so fixed as to be perfectly free and detached, that it may vibrate without interruption.

In witness whereof, I, the said Joseph Manton, have hereunto set my hand and seal, this Thirtieth day of June, One thousand eight hundred and twelve.

JOSEPH (L.S.) MANTON.

ALEXANDER.

AND BE IT REMEMBERED, that on the Thirtieth day of June, in the year of our Lord 1812, the aforesaid Joseph Manton came before our said Lord the King, in His Chancery, and acknowledged the Specification aforesaid, and all and every thing therein contained and specified, in form above written. And also the Specification aforesaid was stampt according to the tenor of the Statute made for that purpose.

Inrolled the Thirtieth day of June, in the year of our Lord One thousand eight hundred and twelve.

LONDON:
Printed by GEORGE EDWARD EYRE and WILLIAM SPOTTISWOODE,
Printers to the Queen's most Excellent Majesty. 1854.

A. D. 1812. April 30. Nº 3558.
MANTON'S Specification.

(1 SHEET)

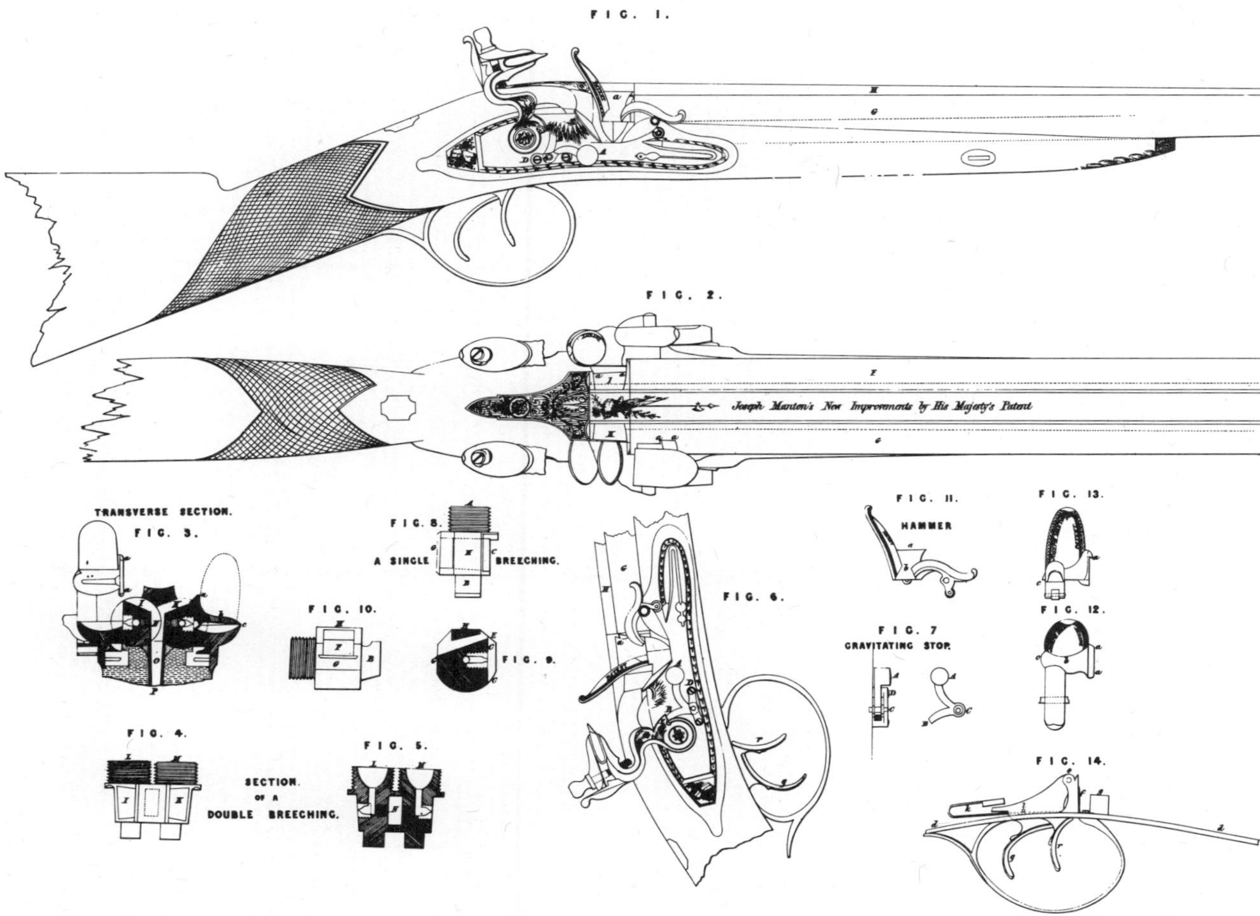

The marked drawing is coloured.

Malby & Sons. Lith.

London: Printed by George Edward Eyre and William Spottiswoode, Printers to the Queen's most Excellent Majesty. 1854.

A.D. 1814 . . . . N° 3833.

Apparatus for Discharging Fire-arms by Means of Compressed Air.

PAULY'S SPECIFICATION.

TO ALL TO WHOM THESE PRESENTS SHALL COME, I, JEAN SAMUEL PAULY, of Little Charlotte Street, in the Parish of Saint George, Hanover Square, and County of Middlesex, Engineer, send greeting.

WHEREAS His most Excellent Majesty King George the Third did, by His Letters Patent under the Great Seal of the United Kingdom of Great Britain and Ireland, bearing date at Westminster, the Fourth day of August, in the fifty-fourth year of His reign, give and grant unto me, the said Jean Samuel Pauly, my ex̃ors, adm̃ors, and assigns, His special licence, full power, sole privilege and authority, that I, the said Jean Samuel Pauly, my ex̃ors, adm̃ors, and assigns, should and lawfully might during the term of years herein mentioned, make, use, exercise, and vend, within England, Wales, and the Town of Berwick-upon-Tweed, my Invention of "IMPROVEMENTS IN THE CONSTRUCTION AND USE OF FIRE-ARMS;" in which said Letters Patent there is contained a proviso that if I, the said Jean Samuel Pauly, shall not particularly describe and ascertain the nature of my said Invention, and in what manner the same is to be performed, by an instrument in writing under my hand and seal, and cause the same to be inrolled in His Majesty's High Court of Chancery within six calendar months next and immediately after the date of the said Letters Patent, that then the said Letters Patent, and all liberties and advantages whatsoever thereby granted, shall utterly cease, determine, and become void, as in and by the same (relation being thereunto had, may more fully and at large appear.

NOW KNOW YE, that in compliance with the said proviso, I, the said Jean Samuel Pauly, do hereby declare that the nature of my said Invention, and manner in which the same is to be performed, are particularly described as follows, that is to say:

By the rapid action of a syringe or instrument of the like description, or by the more rapid action of a syringe or instrument as aforesaid than has heretofore been commonly used, I do put the atmospheric air into or upon gunpowder or other combustible material within the chamber or barrel of a musquet, fowling-piece, or other fire-arm or instrument intended to contain or allow of a charge of explosive powder of any kind or description; and I do prevent the atmospheric air, by a closure of the nature of a valve, or by an impediment produced by crushing a small portion of the powder in or very near the hole of communication or touch-hole of the said musquet, fowling-piece, or other fire-arm or instrument, from escaping before the condensation shall have become sufficient to afford the heat of ignition or giving fire; and when required I do produce the same effect by closing the hole of the syringe by a small slider with an hole therein, which I do introduce between the touch-hole aforesaid & hole of the syringe, and do move the said slider by an oblique stroke of the piston or otherwise, so as to cause a direct communication nearly at the time or very nearly when the stroke terminates. And, farther, I do cause the ignition, explosion, and discharge of such of the said musquets, fowling-pieces, or other fire-arms or instruments intended to contain or allow of a charge of explosive powder of any kind or description, by setting fire to the said charge by air forced upon the same, in manner aforesaid. And, moreover, with regard to the aparatus needful and expedient for carrying my said improvements into effect, which may be greatly varied according to convenience, or the circumstances under which the same may be required to be done, and with regard to any other said improvements, the same may be more particularly understood from the

Drawings hereunto annexed, and the explanations thereof. Fig. 1 A, is the vertical section of a gun, of which *g*, *g*, represents the breecth pin with part of the chamber having its touch-hole behind, and opposite to *b*, *b*, an air syringe, having its minute perforation opposite the touch-hole, and its piston *b*, *b*, driven by the tumbler *d*, which receives its force from the main spring when discharged by the trigger *m*, fitted up as usual, and the machinery is supported by and in connection with the lock plate *a*, and bridle *e*. Fig. 2, shews the same parts distinguished by the same letters, and will be sufficiently understood from the Drawings, excepting the piston *b*, *b*, where the fore part is seen perforated in the upper sketch, and closed by a valve opening outwards, but kept shut by a weak interior spring in the lower sketch, and the effect of the said valve is to admit the external air during the time of cocking the piece; and Fig. 1 A, is a view of the upper part of the same gun which has been drawn as double barrelled; *n*, represents the cock. Fig. 3, represents the method of charging at the breech by means of a joint at the inner end of the barrel, which is perforated all through, and is provided with cartridges attached to or contained in a cup or cap of wood, metal, or other fit material having the touch-hole therein, as shewn at *h*, Fig. 3 A, and at *f*, Fig. 2, one of which, when charged, is to be put into the breech of the piece; and the same kind of cap may be attached to the piece to open and shut, and used with a common cartridge, and the said cartridge or piece is provided with a small touch-hole, and is discharged by the means or method herein-before described at such time as the barrel of Fig. 3, has been restored to the proper right line direction thereof. K, shews a sliding piece for preserving the said direction, which may be done by a variety of other methods in common use for other purposes. Fig. 4, shews a method of charging for the use and application of my means of giving fire also by the breech. *i*, represents a leaver piece which lifts up and down, containing the air syringe and its aparatus, and the other parts of the lock are fixed along with the barrel and stock *u*. When *i* is lifted up, the charge of Figure 3, may be put in at *h*; and upon restoring it to its ordinary position, wherein *i* and *p* are locked together, the whole aparatus becomes united, and the discharge may be made as herein-before described; and in both constructions wherein Fig. 3, is used the parts are made to come together in the manner of a curve lined joint, so that Fig. 3, is very securely held in its place. Fig. 6 *a*, is the profile, and Fig. 6 *b*, the plan of a cannon to be discharged by condensed air, and to be charged at the breech; and the same numeral figures everywhere denote the same things. 1, 1, 1, 1, denote two side pieces which embrace the cannon, and are applied upon the hinder trunions 5, unto which the same are attached by the screw 6; and, moreover, the screws and bolts 10, 10, 10, 10, pass through both side pieces, and connect the same together with great sollidity. 4, is a breech piece capable of being firmly applied to the breach end of the barrel, against which it fitts with great precision. 2, is a powerful screw which passes through the side pieces, and is so attached to the breech piece Fig. 4, that it is not only capable of giving severe pressure thereto, but likewise of withdrawing the same to a small distance to releive the friction of the curve lined joint. 7, 8, represent a syringe piston for giving fire to the powder or other part of the charge by compressed air. In Fig. 6, are seen the cannon with its side pieces and other connected parts raised and the bore open, into which the cartridge or charge 9, Fig. 6 B, which is of the same kind or description as that shewn in Fig. 3 *b*, is shewn, and afterwards when the side pieces have been returned to the situation in which the apertures of the syringe and of the cartridge are accurately in the same line, and the syringe is charged or drawn back, the piece is ready to be fired by the condensed air urged by a stroke or pressure at 7, or otherwise. And the said syringe may be constructed and fitted up in various ways, but I do prefer making the form shewn in *b*, Fig. 2, and that the stroke or pressure be given by the action of a strong helical spring, and the holding up or discharging of the same be done by a detent or catch, as is well known. Fig. 7, shews the method of placing the barrells and cartridges to be used in a considerable number for quick firing, according to my Invention.

In witness whereof, I, the said Jean Samuel Pauly, have hereunto set my hand and seal, this Third day of February, in the year of our Lord One thousand eight hundred and fifteen.

JEAN SAMUEL (L.S.) PAULY.

A.D. 1814. August 4. No. 3833.
PAULY'S SPECIFICATION.

(1 SHEET.)

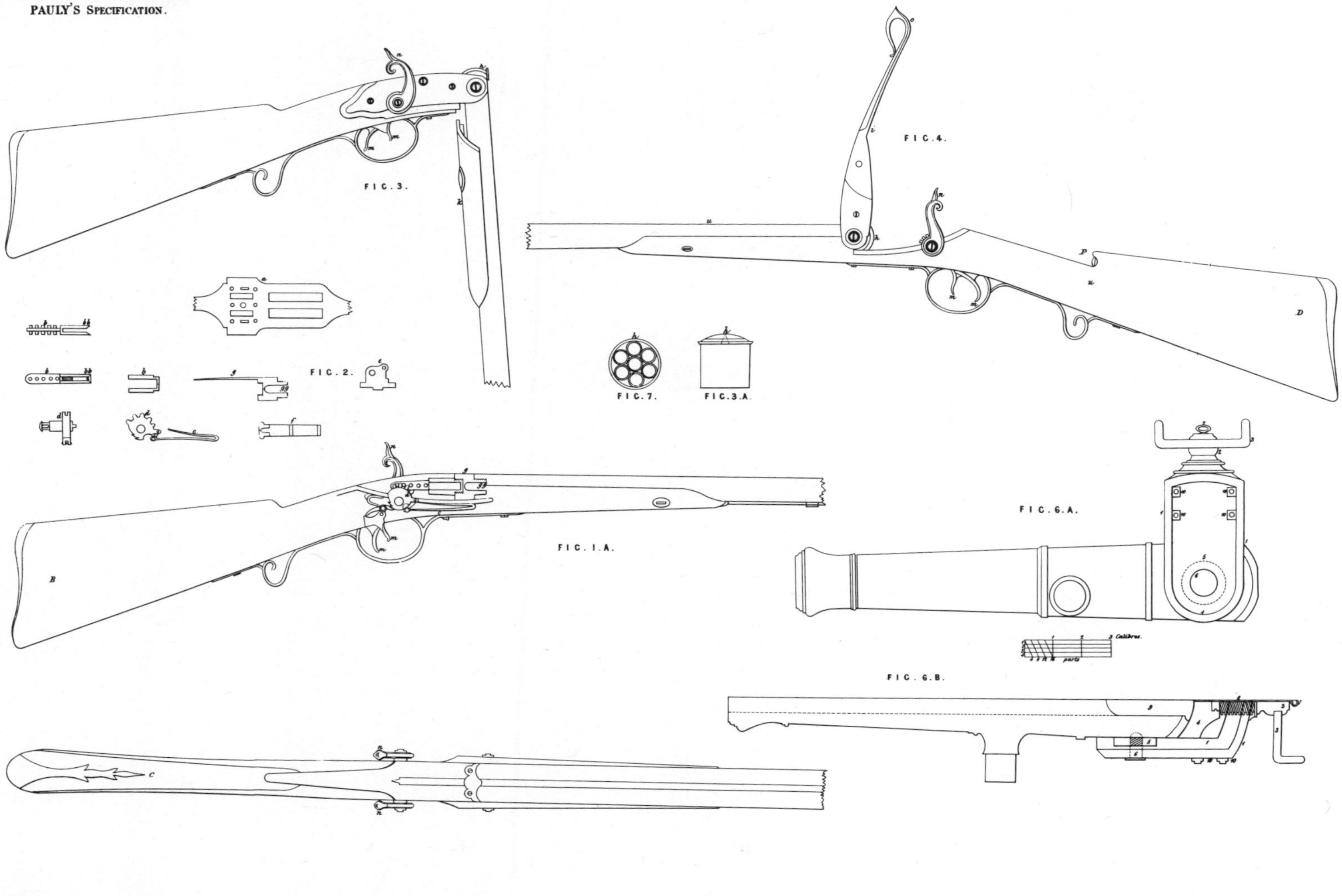

The enrolled drawing is not colored.

Malby & Sons, lith.

London: Printed by George Edward Eyre and William Spottiswoode, Printers to the Queen's most Excellent Majesty. 1854.

A.D. 1815 . . . . N° 3942.

Pans and Hammers of Fire-arms.

MANTON'S SPECIFICATION.

TO ALL TO WHOM THESE PRESENTS SHALL COME, I, JOHN MANTON, of Dover Street, Piccadilly, in the County of Middlesex, Gun Maker, send greeting.

WHEREAS His most Excellent Majesty King George the Third, by His Letters Patent under the Great Seal of the United Kingdom of Great Britain and Ireland, bearing date at Westminster, the Twenty-first day of July, in the fifty-fifth year of His reign, did, for Himself, His heirs and successors, give and grant unto me, the said John Manton, my ex̃ors, adm̃ors, and assigns, His especial licence, full power, sole priviledge and authority, that I, the said John Manton, my ex̃ors, adm̃ors, and assigns, and every of us, by myself and themselves, or by my and their deputy or deputies, servants or agents, or such others as I, the said John Manton, my ex̃ors, adm̃ors, or assigns, should at any time agree with, and no others, from time to time and at all times thereafter during the term of years therein expressed, should and lawfully might make, use, exercise, and vend my Invention of "**AN IMPROVEMENT IN THE CONSTRUCTION OF HAMMERS AND PANS TO THE LOCKS OF ALL KIND OF FOWLING PIECES AND FIRE-ARMS**" within that part of His said Majesty's

United Kingdom of Great Britain and Ireland called England, His Dominion of Wales, and Town of Berwick-upon-Tweed, in such manner as to me, the said John Manton, my exors, admors, or assigns, or any of them, should in my or their discretion seem meet; in which said Letters Patent is contained a proviso, that if I, the said John Manton, should not particularly describe and ascertain the nature of my said Invention, and in what manner the same is to be performed, by an instrument in writing under my hand and seal, and cause the same to be inrolled in His said Majesty's High Court of Chancery within two calendar months next and immediately after the date of the said Letters Patent, that then the said Letters Patent, and all liberties and advantages whatsoever thereby granted, should utterly cease, determine, and become void, anything therein-before contained to the contrary thereof in anywise notwithstanding, as in and by the said Letters Patent, relation being thereunto had, will more fully and at large appear.

NOW KNOW YE, that in compliance with the said proviso, I, the said John Manton, do hereby describe and ascertain the nature of my said Invention, and in what manner the same is to be performed, in manner following, that is to say:

My Invention consists of a ridge or projection along the seat of the hammer and a projecting ridge or division along the centre of the bottom of the pan, from the touch-hole end to the opposite end, which ridge and division I exemplify by the Drawings in the margin of these Presents, and further illustrate by the following explanation (vizt):— The seat of the hammer, instead of being flat or hollowed in the usual manner, is made with a ridge or projection, continued from one end to the other, as shewn by the Drawing marked A, which ridge or projection may be made in an angular or prismatic form, or curvilinear, or may be composed of two curvilinear surfaces, either of which forms will answer the purpose, though I prefer the angular or prismatic form. The ridge or projection in the seat of the hammer may be formed from the solid part of the seat of the hammer, or may be made separate, of iron, steel, or any other metal or metals fit for the purpose, and fastened on with screws, rivetts, or dovetails, or any other way the workmen may prefer. The height of such ridge or projection must be varied, according to the depth of the pan, leaving a space between such ridge or projection, and the projecting division in the pan sufficient to prevent them coming in contact, as represented by the Drawing in the margin marked B. The space or interval between the two ought to be no more than is sufficient to permit the grains of powder to pass. The division in the pan is represented by the Drawing in the margin marked C; such division is made in the centre of the bottom of the pan, from the touch-hole end of the pan to the opposite end, and may be formed out of the solid, or may be made of iron, steel, or any other metal or metals fit for the purpose, and may be fitted in after the pan is formed, by cutting a dovetail notch in the centre of the bottom of the pan from end to end to receive it. The sides of the notch should be closed with a set punch after the division is fitted, in to prevent it moving; this division should be about one sixteenth of an inch in height, and about the thirtieth of an inch thick, and it should extend the whole length of the inside of the pan; but I state these proportions by way of exemplification only; and as the ones which I have found convenient for others will also answer the purpose, the pan may be made either oval or straight, the latter I prefer. The principal use of the ridge or projection in the seat of the hammer is to divide the priming in the pan, and thereby forming two angles, or a curve in the prime, produce a greater surface of powder in the pan with the same or less quantity of powder than if the priming lay flat in the usual way in a pan of the same dimension, which will increase the quickness and certainty of ignition. The use of the projecting division in the pan is, to prevent the powder in the pan from shifting from one side of the pan to the other when the muzzle of the gun is elevated or depressed. To give these improvements the fullest effect it is necessary to have a hole in the ridge or projection on the seat of the hammer somewhat larger than the touch-hole, which hole must be made opposite the touch-hole, pointing downwards, and coming out through the centre of the ridge or projection about one eighth or one tenth of an inch from the touch-hole end of the pan, as shewn by the Drawing in the margin marked A, so as to allow the powder to pass freely from the touch-hole into the pan when compressed with the wadding; but I do not claim this hole as a part of my said Invention.

And I further do hereby declare that the above Drawings and description are given by me for the illustration of my said Invention and the assistance of workmen using it, but that the particulars thereof may be altered and varied; and that my said Invention does consist in the convexity, projection, or ridge on the seat of the hammer, and the division of the pan in whatever way such convexity, projection, or ridge may be made on the seat of the hammer, and such division in the pan only, so that there may be two spaces left in the pan for the powder to be in, which spaces communicate with each other, and an interval left between the projecting part of the pan and the projecting part of the seat of the hammer.

In witness whereof, I, the said John Manton, have hereunto set my hand and seal, the Thirty-first day of July, in the year of our Lord One thousand eight hundred and fifteen.

JN° MANTON. (L.S.)

JEKYLL.

AND BE IT REMEMBERED, that on the Thirty-first day of July, in the year of our Lord 1815, the aforesaid John Manton came before our said Lord the King, in His Chancery, and acknowleged the Specification aforesaid, and all and every thing therein contained and specified, in form above written. And also the Specification aforesaid was stampt according to the tenor of the Statute made for that purpose.

Inrolled the Thirty-first day of July, in the year of our Lord One thousand eight hundred and fifteen.

LONDON:
Printed by GEORGE EDWARD EYRE and WILLIAM SPOTTISWOODE,
Printers to the Queen's most Excellent Majesty. 1854.

A.D. 1815. July 21. N° 3942.

MANTON'S Specification.

(1 SHEET)

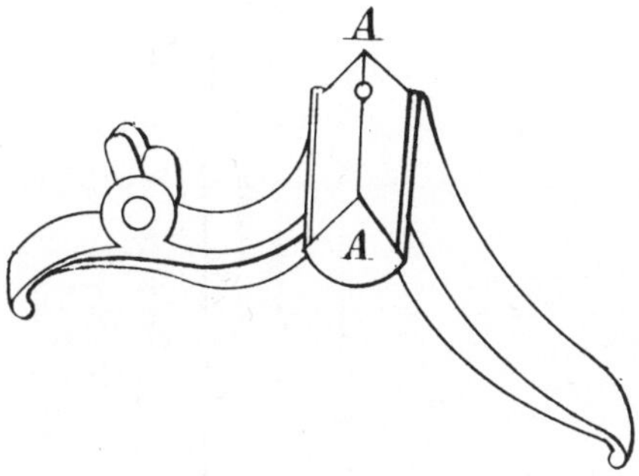

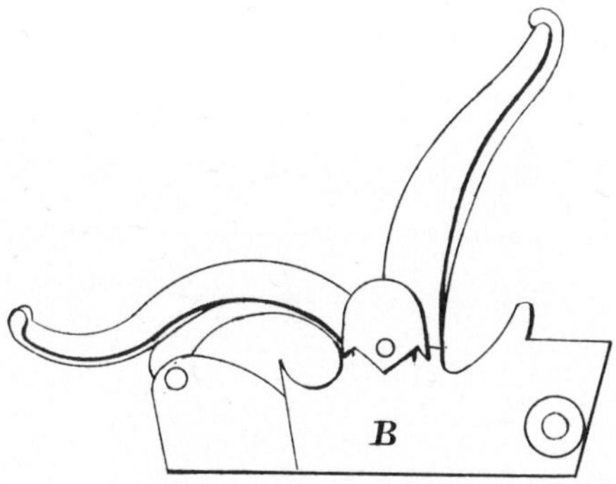

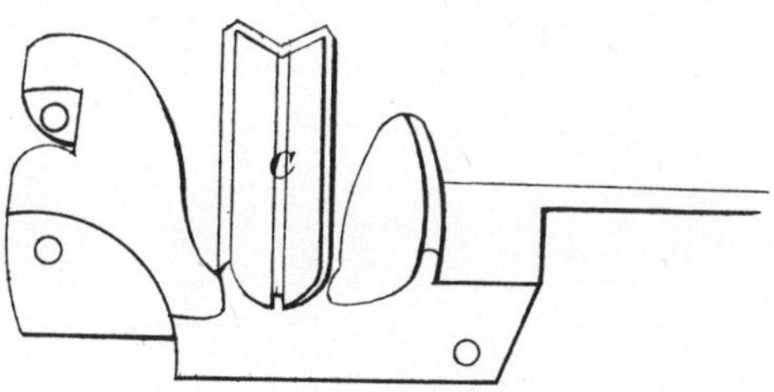

The enrolled drawing is colored.

Malby & Sons, lith.

London: Printed by George Edward Eyre and William Spottiswoode.
Printers to the Queen's most Excellent Majesty. 1854.

A.D. 1816 . . . . N° 3985.

Lock for Fire-arms, and Apparatus for Perforating Horses' Hoofs in Shoeing.

MANTON'S SPECIFICATION.

TO ALL TO WHOM THESE PRESENTS SHALL COME, I, JOSEPH MANTON, of Davies Street, Berkeley Square, in the Parish of St. George, Hanover Square, in the County of Middlesex, [illegible] Maker, send greeting

WHEREAS I did, by petition, humbly represent to His present Majesty King George the Third, that I had invented "**CERTAIN IMPROVEMENTS IN THE CONSTRUCTION AND USE OF CERTAIN OF THE PARTS OF FIRE-ARMS, AND ALSO IN THE SHOEING OF HORSES,**" His Majesty was pleased to issue His Letters Patent under the Great Seal of Great Britain, bearing date at Westminster, the Twenty-ninth day of February, in the fifty-sixth year of His reign, which Letters Patent do give and grant unto me, the said Joseph Manton, and to my ex̃ors, adm̃ors, and assigns, or to my and their servants or agents, His said Majesty's special license to make, use, exercise, and vend my said Invention within England, Wales, and the Town of Berwick upon Tweed, and to have and enjoy the whole profit and advantage arising by reason of the said Invention during the term of fourteen years from the date of the said Letters Patent, by which His Majesty also requires and strictly commands all His subjects whatever, within England, Wales, and the Town of Berwick upon Tweed, that during the continuance of the said term of fourteen years they shall not, either directly or indirectly, make, use, or put in practice my said Invention, or any part of it, or in any ways counterfeit, imitate, or resemble the same; the said Letters Patent also contain a proviso, obliging me, the said Joseph Manton, to paiticularly describe and ascertain the nature of my said Invention, and in what manner the same is to be performed, by an instrument in writing under my hand and seal, and cause the same to be inrolled in His Majesty's High Court of Chancery within six calendar months after the date of the said Letters Patent.

NOW KNOW YE, that in compliance with the said proviso, I, the said Joseph Manton, have executed this instrument, by which I do declare that my improvements in the construction and use of certain of the parts of fire-arms are applicable to and may be substituted for those parts of fire-arms which are for the purpose of fireing the gunpowder or discharging the piece. A piece constructed according to my improvements has no pan or magazine to contain powder for priming, nor any hammer and flint, but instead thereof, that part of the cock which holds the flint in common locks is made with a proper aperture to receive and hold a small tube, which contains a minute quantity of some of those substances which will produce fire and explosion when subjected to a sudden blow or pressure, and which are called detonating or fulminating substances or powders. To fire off or discharge such a piece the trigger must be drawn as usual, and the spring of the lock will throw the cock containing the tube suddenly forwards, by which the perforated end of the small tube which contains the detonating substance is made to strike suddenly on that part of the piece in which is the opening of the touch hole or passage leading to the charge in the barrel, so that the perforated end of the tube will cover the opening of the touch hole, and by the blow or concussion occasioned by the sudden striking of the tube as aforesaid, a pressure is caused to act suddenly upon the detonating substance with sufficient force to explode it, and the flame produced by the explosion issuing with violence from the perforated end of the tube passes through the touch hole or passage into the barrel of the piece and give fire to the charge of gunpowder contained within. When the piece is reloaded the said tube which contained the detonating substance can be readily detached from the cock and replaced by another, which is ready charged with the detonating substance; a sufficient number of such small tubes must be provided and kept ready charged for use. Now I wish it to be clearly understood that I do not make claim to the Invention of discharging fire arms by detonating or fulminating powder or substances, the same having been long since practised; but my Invention consists in the improvements of the mechanism for the production and application of the fire which are herein described. The advantages of a piece constructed in this manner are, that it fires with great quickness and is less liable to miss fire than guns with pans or priming magazines, and the perforated end of the tube and the touch hole may be covered with as much bees' wax or any such substance as will completely exclude damp air, and such covering of wax will form no impediment to the fire passing into the charge; such a piece is also much safer to use, being less liable to accidental discharges whilst loading or whilst carrying or handling in a loaded state, because the tube or tubes being removed from the cock or cocks there is nothing to produce fire.

This is attained by so constructing the aforesaid tube which contains the detonating substance, and also the cock to receive the said tube, that the said tube can be readily detached from the cock or applied thereto for use.

My improvements in the shoeing of horses consist in the application of a machine or instrument to bore holes in the hoof of the horse's foot for the reception of the nails, pins, or screws by which the iron shoe is to be fastened on.

The shoe is to be made with holes and countersinks in it and fitted to the hoof in the usual manner, and being applied in its place the tool or instrument is to be placed flat against the shoe to bore through the holes and countersinks thereof and penetrate into the hoof, or may be placed upon the hoof after the situation of the holes are marked; and its construction is such that it will be certain to enter the hoof in a direction perpendicular to the flat surface of the shoe or hoof until the borer comes out in the inclined or sloping part of the hoof; by this means the nervous or feeling part of the horse's foot will not be pricked, and the nails, pins, or screws, when inserted in their places, will follow the same direction. For a more complete description of the construction of my improvements I have hereunto annexed a Drawing, of which Figures 1, to 6, represent a lock for fire arms, constructed according to my improvements in such a manner that it may be applied to a gun, pistol, or any other piece to which a gun lock with a flint is or can be applied. The Figures are of the full size, and those parts which are of my Invention are marked with numeral characters for reference, but those parts which are marked with the letters of the alphabet are such as have been before used, and to which I make no claim, but are necessarily introduced to shew their connection with the parts of my Invention. A, B, Figures 1, 5, and 6, is the lock plate; C, D, 10, the cock. The tumbler, main spring, seer, and other parts behind the lock plate, which actuate the cock, may be made in the usual manner, or in any other improved manner, such being no part of my present Invention.

1, Fig. 2, is the detached tube to contain the detonating substance, and which is made up in the form of a small pellet *b*, and fitted to the bore of the tube; *e*, is a cylindrical wire fitted into the tube, and when it is in its place one end bears upon the pellet *b*, as shewn in Fig. 3, whilst the other end projects beyond the end of the tube 1, and being struck with a sudden blow will press upon the pellet *b*, and explode it; the end of the tube 1, is perforated with a small hole *f*, through which the fire issues; a number of such tubes, each with its wire *e*, must be provided and kept charged ready for use, in which state they will be like Fig. 3; and the hole at the end *f*, is to be stopped up with soft wax to secure the detonating substances from damp air, but that is not necessary in fine weather, when the tubes are carried in a bag or in the pocket; this wax will make no sensible impediment to the fire passing through the hole *f*. The wire *e*, must be accurately fitted to the bore of the tube, and must be slighly waxed with soft wax, to prevent its falling out of the tube; two small notches are made in the wire, near its end, for the convenience of taking hold of it to draw it out of the tube. To apply these detached tubes to use the upper part 10, of the cock has a cylindrical hole bored into it to receive and exactly fit the tube 1, when it is inserted as in Fig. 1; the end of the hole is plugged up by a screw 4, the end of which is in contact with the projecting end of the wire *e*, as shewn in the section Fig. 4. The tube, when inserted, is held into its place by a tooth projecting from the spring 5, and entering into one of the notches made in the tube at 6, 6, Fig. 3; this prevents the tube being withdrawn, but by turning the tube round on its axis the cylindrical surface of the tube between the notches 6, 6, applies itself to the tooth of the spring and presses it back, and then the tube can be drawn out for the convenience of turning the tube round with the finger and thumb; the end at 1, has three projecting rims which are milled or made rough.

The small screw 8, which holds the spring 5, in its place, enters into the plug screw 4, sideways, and prevents its being withdrawn or getting loose. When the trigger of the piece is drawn the motion of the cock carries the end of the tube 1, down in the dotted lines Fig. 1, and its end *f*, applies upon the surface 12, in which is the opening of the touch hole or passage leading to the charge, as shewn at 16, in Fig. 4, so that the small hole in the end *f*, of the tube will coincide with the opening of the touch hole or passage. By striking upon the surface 12, the motion of the tube 1, is arrested with a sudden stop, and by the motion of the cock the plug screw 4, presses upon the projecting end of the wire *e*, and communicates a sudden blow to the pellet of detonating substance, which causes it to explode; the flame thus produced blows through the small hole *f*, in the end of the tube into the touch hole or passage 16, and gives fire to the charge. The construction of the touch hole or passage is fully explained by Fig. 6, which is a cross section of the lock and breech through the touch. E, represents the solid of the breech, and F, the centre hole or passage to communicate with the charge in the barrel; from this hole the touch hole *a*, is carried sideways, and if the lock of the gun was of the common kind the pan for the priming would be placed against the hole; but when constructed according to my improvements, that part of the lock plate A, B, which applies against the touch hole *a*, has a hole or passage H, through it, forming a continuance of the touch hole. The outside of the hole H, is plugged up by a platina screw with a smaller hole through the centre of it, opening to the air at 13, from the passage H; a small touch hole or passage proceeds sideways in such a direction that it will exactly coincide with the axis of the tube 1, when the lock is discharged, as in Fig. 4, and 5; the surface 12, in which the touch hole opens must be fitted to the end of the tube, so as to make a good joint, and the hole in the cock for the reception of the tube 1, should be so directed that the perforation in the end *f*, of the tube will coincide with the touch hole or passage 16, leading to it. But to make the communication of the fire certain, even if the position of the tube should not be exactly correct, the end *f*, of the tube is hollowed out to a cup much larger than either the perforation through the end or the opening of the touch hole. The small perforation 13, through the center of the plug, which is screwed into the end or opening of the passage H, communicates from the touch hole to the open air and is made to allow the escape of part of the fire produced by the explosion of the detonating substance.

By this means I can charge the tube with rather more of the detonating substance than is absolutely necessary to produce sufficient fire to explode the gunpowder in the barrel, and this will render the explosion of the detonating substance by the concussion more certain than if a less quantity was used, and at the same time the quantity of flame thrown through the touch hole into the charge of powder will not be so great as to dislodge or give motion to the charge before the powder is set on fire. The perforation at 13, should be of such size that the grains of the gunpowder which is intended to be used in the gun will not run through it; the said perforation 13, also allows the air to escape from the barrel when the charge is rammed down in loading the gun. As the violence of the blast from the explosion of the detonating substance renders it liable to penetrate and enter into the joints between the parts, I make a circular channel in that part of the inside of the lock plate A, B, which applies to the breech E, Fig. 6, which channel surrounds that part which fits against the touch hole *a*. The section of this channel is represented by *m*, *m*, Fig. 6, and from it four small holes are drilled through the solid of lock plate and open on the outside, as shewn in Fig. 5; by this means, if any flame or vapour escapes at the fitting between the lock plate and the part of the breech immediately surrounding the touch hole *a*, it will be received in the circular channel *m*, and escape through the four small holes into the open air, without being so liable to penetrate beneath the lock plate into the works of the lock as it might otherwise do.

Figures 7, and 8, represent a double barrelled gun constructed purposely for the application of my improvements; in this the tube 1, and the cock C, D, 10, are made just the same as the former, but the opening of the touch hole or passage 16, which leads to the charge in the barrel, and on which the end of the tube strikes, is made in the breech of the barrel, instead of being made in a part of the lock plate. For this purpose the breach screws of the barrels are constructed as shewn in Figures 9, 10, and 11; the dotted line N, Fig. 10, is the concavity which forms the bottom of the barrel, and on which the charge of the powder lays; *n*, is the passage leading from it; *m*, is a cross passage, which at the out side end is plugged up with a screw having a small perforation, as shewn at 13, Fig. 7, to permit the escape of the superfluous fire, as before described, but too small to allow the powder to run through, as before described. The orifice 16, of the touch hole is made in the surface 12, and the direction of the hole is shewn in Fig. 10, at 16, to explain how it communicates with the cross passage *m*. The surface 12, is so inclined as to make a joint with the end of the tube 1; when the cock is down the direction of the touch hole and the passage is nearly the same in this as in the former, but the passages are made wholly in the breech, instead of being partly in the breach and partly in the lock plate. In a gun thus constructed the locks do not project sideways so far as in other guns, as is shewn by Figure 8, and the barrels fit down upon the upper edge of the lock plates A, B, as shewn by Fig. 7, so that the barrels can be removed from the stock and replaced without first taking off the locks.

The detonating substance which I use is that which is known to chemists by the name of oxymuriate of potash with a proportion of sulphur; this proportion must be such as is to render its explosive action moderately violent, but at the same time sufficient to fire off the piece by a small quantity, as must be ascertained by trying the piece in which it is intended to be used. The detonating substance of which the strength is so ascertained must be made up with gum water into pellets of proper size to fit the tube, and each containing the proper quantity to fire off the piece.

The preparation of the detonating substance being well known, and being no part of my Invention, need not be described; the wax which I use to wax the wire and stop the tubes with, as before described, is of this mixture, four parts bees' wax with one part mutton suet.

My aforesaid improvements in shoeing of horses is represented on half size in Figures 12, and 13. A, B, C, is the horse shoe of the common form, or in any other approved form; 1, 1, is a flat plate of iron, to which a tube 2, 3, is attached, and to the perforation of the tube a cylindrical pin or bore 4, is well fitted, so that it can either turn round or slide backwards and forwards; on one extremity is a handle 5, to turn it round by, and at the other end is the borer 6, which is to pierce the horse's hoof; the form of this borer may be like an augur, or any other form adapted to bore the horn, and the size of the hole which it will perforate must be less than the nails or pins which are to be inserted into the holes for fastening on the shoe; when screws are to be used for this purpose the holes must be exactly the size of the bottom of the thread of the screw, and the screws being made cylindrical, as at L, will hold very fast. The plate 1, 1, to which the tube is fixed is to be applied flat against the surface of the horse shoe or hoof, and by this means the tube 2, 3, will be held perpendicular thereto and will bore the holes in the hoof in that direction.

In witness whereof, I, the said Joseph Manton, have hereunto set my hand and seal, this Twenty-eighth day of August, in the year of our Lord One thousand eight hundred and sixteen.

JOSEPH (L.S.) MANTON.

Signed, sealed, and delivered by the within named Joseph Manton (being first duly stamped), the interlineations being first made, in the presence of

FRA[s] YOUNG, Inrolment Office.

R[D] A. LEACH, his Clerk.

A.D. [illegible]. February 29. No. 3985.

MANTON'S Specification.

(1 SHEET.)

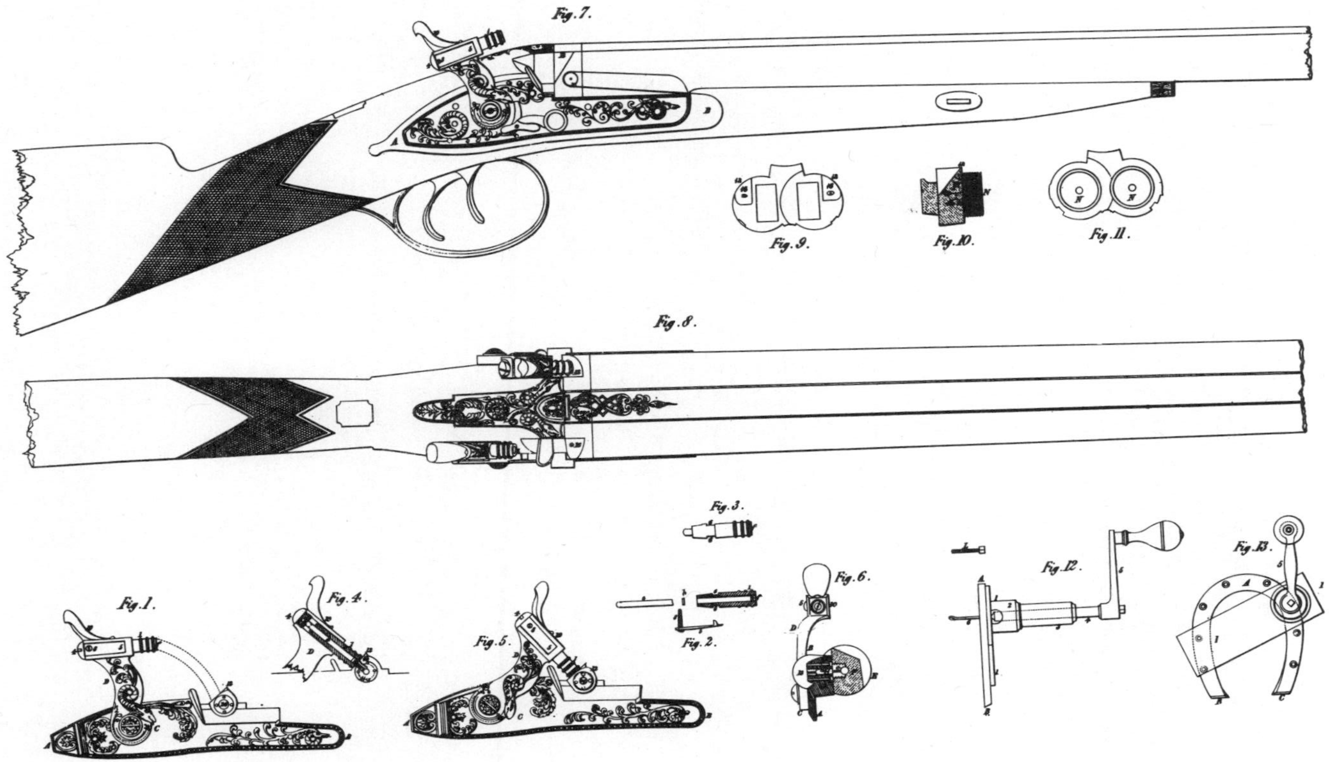

The Enrolled Drawing is Colored.

Malby & Sons. Lith.

London: Printed by George Edward Eyre and William Spottiswoode. Printers to the Queen's most Excellent Majesty. 1854.

A.D. 1816 . . . . N° 4026.

Apparatus and Arrangements for Discharging Fire-arms by means of Condensed Air; also Cartridges applicable thereto.

PAULY'S SPECIFICATION.

TO ALL PERSONS TO WHOM THESE PRESENTS SHALL COME, I, Jean Samuel Pauly, late of Knightsbridge, in the County of Middlesex, Engineer, but now of Kennington, in the County of Surry, send greeting.

WHEREAS, by His Majesty's Royal Letters Patent under the Great Seal of the United Kingdom of Great Britain and Ireland, bearing date the Fourteenth day of May now last past, after reciting that I had by my petition humbly represented unto His Majesty that I had lately invented "Certain Improvements in the Construction and Use of Fire-arms," and obtained His Majesty's Royal Letters Patent, bearing date the Fourth day of August, One thousand eight hundred and fourteen, for exclusively enjoying the benefit of the said Invention; that after considerable application and expence I had invented certain improvements in the said Invention which I believed would be of general benefit and advantage; that I was the true and first Inventor of the said last-mentioned improvements, and that the same had not been made or used by any other person or persons whomsoever, to my knowledge or belief; and that I therefore most humbly prayed that His Majesty would be graciously pleased to grant unto me, my exors, adm̃ors, and assigns, His Royal Letters Patent under the Great Seal of Great Britain, for the sole working, constructing, making, selling, using, and exercising of my said last mentioned improvements and Inventions, and all other benefit and advantage thereof, within that part of the United Kingdom of Great Britain and Ireland called England, His Dominion of Wales, and Town of Berwick-upon-Tweed, and also in all His Colonies and Plantations abroad, for the term of fourteen years, according to the Statute in that case made and provided; His Majesty, being willing to give encouragement to all arts and Inventions which might be for the public good, did, by the said Letters Patent of the Fourteenth day of May now last past, give and grant unto me, the said Jean Samuel Pauly, my ex̃ors, adm̃ors, and assigns, His special licence, full power, sole privilege and authority, that I, the said Jean Samuel Pauly, my ex̃ors, adm̃ors, and assigns, and every of them, by myself and themselves, or by my and their deputy or deputies, servants or agents, or such others as I, the said Jean Samuel Pauly, my ex̃ors, adm̃ors, or assigns, should at any time agree with, and no others, from time to time and at all times thereafter during the term of years therein expressed, should and lawfully might make, use, exercise, and vend my said Invention within that part of the United Kingdom of Great Britain and Ireland called England, our Dominion of Wales, and Town of Berwick-upon-Tweed, and also in all His Majesty's Colonies and Plantations abroad, in such manner as to me, the said Jean Samuel Pauly, my ex̃ors, adm̃ors, and assigns, or any of them, should in my or their discretions seem meet; and that I, the said Jean Samuel Pauly, my ex̃ors, adm̃ors, and assigns, should and lawfully might have and enjoy the whole profit, commodity, and advantage from time to time coming, growing, accruing, and arising by reason of the said Invention, for and during the term of years therein mentioned; to have, hold, exercise, and enjoy the said licence, powers, privileges, and advantages therein-before granted or mentioned to be granted unto me, the said Jean Samuel Pauly, my ex̃ors, adm̃ors, and assigns, for and during and unto the full end and term of fourteen years from the date of the said last-mentioned Letters Patent next and immediately ensuing, and fully to be complete and ended, according to the Statute in such case made and provided;

which said last-mentioned Letters Patent contain a proviso, that if I, the said Jean Samuel Pauly, should not particularly describe and ascertain the nature of the said Invention, and in what manner the same is to be performed, by an instrument in writing under my hand and seal, and cause the same to be inrolled in His Majesty's High Court of Chancery within six calender months next and immediately after the date of my said last-mentioned Letters Patent, then and in such case the said last-mentioned Letters Patent, and all liberties and advantages thereby granted, should utterly cease, determine, and become void.

NOW KNOW YE, that in compliance with the said proviso, I, the said Jean Samuel Pauly, do hereby declare that the nature of my said Invention and improvements, and the manner in which the same are to be performed and used, are set forth and particularly described as follows (that is to say):

My Invention consists of an improved manner of communicating ignition without match or flint and steel, and of priming and loading at the breech of the piece. It applies to all such fire-arms as consist of a barrel or tube so made or constructed as to contain a charge of gunpowder or any other explosive material, which, when ignited, shall be capable of propelling balls, bullets, shot, slugs, or other missile weapons, whether the same fire-arms shall be called guns, pistols, cannon, or by whatever name the same shall be distinguished. And as it will be attended with great difficulty for me to describe my new Inventions and improvements without recurring to and in some measure repeating the Specification of parts contained in my former Specification relating to the said Letters Patent of the Fourth day of August, One thousand eight hundred and fourteen, I shall, whenever I find it necessary, make such repetitions, observing now only and generally that those matters and things which are now and appertain particularly to the said Letters Patent of the Fourteenth day of May now last past are designated and particularly mentioned in a schedule or list thereof which I have placed at the end of this present Specification.

My method of producing fire or ignition to the charge of gunpowder or other explosive material is by the use of the action of condensed air, the condensation of which is effected by the action of a strong spring upon the piston of a small condensing syringe contained within the lock for such purpose, and is operated upon by the trigger in the usual way as the same has been heretofore explained in my former Specification, but with the intervention of various apparatus to be herein-after described, which have now, by repeated trials and experience, been rendered more certain and efficacious than the same then were, as will hereafter appear. The means by which my approved manner of introducing and fixing the charge at the butt or breech end of the barrel, instead of at the muzzle as commonly practised, is to be brought about, and the several parts which are necessary, together with their uses and the relative situations of each, will be clearly seen and understood by the following references to the Drawings hereunto annexed, which in the first Drawing A, are all of half, and in the second Drawing B, all of one eighth of the actual dimensions of the parts they are meant to represent. The same letters of reference are used throughout to express the same parts.

N° 1, is a general side view of part of a gun, in which A, represents the wooden stock or butt, of the usual form and construction. The side plate is supposed in this Figure to be taken off, to shew the interior construction of the lock. B, C, shews a part of the barrel so formed as to load at the breech or butt end C, which in this instance is turned or formed into a portion of a globe, fitting in the nature of a ball and sockett joint into the breech piece D, which contains the barrel of the syringe aforesaid. The particular form of this piece is more distinctly seen at N° 4, where the curved part which joins to the butt of the barrel is seen, and the curved dotted line shews the depth of the socket. The dotted lines *a*, *a*, shew the section of the syringe, and *b*, *b*, a section of a plug of steel or other metal screwed into the piece D, so as to form a bottom to the syringe, and having a small hole as at *e*, (which I call the touch-hole,) to permit the ignition of the priming, when the air in the syringe becomes sufficiently condensed. This hole may be plated in the center of the syringe or not, but must in all cases be opposite to the hole, for the priming to be hereafter described. *c*, is a screw hole, there being another opposite to it, which this piece may be attached to the side plates of the lock, but those holes must not open into the syringe. N° 5, is the plug or piston working in the barrel of the syringe, in which *f* is a solid cylinder or plug of steel or other metal, turned with sufficient accuracy to fit air-tight into the bore of the syringe, which it will do with a little oil or soft grease; *g*, is a tooth rack, by which the plug *f*, is moved backwards and forwards by the action of the toothed tumbler, N° 8. *h*, in N° 5, is a small stud or handle, by which N° 5, can at any time be taken out of the syringe of N° 4, to clean or grease it. *d*, in N° 4, is a small hole, such as is usually placed at the top of condensing syringes, to admit the charge of air when the piston is drawn up. The lock here set forth or drawn consists of N° 10, a bottom plate of metal seen edgewise, and N° 6, the same seen in plan or from above, and which plate screws on to the under part of the stock by the screw holes *i*, *i*. K, N° 10, shews the cocks or blocks in which the toothed tumbler N° 8, turns upon its pivots *l*, *l*, one of which extends through the side plate, and ends in a square or polygonal shape, as may be seen at M, N° 9, (which is an end view of N° 8, to shew its thickness,) in order to carry the cocking piece, N° 16, the end E, of which stands up for the purpose of cocking or half cocking the piece. The tumbler N° 8, is driven, urged, or moved by a common main spring and swivel, the form of which is apparent from the top view of it N° 12, and the side view N° 11, as well as from the view of it placed in its proper situation at F, in N° 1, where the spring is seen held in its place by its two studs *n*, *n*, lodged in holes made to receive them in the standards *o*, *o*, N° 6, and 10. This spring requires to be of considerable force or strength; for the perfection of my mode of ignition depends upon the power and sudden velocity with which the air is condensed. *k*, in N° 13, is a side view, and *p*, a plan or top view of *k*, in N° 10; and *q*, is a top view of the stop to be screwed into the holes *s*, *s*, of N° 6, for the purpose of receiving the blow of the flat face *r*, of the tumbler N° 8, and thereby stopping it when the gun is discharged. The operation of stopping is also assisted by the snail or spiral projection *t*, seen in N° 16, which is an inside view of the cocking piece; the flat surface of *t*, is made to stop either against a projecting part of the under plate, or against any other stop placed to receive it, and becomes necessary from the strength of the spring, to prevent this piece from breaking. G, and H, N° 14, are top views or plans of top plates, which may be applied to the upper part of the gun stock, to strengthen the same. H, is immoveable, and fastened by receiving the ends of the screws, which pass through the bottom plate N° 10, and 6; G, is moveable, either in a vertical or horizontal direction, upon a pivot or hinge at the end nearest the letter G, so as to form a door or opening to get at the syringe and piston, for the purpose of cleaning or greasing them. This door G, is fixed down by a latch, catch, or turnbuckle. A side view of the same is seen at G, and H, in N° 1.

Having so far described my lock, and the means of producing ignition to the charge of gunpowder, I shall in the next place describe my cartridge or cartouche, as I construct it, for loading all small or portable fire-arms at the butt end of the barrel. It is shewn at N° 7, where *w*, represents that part which I nominate the rosette or cap of the cartridge, which may be formed of box or other hard wood, or of paper mache, or lead, or some plastic or yielding substance, turned or cast in a mould, as pressed or stamped, to such size and form, and made so much smaller than the inside of the barrel of the gun as to permit a paper or other case, as *y*, to be glued, tied, or fixed upon the shoulder W, to contain the charge of gunpowder, with the ball, shot, or other missile weapon in it, as indicated in the Drawing, and which must altogether fit the barrel as nearly as possible, but so as to permit it to enter the same without difficulty. *x*, *x*, is a small projection or shoulder on this rosette, which is turned to a conical or other form, fitting a rebate of a similar shape, which is made in the butt end of the barrel, for the purpose of not only stopping or preventing the cartridge or charge from slipping down the barrel, but of forming a close and air-tight joint, when pressed into its situation, by the return of the barrel into its right lined direction, or other contrivance, to be hereafter described. *v*, N° 7, shews a small space or chamber in the rosette of the cartridge, which diminishes into the small hole *e*, which is rammed with a small portion of gunpowder or other explosive material, in the making of the cartridge, before it is charged with its contents of explosive powder, ball or shot, so that when the gun is loaded with such a cartridge as aforesaid, the small hole *e*, in the same, will present itself to and in immediate contact with the small hole *e*, in the end of the syringe before described. And I find that if a small piece of thin paper is glued, pasted, or affixed upon the rosette end of the cartridge over the small hole *e*, it will effectually prevent the bruised gunpowder or explosive material from escaping or being injured by damp, and will not be detrimental to the ignition and discharge. N° 15, represents another cartridge, in which the rosette is made conical or tapering, instead of with a bead or shoulder, to fit the conical or other shaped rebate before described, to be formed in the butt end of the barrel. This form of cartridge, however, requires that the inside of the butt end of the barrel should be made of the same shape to receive it, and it is forced into an air-tight bearing likewise by the return of the barrel, which forces or presses it against the end of the syringe piece or breech. A small stud or nutt is turned, as at *u*, *u*, in both rosettes, for the greater facility of taking them out of the barrel after a discharge. N° 17, is the same gun as N° 1, but with the side plate L, L, N, fixed on, and in a position to introduce one of my aforesaid cartridges at the breech or butt end; and N° 18, shews the same thing as applying to a pistol. In these the barrel B, C, is equipped at its two sides near its breech, with two strong trunnions or gudgeons, as at K, and which may be better seen at K, K, N° 3, which is a top view or plan of such a barrel. These trunnions or gudgeons pass through two opposite holes in the strong plates L, L, which are well secured to the wooden stock as well as to each other, by strong screws, as will appear in the Figures. The barrel by this construction is capable of being turned out of its right-lined direction, as indicated in N° 17, and by this means the bore at the breech becomes exposed and ready to receive its charge. Z, in N° 3, is a small metal door or lid, with a hole in the middle, to confine the cartridge, and makes a more secure joint. This may be used in all constructions of my small fire-arms, or may be altogether dispensed with, though I prefer its use for rifles, carbines, and pistols. The barrel, when loaded, has only to be brought back again into its right-lined direction, in which it is firmly retained by sliding the ring M, over the projecting ends N, of the side plate L, L, or in any other safe and convenient manner. N° 2, represents a front view of this ring as applying to an octagonal barrel. O, O, is a slide and stop, to prevent the ring M from being removed from the barrel, or being lost. N° 19, is an entire top view of the side plates L, L, in which I, I, are strong studs or stops to form an abutment for the back of the syringe piece or breech D, N° 4, which appears in this view fixed between the cheeks, to shew the manner in which this piece is held to its bearing, and secured between the said cheeks or side plates L, L, by the two screws *c*, *c*. N° 20, is a general view of another of my improved guns, in which it will be seen that the same parts as have been hitherto described occur, and to which the same letters of reference are applied. In this gun the barrel does not bend to receive its charge, as in N° 1, 17 and 18, but maintains constantly its right-lined position, it having a false or moveable breeching for the purpose of charging it, and which contains my said improved apparatus for producing ignition, the several parts of which will be evident from an inspection of the Figures and the discription which has already been given of them. N° 21, is the breech or piece containing the barrel of the syringe. The only difference in these is, that the dovetail *p*, in this, affixes to the top moveable plate N° 22, instead of to the bottom one, N° 10, as in N° 4. The bottom plate, N° 23, carries the tumbler, main spring, trigger, and guard, as may be distinctly seen at N° 20, and is drawn separately in profile or side view at N° 24. The top plate is differently formed to that in the gun N° 1, for when the moveable breeching is employed I give this plate a curve so as to cause it to pass from the upper part of the stock through the middle of the lock to the under side of the barrel, as it may be traced in N° 20, by the letters g^1, d^1, g^1, g^1. Its curved shape is also distinctly seen by the same letters in N° 25, where e^1, is a portion of the toothed tumbler N° 8, shewing through it. N° 26, is a top view or plan of this same top plate, in which the dotted line d^1, d^1, and g^1, g^1, shew where it bends, as at the same letters in N° 20 and 25; and the parallelogram e^1, e^1, shews the opening through which the teeth of the tumbler N° 8, present themselves to work into the teeth of the piston rack contained together with the syringe piece in N° 21, altogether in the moveable breeching P, Q, R, which is shewn open, or in the fit state for loading, in N° 25, and shut or ready for discharging at N° 20.

N° 27, shews a side view, and N° 28, a profile or top view, of one of the side plates or cheeks of the moveable breeching, there being one on each side of the barrel. g^1, shews the holes which fit on to or work upon the trunnions or gudgeons K, K. I, I, are solid studs of metal projecting inwards, and coming immediately behind the breech or syringe piece N° 21, to prevent its being driven from its bearing on the but or breach of the barrel by the force of the explosion. N° 21, is fixed between these two cheeks by screws *c*, *c*, which must not proceed so far as the bore of the syringe; and the whole is united into one piece by the top plate R, S, T, affixed by screws, as will appear in the Figures and by the dovetail B in 21, so that when the finger is put into the loop at R, the whole can be raised upon the center K, as seen at N° 25, for the purpose of receiving a cartridge or charge; and as the piece N° 21, when fixed between the two cheeks Nos 27 or 28, or P, Q, N° 25, slides over the globular or spherical end, butt, or breech of the barrel in the nature of a ball and socket joint, by drawing R down into the position shewn in N° 20, the cartridge becomes compleatly fixed, and the gun when cocked is ready to be discharged. As the syringe piece 21, and its piston, all raise together with P, Q, R, there is no necessity for the door plate G, N° 14, as the rack and piston become compleatly exposed, and can be taken out to be cleaned or greased, when R is in the position shewn at N° 25. N° 29, is a view of the under side of the moveable breeching, with the syringe piece and piston rack as appears when elevated, or in a fit state to be loaded. V, in N° 20, is a steel or other spring screwed to the bottom plate, and having a latch head W, to its upper part, which passes through a notch in the loop R, of P, Q, R, and serves to hold it in its place when pressed down; while, by putting the finger into the loop R, to raise it, it is only necessary to first press upon W, when it becomes immediately disengaged. The top and bottom plates Nos 24 and 26, are firmly united to each other and to the gun stock, by screws passing through both of them, as may be seen in N° 20, by which the necessary great strength and stability is given to this construction of guns.

Having so far explained how my said Inventions or improvements apply to the construction and use of small fire-arms, I shall in the next place proceed to describe how the same, or parts or modifications of the same, may be used and applied to cannon, mortars, howitzers, carronades, and all those varieties of large fire-arms commonly known or included under the general name of ordnance. The mode of producing ignition in these is the same as in small arms. N°s 40. 41, 42, and 43, drawn half size, shew different constructions of locks applicable to ordnance. In all these *a* represents the bottom of the syringe with its touch hole; *b*, its plug or piston; *c*, the trigger by which it is discharged; and *d*, the cocking piece. In N°s 42 and 43. I have shewn examples of the use of spiral springs for driving the piston; and in N° 40, an example is given of common musket springs. After the explanations herein-before given of small locks, the construction of these will be sufficiently evident from inspection of the Figures in N° 42. *e*, is a small oblong metal cup or pan, to receive a rosette of hard wood, paper mache, or other material, made to fit it, containing rammed gunpowder or other explosive material, (a section of which is seen at *h*,) and is to be turned down upon the touch-hole of the cannon, in order to effect the discharge, by moving the whole syringe and lock half way round. The same thing is seen at *f*, in N° 41, in which Figure *g*, is a hinge by which the pan may be opened to prime it, and when shut down is likewise brought over and fixed upon the touch-hole of the cannon. In all syringes an aperture must be made at the top, to admit the air when the piston is drawn up; and the manner of applying this, or any of my said contrivances for the purpose of producing ignition in ordnance, will fully appear in the progress of my description of a cannon which is capable of receiving its charge at the breech, for which I have several contrivances, as shewn in Drawing B, in which the Figures are drawn one eighth of the actual dimensions they are meant to represent, as follows:—N° 47, is a top view or section, and N° 46, a side view of a six-pounder gun on nearly the same construction as that described in my former Specification, but considerably improved. In these Figures A, is the body of the gun, which is bored quite through; B, its trunnions by which it is to hang on its carriage; and C, N° 47, the moveable breeching. This breeching is supported by and hangs between the two strong cheeks or side plates

D, D, which are to be made of wrought iron, or other sufficiently strong and tenaceous metal. These cheeks are held together in the form seen at N° 47, by the strong back piece E, which has four or more metal tenons F, F, (seen also in the front view of this piece N° 48,) with screws and nuts upon the ends of them, which pass thro' the holes or mortices G, G, N° 49, which is a separate side view of one of these side plates or cheeks; and N° 50, is a section or top view of the same. The whole of the strong frame thus constituted is made to turn or move upon the trunnions H, H, in N°s 46, 47, 49, 50, and 51; so that when the gun is loaded, and in a state to be fired, it appears like N° 46, but when the moveable breeching is turned up for the purpose of loading or charging it appears as drawn at N° 51. It may so far appear to be similar to the cannon described in my former Specification; notwithstanding which it will, upon examination, be found, not only much more safe, but more manageable also, for by the cheeks or side plates being extended so as to form the hooks I, which lay hold of the hanging trunnions B, B, afford great additional strength, and weights may be added to the ends I, I, to balance the breach and other hinder parts. The manner of fixing in the back piece E, is also much more efficaceous, and it is much strengthened by the ledges or projections K, K, which are parts of the side plates, and project immediately behind the back piece E. The hooks I, I, are so formed at the part where they come into contact with the hanging trunnions B, B, that by their wedge-like form the more they are pressed up the closer they will bring the side plates to the sides of the gun, which is likewise facilitated by the trunnions being made upon the side plates instead of upon the gun. as mentioned in my former Specification. On this account the screws, which were therein described as necessary to fix the trunnions H, now become useless. The moveable breeching is square, and is seen in front view at N° 52, in which *a, b, c, d,* are notches or grooves made to slide upon projections suiting them in the insides of the side plates. N° 53, is a side view of the moveable breeching, in which *e,* is the recess into which the end of the strong screw L, N° 46, 47, and 51, goes, for the purposes of forcing up the breeching, after loading and withdrawing it. When it is to be removed for charging the piece, this screw is worked by its double-ended lever, and attached to the moveable breeching by the screws or pins *f*, *f*, N°s 47, 52, and 53, passing into the groove or notch turned at the end of the screw L, N° 47, to receive it, or in any other convenient manner, so that it will draw back when the screw is withdrawn. N° 54, is a separate view of this screw, and N° 55, the back part E, which it passes into and works thro', which is also a section of N° 48. Provision must be made at the butt end of the bore of the cannon for the culot or cartridge stopper, which will be hereafter particularly described as being essential to the perfection of my said Invention. The cheeks and moveable breeching are likewise stopped when drawn down by the projecting pieces M, cast upon the gun, as at M, N°s 46 and 51, and this also serves to attach the elevating screw to in the ordinary manner when the gun is mounted on its carriage. To apply my mode of ignition to a gun of this description, I introduce the syringe or lock, as before described, into the said moveable breeching; and to do this with sufficient accuracy it is preferable to construct the syringe in a separate plug or piece of metal seen at *g*, in N° 56, and *h*, in N° 53. I make this plug circular, in order that it may be screwed into the said breeching, and I cause the necessary bore for the syringe to be made quite through it, and I afterwards close one end of the same by screwing in a small plug of metal seen at *h*, in N° 56, and also in 57, which is another view of the same thing, together with the piston, spring, trigger, &c., as before described. The touch-hole of the syringe is in this case made at the side instead of the end, and a priming or rosette rammed with gunpowder is to be introduced against this touch-hole in the culot or cartridge stopper, to be afterwards described, before the same is screwed up for firing. The end of the syringe to which the trigger is affixed projects a small distance through a hole in one of the side plates, which is made oval or oblong, as at Q, in N°s 49, 47, and 46, for the purpose of letting the projecting end of the syringe slide with the moveable breeching. It will be needless for me to describe the manner in which my lock and improved method of communicating ignition are to be applied to the other constructions of guns which I am about to describe, since the examples above given will throw sufficient light upon this subject, and

I shall therefore only observe, generally, that the syringe must be so placed in the solid metal, or in the moveable breechings, to be hereafter described, that it may not be in the way, that it may be easily got at to be cleaned, and that the priming of gunpowder, either contained in rosettes or not, may be so placed and fixed with ease and expedition as to come in all cases between the touch-hole of the syringe and the charge of gunpowder to be ignited. With a view to avoid the expence of manufacturing and to diminish the weight of the gun already described, I adopt the mode of leading at the breech, which is shewn by N°s 58 and 59, where the cannon differs but little in appearance from the ordinary construction, it being only made so much longer as to admit the cartridge or charge to be admitted or put into its bore by an opening on one of its sides near the breech, which is afterwards closed by a solid door or stopper of the same materials as the gun, or of iron, brass, or other suitable material, which turns upon a hinge at the side of the gun, and is pressed up to its bearing, and there firmly held by means of a strong screw of metal passing through and working in the breech end of the gun by means of a double or single lever or handspike, as may be most convenient. In these Figures, A, B, C, represent the gun; the dotted lines D, D, its calibre or bore; E, E, E, the space or cavity for introducing the cartridge or charge; H, F, I, the solid door or stopper turning upon the hinge F; and the dotted lines G, K, shews its sweep, or the direction in which it moves; when this door is shut or returned into the cavity E, E, E, it must fill it, and fit as correctly as possible; L, is the strong metal screw working through B, the breech end of the gun, by means of the lever or handle M, M; and the end *n* of this screw must finish in a conical point, to fit into a similarly formed cavity *o*, in the solid door, stopper, or moveable breeching F, I, H. so that whenever that door or moveable breeching is returned into its place, and screw L is forced or pressed upon it, it will not only push it forwards towards the muzzle, but will draw it inwards at the same time, and prevent its opening until the screw is again withdrawn. Fig. 59, is an external view of the gun as it appears when this solid door or stopper is shut. In this Figure the same letters are used as in the last to express such of the parts as are visible. N, is a ring for the greater convenience of opening and shutting this door or moveable breeching. The fixed part of the hinge is cast upon the gun, and the moveable part upon the door, which turns upon the strong pin or screw F. Figures 60, and 61, represents a construction of a gun having a side door, as in the last example, but which has the advantage of being sponged, loaded, or examined from the breech end. The gun in this case must be bored quite through, and of larger dimension at the breech end, as at E, E, in order to introduce the cartridge and culot; and the door or moveable breech, instead of opening in a longitudinal direction, opens in a transverse one. The particular form of this door is seen at X, in Fig. 61, and X, in Fig. 62, as it appears when open; Fig. 62, being a transverse section of the gun in that part in which the door or moveable breeching is situated. Its position in the gun itself is seen at Z, in Fig. 60, where *a*, *b*, *c*, *d*, represents the door or moveable breeching; *e*, *f*, the hinge or part upon which it turns for the purpose of loading. This door is so adjusted at its hinge that its flat side *a*, *b*, which comes nearest to the muzzle of the gun, does not rub or touch against the cavity or recess made to receive it, but only against the culot or cartridge stopper, which is placed with the cartridge within the bore, as will be hereafter described. The thickness of this door or moveable breech should in all cases be made equal to the bore of the gun to which it is to be applied; and its back part *c*, *d*, should be made in a slight degree like a wedge, so that the more it is pressed down the more it will press forward or towards the muzzle of the gun, in order that it may effectually secure the culot or cartridge stopper. In Figures 61 and 62, at X, it will be seen that teeth or cogs are cast or formed on the outside of this door, and which are acted into by similar teeth in a wheel or pinion which works into them, and is moved by a winch projecting behind the breech of the gun, by means of which the door is forcibly driven upon the culot, to fix the same with the cartridge in the gun; and should the door at any time become set fast by friction, rust, or other means, the mechanical power gained by this rack and pinion will overcome the resistence. In this construction of gun it is necessary that the door should be a little larger than the bore of the same, and that a rebate should be made for it to fit into, as at *g*, *h*, *e*, Figure 62. In all other parts the door may be made less than its recipient cavity, to prevent friction, but the curved lines of its sides must be parrellel with each other, as at *g*, *e*, and concentric with the centre of the hinge *c*, upon which the same turns. *i*, *i*, in Figure 61, is a front view of the toothed edge of the door X; and a side view of the same is seen at *i*, *i*, in Figure 62. The manner in which my syringe locks may be applied to the guns Nos 58, 59, and 62, will be seen at Y, in those Figures; and in these the priming or rosette may be applied to the end of the syringe in these several doors, instead of applying it in the culot.

Having described the various forms and constructions which I give to cannon or other large arms, for the purpose of loading or charging them at the breech or butt end, I shall in the next place describe my culot, plug, or cartridge stopper, which is an essential part of Invention, which must be used with any or all the constructions of cannon which have herein-before been described or mentioned by me, when the same are to be fired or discharged. The great obstacle to the use of large arms loading at the breech has hitherto been the difficulty of producing an air-tight or perfect closure between the moveable breech and the body or remainder of the gun; and if the effort of the explosion can at all make way into this joint or closure, the construction becomes dangerous, inasmuch as the opening will become larger, and eventually the moving breeching may be blown away or torn from its situation. This effect will entirely be prevented by the adoption of my culot, plug, or cartridge stopper, which is so placed in the gun as to come between the charge of gunpowder and the moveable breeching in all cases, and is formed of lead, copper, or such other ductile materials as will give way to the explosive force of the charge, and so formed and situated that by yielding or giving way it will completely and effectually cover and close up the joint or joining between the moveable breeching and the gun itself. Some of the manners in which this is to be performed and placed will be seen in the Drawing A in which N° 63, is a general section of one of these culots or cartridge stoppers, drawn half of their size in dimensions for a six-pounder; and N° 64, a front view of the same. The external edge *a*, *a*, *b*, *b*, of this must in all cases accord with the form of the butt or chamber end of the barrel, so that if that is made with a rebate or rebates, or conical or cylindrical, this part of the culot or cartridge stopper must be the same, and they must fit each other pretty accurately. The back part of this piece *b*, *b*, must likewise be made flat, convex, or concave, to suit the form of the moveable breeching which is to come against it; and the front part *a*, *a*, must be hollowed out, or so cast as to leave a recess *c*, *c*, into which the cartridge may fit. For my guns the cartridge case will be best made of sheet iron, copper, tinned iron, or other thin metalic plating capable of holding the gunpower, with or without a ball, or as *d*, *d*, *e*, N° 63, and is completely detached and separate from the culot, but to which it may be attached instantaneously by pushing the part *d*, *d*, into the recess *c*, *c*, by which operation the small metalic piece *f*, *g*, enters the cartridge by a hole *h*, left or made in the end plate *d*, *d*, of the cartridge, but stopped up by pasting or otherwise fixing a piece of paper, parchment, or other material capable of retaining the gunpowder over it, but which becomes broken by the point of *f*, *g*, when they are pressed together. The end of the metalic piece *f*, *g*, is hollowed out at *g*, so as to form a cup or cavity for one of my primings or rosettes, hereinbefore described, and as may be seen at *r*, and which will come into immediate contact with the touch-hole of the syringe when the gun is in a state fit to be discharged, and the fire from the rosette at *g*, is conducted into the gunpowder through *h*, by a hole drilled through the piece *f*, *g*, but not in a right-lined direction, but in a right angle, since I find it better by experience to carry the hole from *g*, nearly to *f*, and then to carry it into the gunpowder through the side of the piece *f*, *g*, as at I, I, I, where this hole is marked in. The hole *h*, in the end *d*, *d*, of the cartridge is not circular, but oval, or with two wings or latteral openings, as at *k*, *k*, N° 66; and the piece *f*, *g*, N° 63, has two pins or projections, as at *m*, *m*, which may pass into *k*, *k*, N° 66; and then by turning it partly round the the cartridge becomes completely fixed or held upon the culot or cartridge stopper, and the two together may be very conveniently held to introduce the cartridge, and withdraw its case, after the gun has been discharged, by the semi-circular handle *l*, *l*, turning on the pins or rivets *n*, *n*, and so adjusted as to fit into the countersunk or hollow circle *o*, *o*, *o*, made or turned in the back of the culot to receive it, by which means it gets completely out of the way of the moveable breeching whenever it may be pressed upon it. N° 65, is a front view of my culot or cartridge stopper, in which the touch-hole piece *f*, *g*, with its projecting arms *m*, *m*, to lay hold of the parts *k*, *k*, of the cartridge, are seen. N° 67, is another construction of my culot, in which the back part *a*, *b*, *a*, *b*, may be made of cast iron or other metal; but in this the end of the cartridge is made eliptical or spherical, and the space between them, as at *q*, *q*, is fitted up with lead or other soft and ductile material to receive the first effort of the charge; and in all cases in which it can be used I recommend a collar of leather *s*, *s*, to be fixed round the call, as seen at N° 68, to prevent the windage which usually occurs, particularly in large arms.

In witness whereof, I, the said Jean Samuel Pauly, have hereunto set my hand and seal, the Twelfth day of November, in the year of our Lord One thousand eight hundred and sixteen.

THE SCHEDULE ABOVE REFERRED TO.

The union of the syringe with the breech piece, as in Nos 4 and 21, Drawing A, and the form now given to it, and the manner of fixing it between the said plates.

The form and construction of N° 5, in Drawing A.

The door G, N° 14, in Drawing A, by which the syringe and piston can be got at and kept in order with great facility.

The form and manner of stopping N° 8 or 9, by the application of *q*, N° 13, in Drawing A.

The improved form of the rosettes for the cartridges of large and small arms, as at Nos 7 and 15, Drawing A; and *r*, N° 63, and *h*, N° 42, Drawing B; and the application of paper fixed over them to confine the rammed powder.

The improved form of locks shewn in Nos 40, 41, 42, and 43, Drawing A, and N° 57, Drawing B.

The improved construction of side plates to ordnance in N° 49, and 50, and as shewn in Nos 46, 47, and 51, in Drawing B.

The guns shewn at N° 58, and 59, 60, 61, and 62, in Drawing B.

The culot or cartridge stopper, shewn at Nos 63, 64, 65, and 67, in Drawing A.

J. SAMUEL (L.S.) PAULEY.

Signed, sealed, and delivered in the presence of

J. FRISARD, 31, Gt Winchester Street, London.
JOHN CROSLAND, Broad Street, London.

SMITH.

AND BE IT REMEMBERED, that on the Twelfth day of November, in the year of our Lord 1816, the aforesaid Jean Samuel Pauley came before our said Lord the King, in His Chancery, and acknowledged the Specification aforesaid, and all and every thing therein contained and specified, in form above written, thro' the interpretation of James Frisard, the said interpreter being first sworn that he would truly interpret the acknowledgment to the said Jean Samuel Pauley. And also the Specification aforesaid was stamped according to the tenor of the Statute made for that purpose.

Inrolled the Twelfth day of November, in the year of our Lord One thousand eight hundred and sixteen.

LONDON:
Printed by GEORGE EDWARD EYRE and WILLIAM SPOTTISWOODE, Printers to the Queen's most Excellent Majesty. 1854.

A.D. 1816. May 14. No 4026.
PAULY'S SPECIFICATION

(2 SHEETS)
DRAWING A.

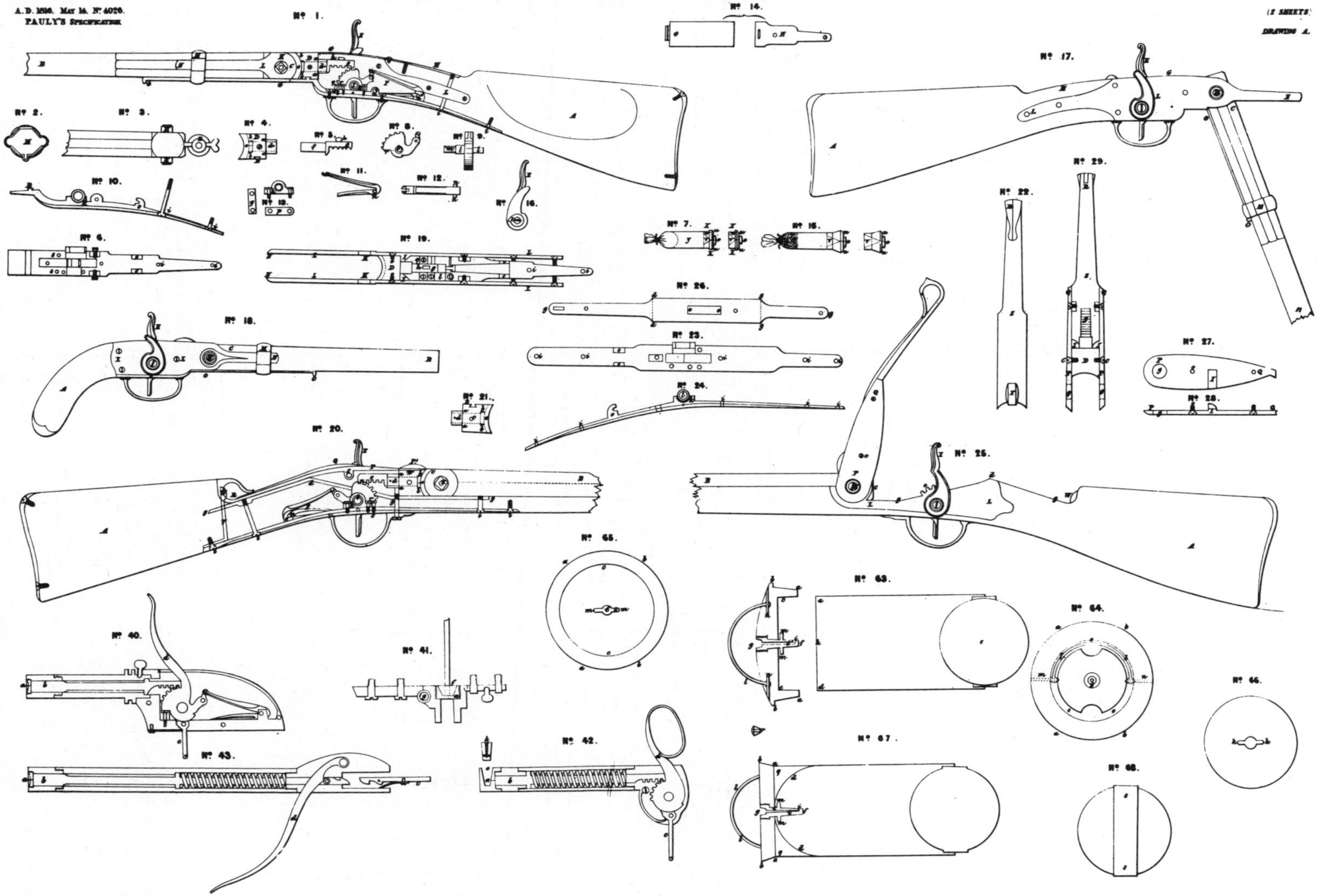

The scrolled drawing is not colored.

Malby & Sons, Lith.

London: Printed by George Edward Eyre and William Spottiswoode,
Printers to the Queen's most Excellent Majesty. 1854.

A.D. 1816 . . . . N° 4054.

Pan for Locks of Fire-arms.

NOCK'S SPECIFICATION.

TO ALL TO WHOM THESE PRESENTS SHALL COME, I, Samuel Nock, of Number One hundred and eighty, Fleet Street, in the City of London, Gun Maker, send greeting.

WHEREAS His Royal Highness the Prince Regent, in the name and on behalf of His Majesty, did, by Letters Patent under the Great Seal of Great Britain, bearing date at Westminster, the Twelfth day of August, in the fifty-sixth year of the reign of His Majesty King George the Third, give and grant unto me, the said Samuel Nock, his especial licence, that I, the said Samuel Nock, during the term of years therein mentioned, should and lawfully might exclusively use, exercise, and vend, within England, Wales, and the Town of Berwick-upon-Tweed, my Invention or "Improvement in the Pan of the Locks of Guns and Fire-arms;" in which said Letters Patent there is contained a proviso, obliging me, the said Samuel Nock, by an instrument in writing under my hand and seal, to cause a particular description of the nature of my said Invention, and in what manner the same is to be performed, to be inrolled in His Majesty's High Court of Chancery within two calendar months after the date of the said recited Letters Patent.

NOW KNOW YE, that in compliance with the said proviso I, the said Samuel Nock, do hereby declare that my said Invention is described in the Drawing hereunto annexed, and the description hereunder written. In witness whereof, I, the said Samuel Nock, have hereunto set my hand and seal, the Eleventh day of October, in the fifty-sixth year of the reign of His Majesty King George the Third, and in the year of our Lord One thousand eight hundred and sixteen.

The following is a Specification of my Patent for an Improvement in the Pan of Locks of Guns and Fire-arms :—The improvement consists in a hollow tube or chamber in the pan of the lock opposite to the touch-hole of the barrel, and lying in such a direction as that a straight line passing through the centre of such tube or chamber from one end to the other would enter the touch-hole of the barrel. This tube or chamber in its length occupies about three fourth parts of the length of the pan, leaving the remaining part of the pan between it and the touch-hole of the gun a free space, or it may occupy in its length the whole or any part of the length of the pan. That end of the tube or chamber nearest the touch-hole of the barrel is open, the other or exterior end is closed or stopped up, either by being left or made solid, or by a screw plug or slide applied to such exterior end, and which screw plug or slide, if capable of being removed or taken out, will facilitate the cleaning out of the tube or chamber when foul. The tube or chamber can be made of such proportionate size to the pan as to be wholly or partly covered with the gunpowder, and to leave either the whole or part of the surface of the pan to be covered with gunpowder. A small hole or vent resembling a touch-hole is made in the tube or chamber near to its exterior end, whereby a more ready communication between the gunpowder on the surface of the tube or chamber and the gunpowder within it is obtained, though such hole or vent is not absolutely necessary. This tube or chamber may be shortly described as resembling a small cannon fixed in the hollow of the pan, opposite to and firing straight into the touch-hole of the barrel. This improvement may be applied to all sorts of guns and fire-arms, but it is more peculiarly adapted to such as prime themselves from the loading or charge put into the barrel. For the more minute particulars the

A. D. 1816, AUG^{T} 12. N° 4054.
NOCK'S SPECIFICATION.

(1 SHEET)

F I G. 2.

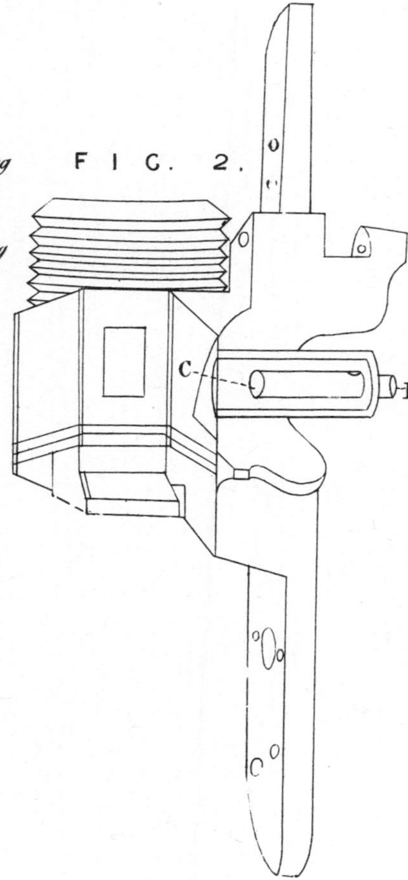

Fig. 2. A View of the improved pan presenting its appearance when the Lock is screwed or fastned to the Barrel shewing the outward or closed end of the Tube or Chamber stopped up by means of a moveable ~~Screw Plug~~ or Slide marked F

C. Is the interior or open end of the Tube placed opposite to or in a line with the Touch Hole of the Barrel.

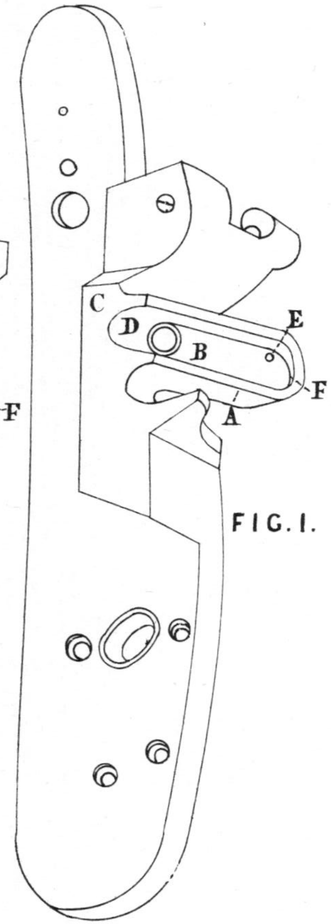

F I G. I.

Fig. 1. Is a view of the Lock Plate with the improved pan applied to it

A. The Pan

B. The Tube or Chamber in the Pan in a line with the Touch Hole of the Barrel

C. The Interior, or open End of the Tube

D. The free or open Space between the Interior End of the Tube and the Touch Hole of the Gun

E. The Hole resembling a Touch Hole on the surface of and into the Tube near its outward or closed End

F. The outward or closed End.

The enrolled drawing is colored

Malby & Sons Lith

LONDON: Printed by GEORGE EDWARD EYRE and WILLIAM SPOTTISWOODE
Printers to the Queen's most Excellent Majesty 1854

details in the annexed Drawing may be referred to. Figure 1, is a view of the lock plate with the improved pan applied to it. A, is the pan; B, the tube or chamber in the pan, in a line with the touch-hole of the barrel; C, the interior or open end of the tube; D, the free or open space between the interior end of the tube and the touch-hole of the gun; E, the hole resembling a touch-hole on the surface of and into the tube near its outward or closed end; F, the outward or closed end. Figure 2, is a view of the improved pan, presenting its appearance when the lock is screwed or fastened to the barrel, shewing the outward or closed end of the tube or chamber stopped up by means of a moveable screw plug or slide marked F. C, is the interior or open end of the tube or chamber placed opposite to or in a line with the touch-hole of the barrel.

SAML NOCK. (L.S.)

Signed, sealed, and delivered by the within-named Samuel Nock, in the presence of

JAS B. BOSTOCK,
Tokenhouse Yard.

AND BE IT REMEMBERED, that on the Eleventh day of October, in the fifty-sixth year of the reign of His Majesty King George the Third, the said Samuel Nock came before our said Lord the King, in His Chancery, and acknowledged the instrument aforesaid, and all and every thing therein contained and specified, in form above written. And also the instrument aforesaid was stamped according to the tenor of the Statute made in the fifty-fifth year of His said Majesty's reign.

JEKYLL.

Inrolled the Eleventh day of October, One thousand eight hundred and sixteen.

LONDON:
Printed by GEORGE EDWARD EYRE and WILLIAM SPOTTISWOODE,
Printers to the Queen's most Excellent Majesty. 1854.

A.D. 1817 . . . . N° 4107.

Barrels of Breech-loading Fire-arms.

SARTORIS' SPECIFICATION.

TO ALL TO WHOM THESE PRESENTS SHALL COME, I, URBANUS SARTORIS, of Winchester Street, in the City of London, Merchant, send greeting.

WHEREAS His most Excellent Majesty King George the Third, by His Letters Patent under the Great Seal of Great Britain, bearing date at Westminster, the Eleventh day of March, in the fifty-seventh year of His reign, did, for Himself, His heirs and successors, give and grant unto me, the said Urbanis Sartoris, my executors, administrators, and assigns, His especial licence, full power, sole privilege and authority, that I, the said Urbanus Sartoris, my executors, administrators, and assigns, and every of them, by myself and themselves, or by my and their deputy or deputies, servants or agents, or such others as I, the said Urbanus Sartoris, my executors, administrators, or assigns, should at any time agree with, and no others, from time to time and at all times during the term of years therein expressed, should and lawfully might make, use, exercise, and vend, within England, Wales, and the Town of Berwick upon Tweed, and also within all His Majesty's Colonies and Plantations abroad, my Invention of "**IMPROVEMENTS IN THE CONSTRUCTION AND USE OF FIRE-ARMS**," in such manner as to me, the said Urbanus

Sartoris, my executors, administrators, and assigns, or any of them, shall in my or their discretions seem meet; in which said Letters Patent there was contained a proviso, obliging me, the said Urbanus Sartoris, by an instrument in writing under my hand and seal, particularly to describe and ascertain the nature of my said Invention, and in what manner the same is to be performed, and to cause the same to be inrolled in His Majesty's High Court of Chancery within six calendar months next and immediately after the date of the said recited Letters Patent, as in and by the same, reference being thereunto had, will more fully and at large appear.

NOW KNOW YE, that in compliance with the said proviso, I, the said Urbanus Sartoris, do hereby declare that the nature of my said Invention, and the manner in which the same is to be performed, is described and ascertained in the plan or Drawing hereunto annexed, and in the manner following (that is to say):

My said Invention of improvements in the use and construction of fire-arms consists of a new construction of barrels with moveable breeches, at which the charge is to be introduced, and for afterwards closing up those breeches with a degree of expedition, certainty, and security which has never before been attained, all which I effect and produce in the manner herein-after particularly set forth and described (that is to say):—The moveable chamber which contains the charge of gunpowder and ball or shot is to be fixed to the band by a particular formation of screw, to be hereafter described, and when the barrel is disengaged from it is capable of turning up into the direction seen at B, in Figure 1, for receiving its charge. For this purpose it turns upon the pin M, as a centre, this pin fixing it in the manner of a hinge to the chamber carriage, no part of which can be seen in Figure 1, but the whole of it is shewn as it appears when taken from the wooden stock at Figure 5, which shews a side view of it, while Figure 3, shews the touch hole side of the moveable chamber (Y, being the touch hole), Figure 4, an end view of the same, as seen from its junction with the barrel; and Figure 6, a top view of the moveable chamber, placed in its proper situation upon the chamber carriage. In all these M, represents the pin upon which this moveable chamber B, turns; and E, E, in Figures 5, and 6, the entire form of the chamber carriage, which is to be let into the wooden stock and secured there in any convenient manner, or as shewn here, by the screws H, K, N, Figure 1, which correspond with the screw holes marked with the same letters in Figures 5 and 6, and this piece is also secured by the two lock screws I, I, passing thro' it in the situations so marked in Figure 5, as well as by the pin L. *b*, in Figures 1, 2, 3, 4, and 6, answers the double purpose of an handle to raise the moveable chamber and a sight to direct the eye in firing. The barrel C, in Figures 1 and 2, is not fixed to the wooden stock, as in common guns, but is made to slide in two rings of metal, one of which is seen at G, and is fixed into its place by the pin L, passing through it, and at the same time through the stock and chamber carriage, as will be more particularly seen in Figure 5, at G, while Figure 7, represents a separate view of this ring, as seen in front, as well as the pin L, by which it is fixed to Figure 5. The further or muzzle end of the barrel is held to the wooden stock by another similar ring, the ends of which may be united or brought together, because they simply slip into a mortise in the wooden stock, in which they are fixed by a metal pin passing through them, or it may encompass barrel and stock together. It is evident then that the barrel, instead of being fixed to the stock, will be capable of not only sliding backwards and forwards, but of turning round upon its own axis within these rings, for which purpose they must be made to fit the outside of the barrel pretty accurately, and should the outside of the barrel be of an octagonal or any other polygonal form, those parts which worked through the rings should be cylindrical. To obtain an action or power over the motion of the barrel I apply a strong handle D, R, Figure 1, the part R, of which is firmly brazed or otherwise fixed upon the barrel C, and which might always remain in the position shewn in this Figure, but as it might be inconvenient when the gun was not in use, I equip this handle with a strong joint and spring, similar in its action to that of a common pocket knife, by which means it may be shut up into the form shewn at Figure 2, all which circumstances are sufficiently obvious from the two last named Figures without further explanation; and it is equally obvious that any other form of lever for moving the barrel may be adopted in my Invention. The motion which may be communicated by this handle to the sliding barrel is, however, regulated by a small metal stud or pin X, Figure 8, projecting from the under side of the barrel and working in the ward or labarynth O, of the chamber carriage, seen in plan in Figure 6, and in front elevation at Figure 5; and if the projecting stud of the barrel X, is supposed to be at O, it is evident the handle can have no effect in turning the barrel round, and that it can only move it in a straight direction from O, towards K; at this point the further progress of the barrel in this direction is stopped, but it is now permitted to turn upon its axis, the stud working in the opening P, so that in this situation the barrel C, can be screwed upon or unscrewed from the moveable chamber B, by the handle D, R, and as soon as it is unscrewed it will slide as far as from P, to O, for the purpose of disengaging itself, in order that the moveable chamber B, may lift up for loading, as in Figure 1; but inasmuch as no common screw can sufficiently secure the barrel and the moveable chamber together without making several turns or revolutions, which would not only be very inconvenient but attended with groat loss of time, I prepare my screw as follows:—I first cut my screws, male and female, in the usual manner, leaving the male screw upon the moveable chamber B, Figures 3 and 6, while the female screw is formed in the barrel C, or else I make them in the reverse manner, by placing the female screw in the chamber piece B, while the male screw is upon the barrel C; and I prefer a screw the threads of which are from 8 to 10 in an inch. Having finished my screw in the ordinary manner, I divide its circumference into a certain number of equal parts, suppose 8, and then I file or otherwise cut away every alternate division of the screw down to the cylinder upon which it is formed, as in the Figures 1, 3, 4, and 6, in all of which the screw has been divided into 8 parts, so that four spaces and four portions of screw occur, as will be distinctly seen at Figure 4, where *a*, *a*, *a*, *a*, show the four projecting pieces of screw, and *c*, *c*, *c*, *c*, the four spaces which have been cut out, and the same letters indicate the same parts in Figs. 1, 3, 4, and 6, The female screw is treated in precisely the same manner, having a similar number of spaces and projections left in it; and now, of course, since the stud upon the barrel before-mentioned, by working in the metal guide O, Figure 6, prevents it from moving in any other directions than backwards and forwards, so if after putting a charge into the chamber B, Figure 1, it is turned down upon its centre M, in the direction of the dotted line *d*, *d*, and the barrel C, is then pulled towards it by the handle D, R, the projections in the one screw will pass through the spaces in the other and allow the barrel to come immediately up to the shoulder on the chamber; and now the eighth part of a turn of the barrel by the handle D, R, will completely bring all the divided parts of each screw into action together, and the one part will be forcibly drawn or screwed towards the other, so as firmly to unite the barrel and the breech when the hole is ready for firing, and will appear as in Figure 2, in which the tail D, of the handle is represented as shut up by turning it upon its pivot screw S, which will be more handy for shouldering or carrying the gun. If the barrel is turned an eighth part round in the contrary direction, so as to unscrew it, it may then be pushed away from the moveable chamber, which becomes at liberty, and may be raised as at B, Figure 1, to receive a new charge; and thus the process of charging may be effected in less time and with more certainty than in any other kind of gun. In the above example I have described the screw as being divided into eight equal parts, four of which are alternately cut out and left, and it is almost needless to observe that a greater or less number of such divisions may be used, tho' I find eight the most convenient, because the handle has a less quantity of motion, which must always increase as the number of divisions decrease, and thus if the screw were divided into two parts the handle must move thro' a semicircle, which would be very inconvenient. In all cases, however, the stud upon the barrel must be so placed as by its working in the guide O, Figures 5 and 6, it may bring the space in the screw of one joint against the projection in the other; and the manner in which this stands with regard to the handle and its manner of working will perhaps be better seen in Figure 8, which is a transverse section thro' the barrel, showing the handle and the stud X, which works in O, P, Figures 5 and 6. The part of the handle D, in Figure 1, likewise appears shorter than those in Figures 2 and 8, because it is elevated and consequently foreshortened. It must be understood that I do not claim the exclusive right of loading fire-arms at the breech by means of a moveable piece or pieces, but that my claim is to the particular combination of machinery for the purpose of fire-arms which is above described.

In witness whereof, I, the above-named Urbanus Sartoris, have hereunto set my hand and seal, the Second day of August, in the year One thousand eight hundred and seventeen.

URB. (L.S.) SARTORIS.

Signed, sealed, and delivered by the within-named Urbanus Sartoris in the presence of

JOHN CROSLAND
Broad Street London.
G. W. PARRY,
of the same place.

AND BE IT REMEMBERED, that on the Second day of August, in the fifty-seventh year of the reign of His Majesty King George the Third, the said Urbanus Sartoris came before our said Lord the King, in His Chancery, and acknowledged the instrument aforesaid, and all and every thing therein contained and specified, in form above written. And also the instrument aforesaid was stamped according to the tenor of the Statute made in the fifty-fifth year of His said Majesty's reign.

Inrolled the Fourth day of August, One thousand eight hundred and seventeen.

LONDON:
Printed by GEORGE EDWARD EYRE and WILLIAM SPOTTISWOODE,
Printers to the Queen's most Excellent Majesty. 1854.

A. D. 1817. March 11. N° 4107.

SARTORIS' Specification.

(1 SHEET)

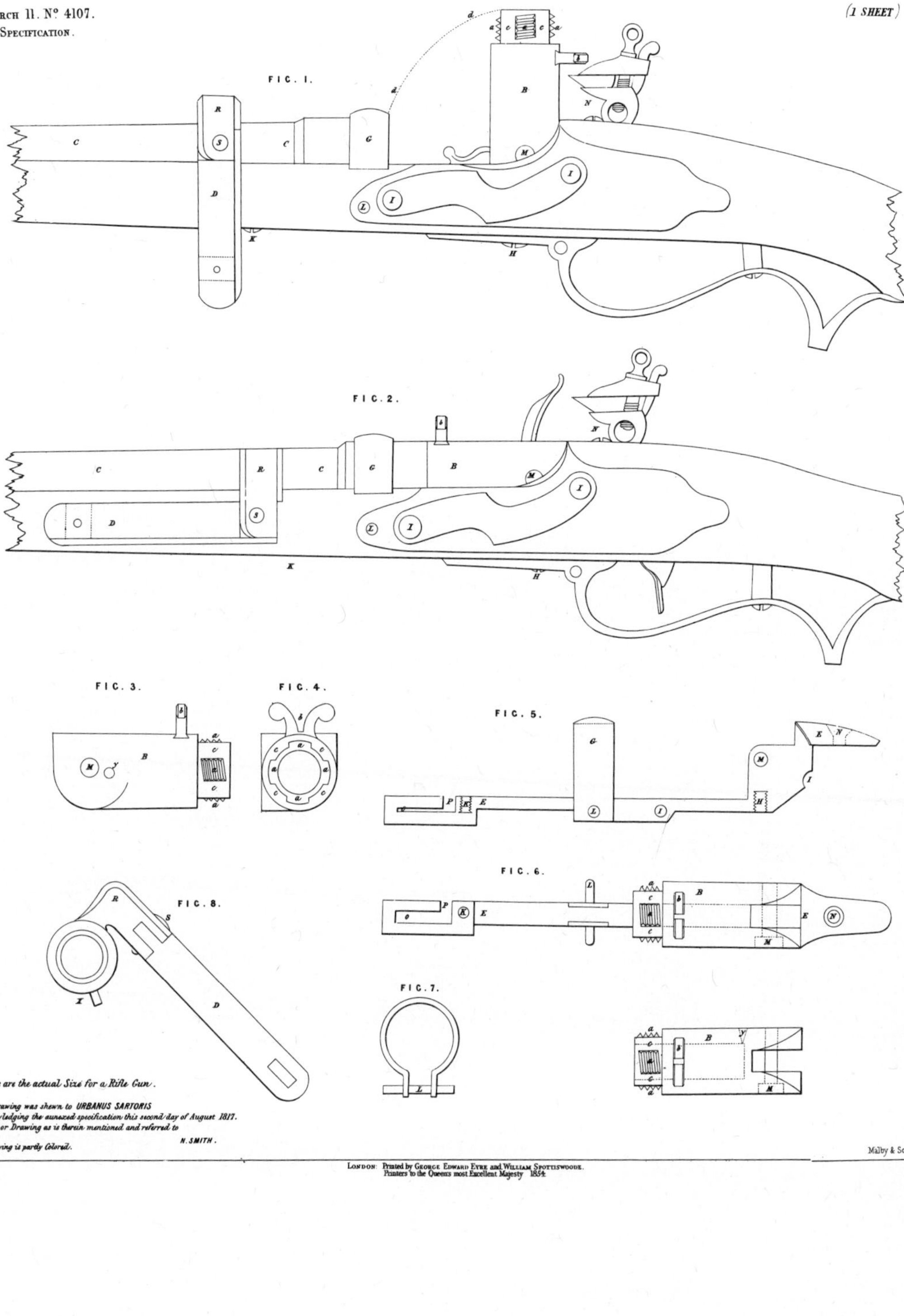

The above drawings are the actual Size for a Rifle Gun.

This Plan or Drawing was shewn to URBANUS SARTORIS at the time of acknowledging the annexed specification this second day of August 1817. and is the same Plan or Drawing as is therein mentioned and referred to

N. SMITH.

The Enrolled Drawing is partly Colored.

Malby & Sons, lith.

London: Printed by George Edward Eyre and William Spottiswoode, Printers to the Queen's most Excellent Majesty. 1854.

A.D. 1818 . . . . Nº 4285.

Locks and Primers of Fire-arms.

MANTON'S SPECIFICATION.

TO ALL TO WHOM THESE PRESENTS SHALL COME, I, Joseph Manton, of Davis Street, Berkley Square, in the Parish of St. George, Hanover Square, in the County of Middlesex, Gun Maker, send greeting.

WHEREAS I did, by petition, humbly represent to His present Majesty King George the Third, that I had invented "**Certain Primers for Fire-arms, and also certain Improvements in the Construction of certain of the Parts of Fire-arms,**" His Majesty was graciously pleased to issue His Letters Patent under the Great Seal of the United Kingdom of Great Britain and Ireland, bearing date at Westminster, the Third day of August, in the fifty-eighth year of His reign, which Letters Patent do give and grant unto me, the said Joseph Manton, and to my executors, administrators, and assigns, or to my or their servants or agents, His said Majesty's especial licence to make, use, exercise, and vend my said Invention within England & Wales, and the Town of Berwick-upon-Tweed, and to have and enjoy the whole profit and advantage arising by reason of the said Invention, during the term of fourteen years from the date of the said Letters Patent; His Majesty also, by the said Letters Patent, requires and strictly commands all His subjects whatsoever, within England and Wales, and the Town of Berwick upon Tweed, during the continuance of the said term of fourteen years, they shall not, either directly or indirectly, make or put in practice my said Invention, or any part of it, or in any ways counterfeit, imitate, or resemble the same; the said Letters Patent also contain a proviso, obliging me, the said Joseph Manton, to particularly describe and ascertain the nature of my said Invention, and in what manner the same is to be performed, by an instrument in writing under my hand and seal, and to cause the same to be enrolled in His Majesty's High Court of Chancery within six calendar months after the date of the said Letters Patent.

NOW KNOW YE, that in compliance with the said proviso, I, the said Joseph Manton, have executed this instrument, by which I do declare that my said Invention is described and ascertained in manner following, that is to say:

My primers for fire arms are small hollow tubes made of very thin metal or other suitable substance. The dimensions of the tubes must be according to the size of the piece to which they are to be applied as primers, but for a musket or fowling piece they should be from half an inch to three quarters of an inch in length, and from one tenth to one eighth of an inch in diameter, and open at both ends. To prepare the primers for use the interior of the tube is filled (or partly filled) with a detonating or fulminating substance or powder, which will explode and produce fire when struck with a sudden blow, which substance is well known, being already in use for discharging fire-arms by percussion. The open ends of the primer are stopped with bees wax to retain the fulminating substance in the tube, and preserve the same from damp. The metal or other substance of which the primer is composed must be thin, because the detonating powder or substance is to be exploded by a blow struck on the outside of the primer, as herein-after described. The primer being supported against a solid support, the tube is crushed or squeezed up by the blow so as to explode the detonating powder or substance within it. My improvements in the construction of certain of the parts of fire arms consist in the form and arrangement of certain parts of the lock, which parts are adapted to receive one of the aforesaid primers and hold it in a proper position to be struck by a projecting part of the cock of the lock, and in which position it will, on exploding or detonating, fire the gunpowder with which the piece is charged. These improvements on the lock are as follows:—In place of the pan of the lock is a piece of solid metal projecting from the lock plate, and made with a flat surface, against which the primer is supported whilst it receives the blow of the cock. In place of the hammer of the lock I apply what I call a primer holder, because it partly covers the primer and holds it by the gentle pressure of a spring against the aforesaid flat surface, in such position that one end of the primer applies to the touch hole of the piece, and the length of the primer is nearly in the line or direction of the touch hole. On the underside of the primer holder is a groove, which enables it the better to hold the primer in its said position, and which permits the primer to be introduced between the said primer holder and the said flat surface. The primer holder has an opening cut through it to expose the middle part of the primer which lies beneath it, and when the cock of the lock is let fly the projecting end thereof passes through this opening and strikes suddenly upon the outside of the primer, with sufficient force to explode the fulminating powder or substance contained in the primer. The flash of fire from the end of the primer communicates fire by the touch hole to the gunpowder contained in the barrel of the gun. To explain the nature of my said Invention more fully, I have hereunto annexed Drawings of a lock of a gun, pistol, or other fire-arm, which lock is constructed according to my improvements and is to be primed with my primers. A, A, is the lock plate seen on the outside; the parts situated behind the lock plate are made in the usual manner. B, is the cock, the end *b*, of which is adapted to strike upon and bruise the primer when the trigger is drawn; 1, is a primer; 2, is the part substituted for the pan of the lock; the surface thereof against which the primer is applied is flat and inclined, so that the end *b*, of the cock will strike down flat against it. The part 2, forms a solid support to the primer when the cock strikes it, as in Fig. 5. 3, is the primer holder which is substituted for the hammer; it turns on a screw pin as a centre of motion, and the spring D, causes it to press gently down on the primer 1; to hold the same steady against the flat surface 2, a groove or notch is cut in the underside of the primer holder 3, to lodge the primer in, and retain it opposite the touch hole; a square opening is made through the primer holder 3, to expose the middle part of the primer, as shewn in the plan Fig. 4, and allow the nose *b*, of the cock to strike thereupon. The tubes for the primers 1, may be made by rolling up small pieces of metal plate or by drawing thin tubes of brass, copper, or other suitable metal, in the manner of wire, as is well known to workmen. The piece is primed when the primer holder 3, is the situation shewn in Fig. 1. The end of the primer 1, is pushed in between the parts 2, and 3, 'till it can go no farther, and then the end of the primer will be close to the touch hole. The groove in the under side of the primer holder 3, is not quite so large as the primer, and therefore when the primer is pushed in it raises the primer holder 3, a little up from the surface 2, and the spring D, holds the primer holder on the primer with a gentle pressure, so as to prevent the primer falling out of its place. When the cock is let fly, as in Fig. 5, the end *b*, strikes on the primer and beats it flat in the middle part, so as to explode the detonating substance contained in it. The fire then flashes from both ends of the primer, as shewn in the Figure, and that end which applies to the touch hole gives fire to the charge of gunpowder in the barrel of the piece; the fire from the other end discharges itself in the open air. When the piece is to be reprimed the primer holder 3, is lifted up with the thumb or finger, as in Fig. 3, to let the old primer drop out, and the primer holder is then shut down, as in Fig. 1, ready to receive a fresh primer. A sufficient number of primers must be kept prepared ready for use. The detonating substance or substances with which I fill my primers are those which are known to chemists by the name of fulminating powders, the preparation of which, being well known and being no part of my Invention, need not be described. I confine my claim to the Invention of primers or detached tubes, such as I have described, in which the detonating substance is enclosed so as to preserve it, and is exploded in the manner herein-before described by striking the outside of the tube; and also to the improved construction of those parts of the lock which receive the primer and hold it in the proper situation to receive the blow of the cock and to communicate fire to the touch hole, as herein-before described.

In witness whereof, I, the said Joseph Manton, have hereunto set my hand and seal, this Second day of February, in the year of our Lord One thousand eight hundred and nineteen.

JOSEPH MANTON. (L.S.)

Signed, sealed and delivered
in the presence of

JOHN FARCY,
37, Howland Street, London.
HENRY GEO. HOLDEN,
Rolls Chapel.

AND BE IT REMEMBERED, that on the Second day of February, in the fifty-ninth year of the reign of His Majesty King George the Third, the said Joseph Manton came before our said Lord the King, in His Chancery, and acknowledged the instrument aforesaid, and all and every thing therein contained and specified, in form above written. And also the instrument aforesaid was stamped according to the tenor of the Statute made in the fifty-fifth year of His said Majesty's reign.

JEKYLL.

Inrolled the Second day of February, One thousand eight hundred and nineteen.

LONDON:
Printed by GEORGE EDWARD EYRE and WILLIAM SPOTTISWOODE,
Printers to the Queen's most Excellent Majesty. 1854.

A.D. 1818. AUG^T 3. N° 4285.
MANTON'S SPECIFICATION.

(1 SHEET.)

FIG. 1.

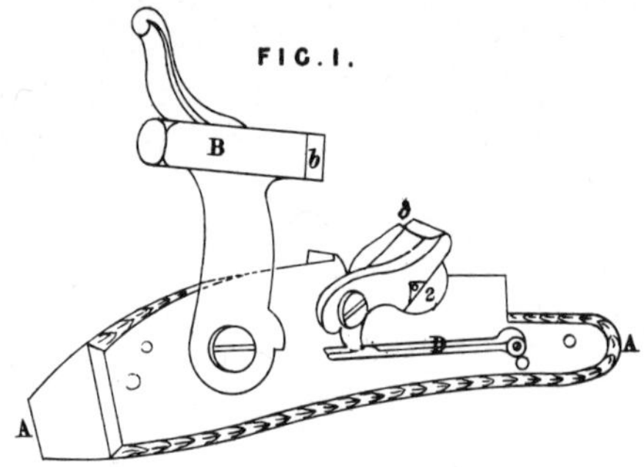

FIG. 2.

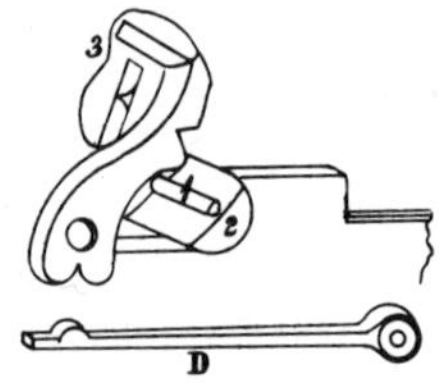

FIG. 3.

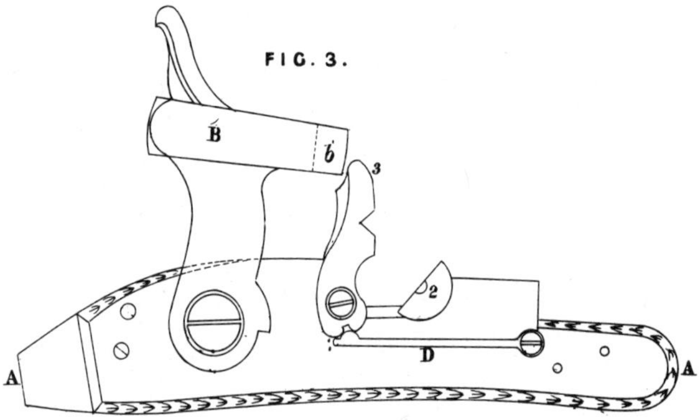

FIG. 4.

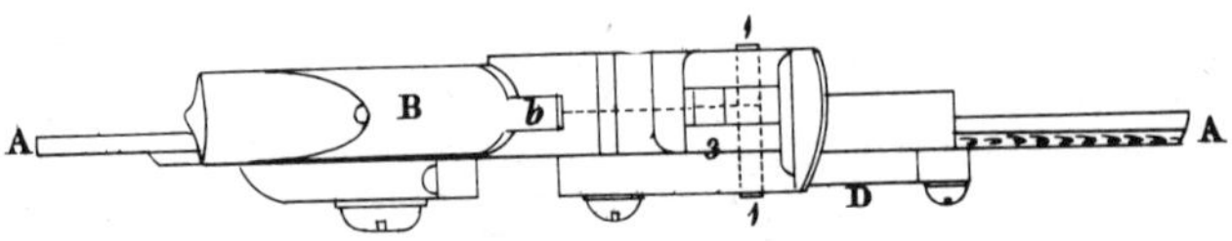

FIG. 5.

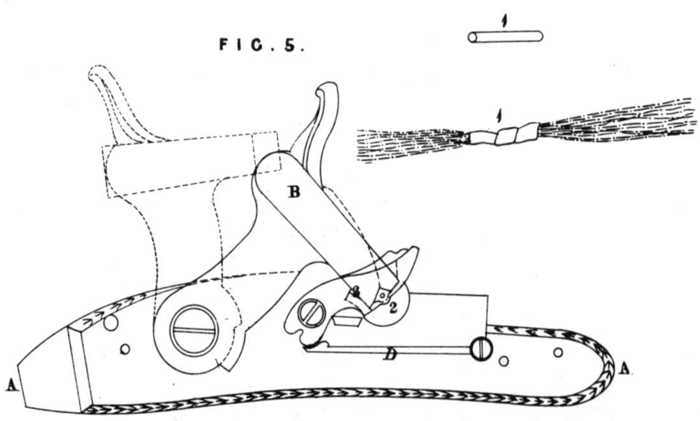

The enrolled Drawing is Colored.

Malby & Sons lith.

LONDON: Printed by GEORGE EDWARD EYRE and WILLIAM SPOTTISWOODE, Printers to the Queen's most Excellent Majesty. 1854.

A.D. 1818 . . . . Nº 4315.

Fire-arm combining Single Barrel with several Chambers, to obtain Succession of Discharges from One Loading.

COLLIER'S SPECIFICATION.

TO ALL TO WHOM THESE PRESENTS SHALL COME, I, Elisha Hayden Collier, late of Boston, in the State of Massachusetts, one of the United States of America, but now of Charter House Square, London, Gentleman, send greeting.

WHEREAS His most Excellent Majesty King George the Third did, by His Letters Patent under the Great Seal of the United Kingdom of Great Britain and Ireland, bearing date at Westminster, the Twenty-fourth day of November, in the fifty-ninth year of His reign, give and grant unto me, the said Elisha Hayden Collier, my ex̃ors, adm̃ors, and assigns, His especial licence, full power, sole privilege and authority, that I, the said Elisha Hayden Collier, my ex̃ors, adm̃ors, and assigns, during the term of years therein mentioned, should and lawfully might make, use, exercise, and vend, within England, Wales, and the Town of Berwick-upon-Tweed, and also in all His Majesty's Colonies and Plantations abroad, the Invention, partly communicated to me by a certain foreigner residing abroad, together with certain additions of my own Invention, of an "Improvement in Fire-arms of various Descriptions, which Improvement is also applicable to Cannon;"

in which said Letters Patent there is contained a proviso, that if I, the said Elisha Hoyden Collier, shall not particularly describe and ascertain the nature of my said Invention, and in what manner the same is to be performed, by an instrument in writing under my hand and seal, and cause the same to be inrolled in His Majesty's High Court of Chancery within six calendar months next and immediately after the date of the said Letters Patent, that then the said Letters Patent, and all liberties and advantages whatsoever thereby granted, shall utterly cease, determine, and become void, as in and by the same, relation being thereunto had, will more fully and at large appear.

NOW KNOW YE, that in compliance with the said proviso of the said Elisha Haydon Collier, do hereby declare that this Invention is described in manner following, that is to say:

The object of the said Invention is to render a gun, pistol, or other fire-arm capable of being fired off several times in succession with only once loading. The manner in which this effect is produced will, however, be better understood by a reference to the Drawings accompanying this Specification, and by the following description thereof, the same letters of reference applying to the same parts in all the Figures.

Fig. 1, is the side view of a pistol formed upon this construction; Fig. 2, an end view of the same; and Fig. 3, a section through the middle of it, viewed on the opposite side to Fig. 1. Fig. 4, is another section similar to Fig. 3, excepting that the situations of several parts of the movements are changed, as will be described hereafter. Figs. 5, and 6, are also introduced to shew the different situations of particular parts of the lock in performing their several offices. Other Figures, from Fig. 7, to Fig. 26, inclusive, are also added to explain the detail of the several parts. A, Figs. 1, and 2, is the wooden stock of the pistol. B, a metal plate forming one side of a frame or case to contain the parts of the lock and other movements, to be particularized hereafter. C, is another metal plate with straps forming the top, end, and bottom of this frame, which is firmly secured to the stock A, by screws. Into the end of the frame C, is screwed the axis D, at the other end of which is an elevated off-set having a hollow made in its upper part, into which one end of the long barrel E, which is open at both ends, is received and secured by a screw. To the top of the frame C, is secured the elevated brace F, having also at is other end a hollow to fit the end of the long barrel E, and to which it is also joined by a screw; and thus the long barrel E, is firmly united to the frame C. Upon the axis D, revolves a number of short barrels or chambers G, arranged in a circular manner, and at equal distances ground the axis. One end of each of these short barrels is closed, and they are of sufficient capacity to contain the charges of gunpowder, shot, or ball intended to be used. These short barrels may either be made of iron, or of any other fit metal or composition of metals, in separate tubes, and afterwards firmly united together, or they may be made out of one solid mass of metal properly bored and shaped into form. A cylindrical hole is made through the centre of this set of short barrels, into the fore end of which is screwed a nut H, having a cylindrical hole in its centre, through which the axis D passes, and into the opposite end of it another nut I, is screwed, with a cylindrical hole also in its centre, fitted to the axis. This nut I, is prevented from becoming unscrewed by means of a steady pin screwed into the joint between the male and the female screw. To the nut I, is screwed a plate or wheel K, having a number of teeth equidistant around it, corresponding to the number of short barrels used, the use of which teeth will be described hereafter. On the axis is fastened one end of a spiral spring L, similar to the main spring of a clock or watch, the other end of which is secured to the inner part of the cylindrical hole within the set of short barrels, and which spring, when wound up, has a tendency to cause the barrels to revolve in a circular direction. A boss or ring M, is also fixed upon the axis D, by means of a pin passed through it, against which boss one end of an helical spring N, placed upon the axis D, presses, the other end of it pressing against the nut H, in the fore end of the set of short barrel G, and thus gives it a tendency to thrust or push outwardly, so as to cause a recess made in the end of each short barrel to fit upon a shoulder made upon the end of the long barrel as they pass it in succession, and thereby to make a close joint with it ready to be discharged. In order to prevent the short barrel from recoiling during the discharge, a bolt O is thrust against it by the action of a tooth P, formed upon the axis of the lock-tumbler, as shewn in Fig. 3; that part of the axis opposite to the tooth P, being also supported by a bracket Q, firmly screwed to the top plate of the frame C, and rivetted to the end plate thereof. Another tooth R, fitted upon the square part of the axis of the tumbler, operates upon one end of a lever S, turning upon its fulcrum at T, to the other end of which is jointed the catch bolt U, so as to cause it to advance and recede alternately, and when advanced to catch or lay hold of that tooth of the wheel K, which may happen to be opposite to it, and by being withdrawn to bring the set of short barrels along with it, until that one which was engaged in or jointed with the long barrel becomes sufficiently separated from it to allow the spiral spring L, to cause it to revolve far enough for the next short barrel in succession to take its place, which it does by the pressure exerted by the helical spring N, against the set of short barrels, and by the catch bolt U, slipping off the tooth of the wheel K, in consequence of the circular motion of the barrels, and thereby permitting the recess in the approaching short barrel to slip upon the shoulder on the end of the long barrel.

Fig. 1, represents the pistol as just discharged, with the catch bolt U, hooked upon one of the teeth of the wheel K.

Fig. 3, is a section of the pistol in this situation, to shew the tooth R, lying ready to act upon the lever S, and pull back the catch bolt U.

Fig. 4, shews the cock a little raised, the tooth R being engaged in pushing forwards the end of the lever S, and threby drawing back the catch bolt U, and the set of short barrels, until that which was united with the long barrel has become disengaged from it. The tooth of the wheel K disengages itself from the catch bolt U, by the action of the spiral spring, and the cock will then be at liberty to be brought to the half cock, and the catch bolt withdrawn so far as to let the next tooth of the wheel R pass by it, and the next short barrel will then be at liberty to engage itself with the long barrel by the action of the helical spring N.

In bringing the cock from the half cock to the full cock, the tooth R depresses the lever S, until it can pass over it, at the same time compressing the spring V, which is attached to the catch bolt U; but when brought to the full cock, as shewn in Fig. 5, the tooth R will have passed over the lever S, and by the reaction of the spring V the lever will be raised, so as to be acted upon by the tooth R, in the opposite direction when the cock is discharged. By the time it has arrived at the situation of half cock the tooth R will have pushed back the lever S, and moved the catch bolt U, forwards, until the spring W (also fixed to the catch bolt) acts against the stud X, and as the cock moves further down the lever S is still more depressed, so that the tooth R, is just ready to pass over it; and when in this situation, the end of the spring W, being pressed against the stud X, which is above the fulcrum T, upon which the lever S moves, it causes the catch bolt U to rise up to the upper side of the cap or cavity through which it passes, at the same time receiving the spring Y, which acts on the top of it, as shew in Fig. 6; the catch bolt is thus lifted up sufficiently high to pass over the before-mentioned tooth of the wheel K. When the cock has descended a little more the tooth R will have passed the end of the lever S, and by the reaction of the spring W, the lever S will be raised to its first situation, as shewn in Fig. 3, and at the same time, by the reaction of the spring Y, the catch bolt U, will be depressed and hooked upon the tooth of the wheel K, ready to again bring forwards the short barrels, as before described. When the cock is quite down, as shewn in Fig. 3, and the discharge may be taking place, it will be observed that the point of the tooth P is depressed below the centre of the axise of the tumbler; and the greater is the effort made by the discharge of the piece against the bolt O, to cause it to recede, so much the more does it prevent the tooth P from rising out of the recess in the end of the bolt O, which is below the centre of the axis of the tooth, the bolt itself being supported upon the brackett Q. When the cock is raised the tooth P is lifted out of the recess in the end of the bolt O, and permits it to slide back by the pressure of the short barrels against it, as shewn in Fig. 4. The sear and other parts of the lock not described, being common to other locks, need not be particularized here; I can apply either the usual hammer or one of the magazine hammers (to be herein-after described) to this lock. The hammer Z, is united to the elevated brace F, (instead of to the lock plate as usual) as shewn in Figs. 1, and 8. Each short barrel has a separate pan *a*, and a sliding cover *b*, fitted to it, moving in dovetailed grooves, and regulated by the spring *c*, so as to move freely, but not too easily, so as to be affected by the recoil in firing the piece. Each sliding cover has also a tooth or stud *d*, upon it, against which the toe *e*, of the hammer acts on shutting the pan, so as to cause the cover to slide back, the hammer itself then covering the priming ready to receive the fire from the flint. In order to use this weapon the short barrels or chambers much each receive a charge of gunpowder, and either shot or ball, as may be required. The pans also must also be filled with prime, and the sliding covers shut upon it. All the short barrels but one may be loaded, and that one, namely, that which is engaged in the long barrel, will be disengaged from it on winding up the spring L, by turning the short barrels round in the contrary direction to that in which they move when fired, the remaining short barrel may then be loaded and primed, and the piece is ready to fire off each charge in succession by merely cocking the lock, shutting the hammer, and discharging the piece.

Fig. 27, is a side view, and Fig. 28, a top view of a gun, fitted up with a lock and its appendages of the construction before described as applied to the pistol, but witl a magazine hammer and one pan only, which serves all the short barrels as they revolve. This magazine hammer *f*, has a chamber *g*, Fig. 31, made at the back of it, to hold a sufficient quantity of gunpowder to fire off the number of barrels required. At the lower part of the chamber is fitted a cylinder *i*, shewn separately at Figure 32, having a number of equidistant cavities *j*, *j*, *j*, around it, and each sufficiently large to receive a supply of gunpowder for a priming. This cylinder communicates with the chamber by the opening *l*, at the bottom of it; and underneath it is another opening *m*, which delivers the gunpowder it has received into the pan *n*; a sliding cover *o*, with a stud and regulating spring in the inside of the chamber, closes it. On one end of the cylinder is fitted a ratchet wheel *p*, Figs. 27, and 30, having teeth corresponding to the number of cavities in the cylinder, and a spring *q*, to retain it in the required situation. This ratchet wheel meets a click *r*, jointed to the elevated brace F, having a spring *s*, acting upon it, and which click causes the cylinder to turn on shutting the pan until the next cavity in succession is brought over the pan, as shewn in Fig. 31, and delivers its priming into the pan, whilst at the same time another cavity is receiving another priming from the chamber, as is likewise shewn in Fig. 31. The toe of the hammer is acted upon by the spring *t*, in the manner represented in Figs. 27, and 30. The pan *n*, is attached to a metal hoop or cylinder *a*, at the end of which is a plate with a hole in the middle of it, to permit the axis and bolts to pass through; this plate is firmly secured to the end plate of the frame C, by means of screws. Within the hoop *u*, the set of short barrels revolves, the interstices between them at that end being filled up or left solid, so as to form a cylindrical surface nearly filling the hoop, so as not to permit the gunpowder to pass out of the barrels through the touch-holes, nor the flame to communicate from one barrel to another. Within this hoop the short barrels are capable of moving in all the various directions, the same as if it was not there.

Fig. 29, represents the set of short barrels G, within the hoop *u*, and the situation of the pan *n*, upon it communicating with one of the touch-holes. Fig. 7, is the set of short barrels viewed from the opposite end in Fig. 2. Fig. 9, shews the axis D, separate. Fig. 10, is the end of the lock frame C. Fig. 11, is a side view of the cock; and Fig. 12, shews its thickness. Fig. 13, is the tooth R, of which Fig. 14, shews the thickness. Fig. 15, 16, & 17, are different views of the bridle of the tumbler, which is situated when in use near the middle of it, between the teeth P, and R. Figs. 18, and 19, are an end and front view of the axis of the tumbler and tooth P. Figs. 20, and 21, are a top and side view of the bolt O. Fig. 22, is a side view of the sear and sear bridle. Fig. 23, is an edge view of the sear. Fig. 24, an edge view of the sear bridle; and Fig. 25, is a side of the lever S, and catch bolt U, with its springs W and V, the dotted lines shewing oter situations of the different parts, as already described. Fig. 26, is an edge view of Fig. 25.

I do not mean hereby to claim as my Invention any such parts of fire-arms as are already known and in use, although such may be here represented as connected with those parts which I do consider as my Invention, for the sake of explanation and the better understanding thereof. But I do hereby claim the new combinations and applications of those parts to produce the different effects herein described; I can also apply any of the parts either to guns, pistols, or other fire-arms, as well as to cannon, under proper modifications thereof.

In witness whereof, I, the said Elisha Hayden Collier, have hereunto set my hand and seal, this Twenty-fourth day of May, One thousand eight hundred and nineteen.

E. H. (L.S.) COLLIER.

STRATFORD.

AND BE IT REMEMBERED, that on the Twenty-fourth day of May, in the year of our Lord 1819, the aforesaid Elisha Hayden Collier came before our said Lord the King, in His Chancery, and acknowledged the Specification aforesaid, and all and every thing therein contained and specified, in form above written. And also the Specification aforesaid was stamped according to the tenor of the Statute made for that purpose.

Inrolled the Twenty-fourth day of May, in the year of our Lord One thousand eight hundred and nineteen.

LONDON:
Printed by GEORGE EDWARD EYRE and WILLIAM SPOTTISWOODE,
Printers to the Queen's most Excellent Majesty. 1854.

A.D. 1818. November 24. Nº 4315.

COLLYER'S Specification.

(1 SHEET)

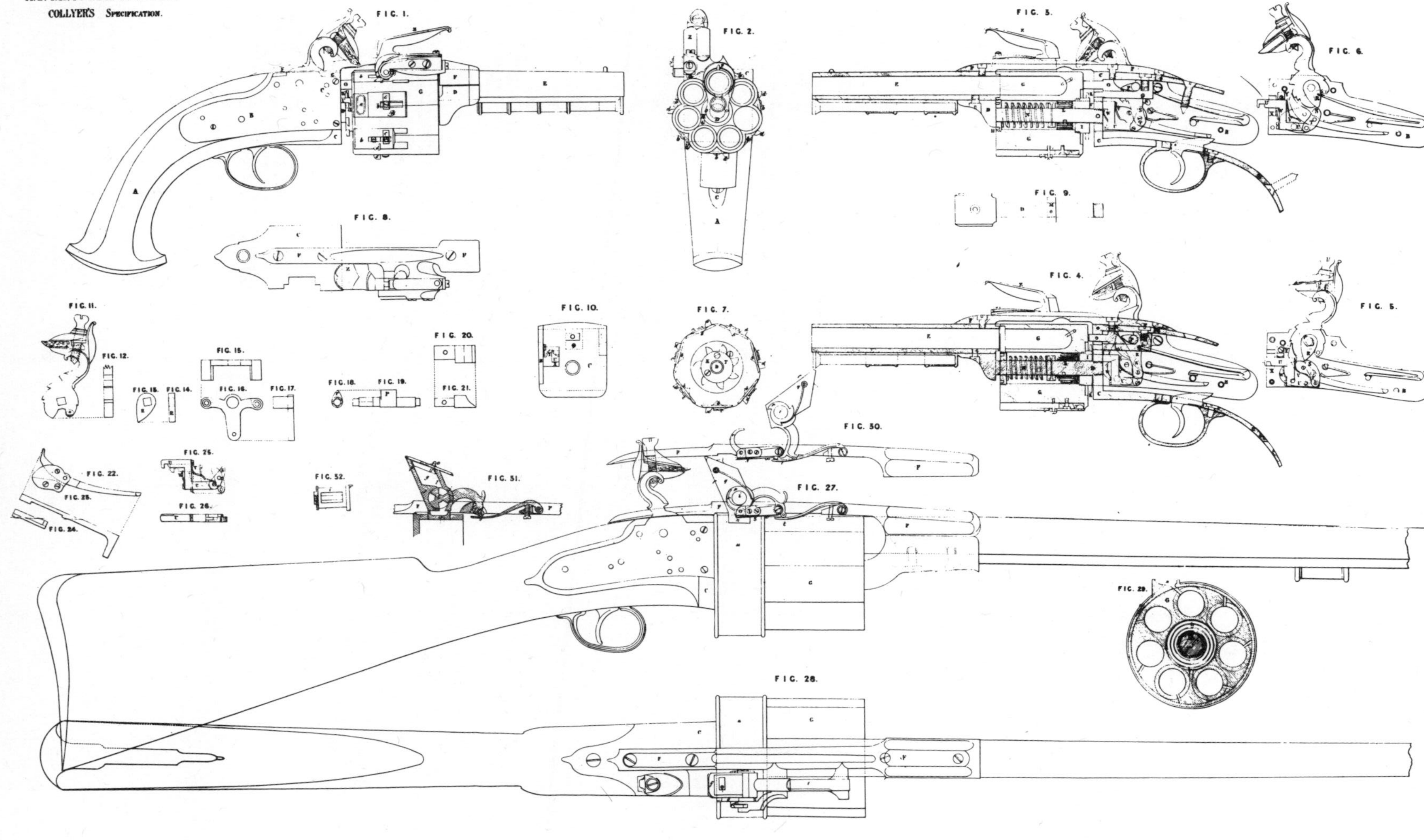

The Enrolled Drawing is colored.

Malby & Sons, Lith.

London: Printed by George Edward Eyre and William Spottiswoode, Printers to the Queen's most Excellent Majesty. 1854.

A.D. 1821 . . . . Nº 4577.

Locks of Fowling Pieces and Fire-arms.

MANTON'S SPECIFICATION.

TO ALL TO WHOM THESE PRESENTS SHALL COME, I, John Manton, of Dover Street, Piccadilly, in the County of Middlesex, Gunsmith, send greeting.

WHEREAS His most Excellent Majesty King George the Fourth, by His Letters Patent under the Great Seal of the United Kingdom of Great Britain and Ireland, bearing date at Westminster, the Thirtieth day of July, in the second year of His reign, did, for Himself, His heirs and successors, give and grant unto me, the said John Manton, my executors, administrators, and assigns, His special licence, full power, sole privilege, and authority, that I, the said John Manton, my executors, administrators, and assigns, and every of us, by myself and themselves, or by my and their deputy or deputies, servants or agents, or such others as I, the said John Manton, my executors, administrators, or assigns, should at any time agree with, and no others, from time to time and at all times thereafter, during the term of years therein expressed, should and lawfully might make, use, exercise, and vend my Invention of "**An Improvement in the Construction of Locks of all kind of Fowling Pieces amd Fire-arms**," within that part of His said Majesty's United Kingdom of Great Britain and Ireland called England, His Dominion of Wales, and Town of Berwick upon Tweed, in such manner as to me, the said John Manton, my executors, administrators, and assigns, or any of them, should in my or their discretion seem meet; in which said Letters Patent is contained a proviso, that if I, the said John Manton, should not particularly describe and ascertain the nature of my said Invention, and in what manner the same is to be performed, by an instrument in writing under my hand and seal, and cause the same to be enrolled in His Majesty's High Court of Chancery, within two calendar months next and immediately after the date of the said Letters Patent, that then the said Letters Patent, and all liberties and advantages whatsoever thereby granted, should utterly cease, determine, and become void, any thing therein-before contained to the contrary thereof in anywise notwithstanding, as in and by the said Letters Patent, relation being thereunto had, will more fully and large appear.

NOW KNOW YE, that in compliance with the said proviso, I, the said John Manton, do hereby describe and ascertain the nature of my said Invention, and in what manner the same is to be performed, in manner following (that is to say):—

My Invention consists in forming the sear of the lock so as to have the back end placed crosswise in the plate instead of lengthwise, which is a great advantage, as it admits of the trigger blade being so formed as to require only a small notch in the trigger plate instead of a long one, as with the common trigger, which lets in much wet, thereby rusting the trigger and the notch in which it acts and impeding its action, and also that of the sear, and consequently causing considerable danger. This Invention I exemplify by the Drawings at the foot of these Presents, and further illustrate by the following explanation (that is to say):—The sear, instead of being formed in the usual manner, as represented by the Drawing marked B, is made in the form represented by the Drawing marked A; and the Drawing marked B, is inserted with a view of shewing the difference between the sear invented by me and the one in general use, the latter of which such Drawing represents. In the Drawing marked C, the situation and position of the sear is shewn, and it is there marked D. The trigger of the lock to which the sear is applied is fitted to the trigger plate in the same manner as a pocket pistol (that is to say) with a blade nearly upright, so as to press upon the shank of the sear, which shank in the Drawing C, is marked E.

In witness whereof, I, the said John Manton, have hereunto set my hand and seal, this Twenty-fourth day of September, One thousand eight hundred and twenty-one.

JN° MANTON. (L.S.)

AND BE IT REMEMBERED, that on the same Twenty-fourth day of September, in the year above mentioned, the aforesaid John Manton came before our Lord the King, in His Chancery, and acknowledged the Specification aforesaid, and all and every thing therein contained, in form above written. And also the Specification aforesaid was stamped according to the tenor of the Statute in that case made and provided.

Inrolled the Twenty-eighth day of September, in the year above written.

J. E. DOWDESWELL.

LONDON:
Printed by GEORGE EDWARD EYRE and WILLIAM SPOTTISWOODE,
Printers to the Queen's most Excellent Majesty. 1854.

A.D. 1821. July 30. No. 4577.

(1. SHEET.)

MANTON'S SPECIFICATION.

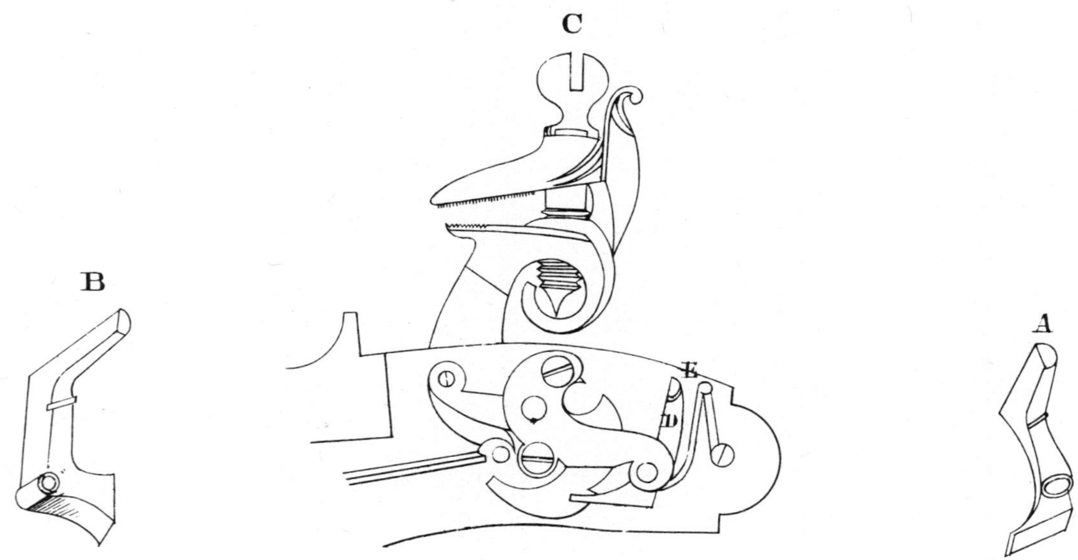

The Enrolled Drawing, is Colored.

Malby & Sons, lith.

London: Printed by George Edward Eyre and William Spottiswoode.
Printers to the Queen's most Excellent Majesty. 1854

A.D. 1821 . . . . N° 4590.

Discharging Fire-arms.

WEBSTER'S SPECIFICATION.

TO ALL TO WHOM THESE PRESENTS SHALL COME, I, William Webster, late of George Court, Princes Street, Soho, in the County of Middlesex, but now of Regent Street, Picadilly, in the said County, Gun Maker, send greeting.

WHEREAS His present Majesty King George the Fourth, in and by His Letters Patent under the Great Seal of Great Britain, bearing date at Westminster, the Fourteenth day of September last, for Himself, His heirs and successors, did give and grant unto me, the said William Webster, His special licence, full power, sole privilege and authority, that I, the said William Webster, my ex̃ors, adm̃ors, and assigns, and every of them, by myself and themselves, or by my and their deputy or deputies, servants or agents, or such others as I, the said William Webster, my ex̃ors, adm̃ors, or assigns, should at any time agree with, and no others, from time to time and at all times thereafter during the term of years therein unexpired, should and lawfully might make, use, exercise, and vend, within England, Wales, and the Town of Berwick-upon-Tweed, and also in all His Majesty's Colonies and Plantations abroad, my Invention of "Certain Improvements in the

MECHANISM OF AND APPERTAINING TO FORSYTH'S ROLLER MAGAZINE FOR THE DISCHARGE OF FOWLING PIECES AND FIRE-ARMS IN GENERAL BY MEANS OF PERCUSSION;" in which said Letters Patent there is contained a proviso, obliging me, the said William Webster, by an instrument in writing under my hand and seal, particularly to describe and ascertain the nature of my said Invention, and in what manner the same is to be performed, and to cause the same to be inrolled in His Majesty's High Court of Chancery within two calendar months next and immediately after the date of the said recited Letters Patent, as in and by the same, reference being thereunto had, will more fully appear.

NOW KNOW YE, that, in compliance with the said proviso, I, the said William Webster, do hereby describe and ascertain the nature of my said Invention, and in what manner the same is to be performed, by the plan or Drawing annexed to these Presents, and the following description thereof, that is to say:

My improvements in the mechanism of the roller magazine described in the Specification of a Patent granted to Alexander John Forsyth (Figures 5, and 6), and inrolled or dated the Eleventh April, One thousand eight hundred and seven, consists of a mode of attaching and detaching the magazine to and from the cylindrical piece or plug which contains the touch-hole, which improvements are shewn in the Drawings hereunto annexed. Figure 1, is a representation of a percussion gun lock with my improvement applied to the magazine. Figure 2, the magazine removed from the lock, in which *a*, is a dove-tailed groove for the purpose of receiving a bolt. Figure 3, is the plug or cylindrical piece before alluded to, but which is found most eligible if made slightly tapering, as in the Drawing. (This alteration, however, in the form of the plug I do not claim.) In this Figure *b*, is a groove formed round the plug for the purpose of receiving the end of a bolt *c*. Figure 4, shews the said bolt detached; and Figure 5, is a section of the plug and the magazine, in which the parts are seen put together. The object of this improvement is that the magazine may be taken off or removed from the plug without the trouble of unscrewing any part of the magazine or its appendages. The bolt *c*, slides in the groove *a*, as above mentioned, and has a spring acting behind it for the purpose of pressing up the bolt against the dovetail, so as to render it sufficiently tight, and at the same time enabling it to slide easily. The end of the bolt intended to fall into the groove *b*, and which I call the nose, may form part of the slider *c*, or be united to the end of the spring behind it, the latter of which I prefer. In order to take the magazine off the plug the bolt is to be sliden down by the thumb nail pressing against the screw head or knob seen in front of the bolt; when the nose of the bolt is thus drawn from the groove of the plug *b*, the magazine may be immediately removed from the plug, and on replacing the magazine upon the plug the bolt is to be slidden up again into the grove *b*, which secures it from falling off. Though I have described this particular manner of securing the magazine upon the plug, yet I do not mean to confine myself precisely to this form or situation of the bolt, as the same object may be effected by placing a spring catch upon the side of the magazine, as shewn at *c*, Figure 6, or in several other modes or modifications of my plan, which consists in securing the magazine upon the plug by means of a bolt catch or catches falling into a groove.

My further improvements in the mechanism of a percussion gun lock consists of a magazine which is contrived to deposit its priming in the act of cocking into a receptical placed on the inside of the cock, which priming is effected every time that the lock is brought to full cock. Figure 7, exhibits the external appearance of a gun lock upon this construction; and Figure 8, a front view of the lock, the magazine, and the receptical, Figure 9, being a section of the same; the respective letters referring to the same parts in Figures 7, 8, and 9. *a*, is a curved piece of metal with a chamber *b* within, containing the detonating powder intended to be employed as priming. *c*, is another piece of metal attached to the inner part of the cock formed to fit the curve of *a*, under which it slides. In this last-mentioned piece a small receptical or chamber *d*, is made to receive the priming when it comes immediately under the chamber *b*, which is the case upon the lock being brought to full cock. *e*, is a stopper placed in the front of the priming chamber, which is held in that situation by its arm *f*, fastened at bottom by a screw to the breast of the cock.

This arm *f*, is a spring in order to permit the stopper to slip away from the front of the priming chamber *d* (as shewn by dots at Fig. 8), when it strikes against the plug *g*, for the purpose of discharging the priming. There are several modes which may be adopted of removing this stopper from the front of the priming chamber so as to explode the priming. Two particularly I adopt; the first of which is by making the front of the stopper, as seen at Figure 8, an inclined plane, which by the descent of the cock strikes against the end of the plug, and being thus displaced admits the end of the plug into the chamber, the percussion of which discharges the powder. The second mode is by placing an inclined plane upon the plate of the lock, as seen at *h*, Figure 7, when, as the cock comes down by the spring arm *f*, acting against the inclined plane *h*, the stopper *e* is forced on one side, and the priming chamber opened to receive the plug *g*, by which the percussion is effected. The curved piece *a*, is kept in close contact with the upper or curved piece *c*, by means of a spring acting against the toe within the lock, as seen a *i*.

This part of the improvement may be further varied by removing the plug *g*, from the breach, and attaching it to the end of the spring arm (a view of which is seen at Figure 10, and a section of the cock and chamber at Figure 11). The operation of depositing the detonating powder in the priming chamber will be the same as above described, there being a slight variation in the form of the entrance of the chamber, as will be seen at Figure 11; the plug is by the spring withdrawn a sufficient distance to allow the priming to descend. When the cock is discharged the plug, striking against the breach, causes the powder to explode, which communicates its fire through the small perforation or touch-hole, as shewn by dots at *k*. Figure 12, is a section of the spring arm and plug detached.

In witness whereof, I have hereunto set my hand and seal, the Fourteenth day of November, in the year of our Lord One thousand eight hundred and twenty-one.

WILLIAM WEBSTER. (L.S.)

A.D. 1821. SEPTEMBER 14. Nº 4590.

WEBSTER'S SPECIFICATION.

(1 SHEET)

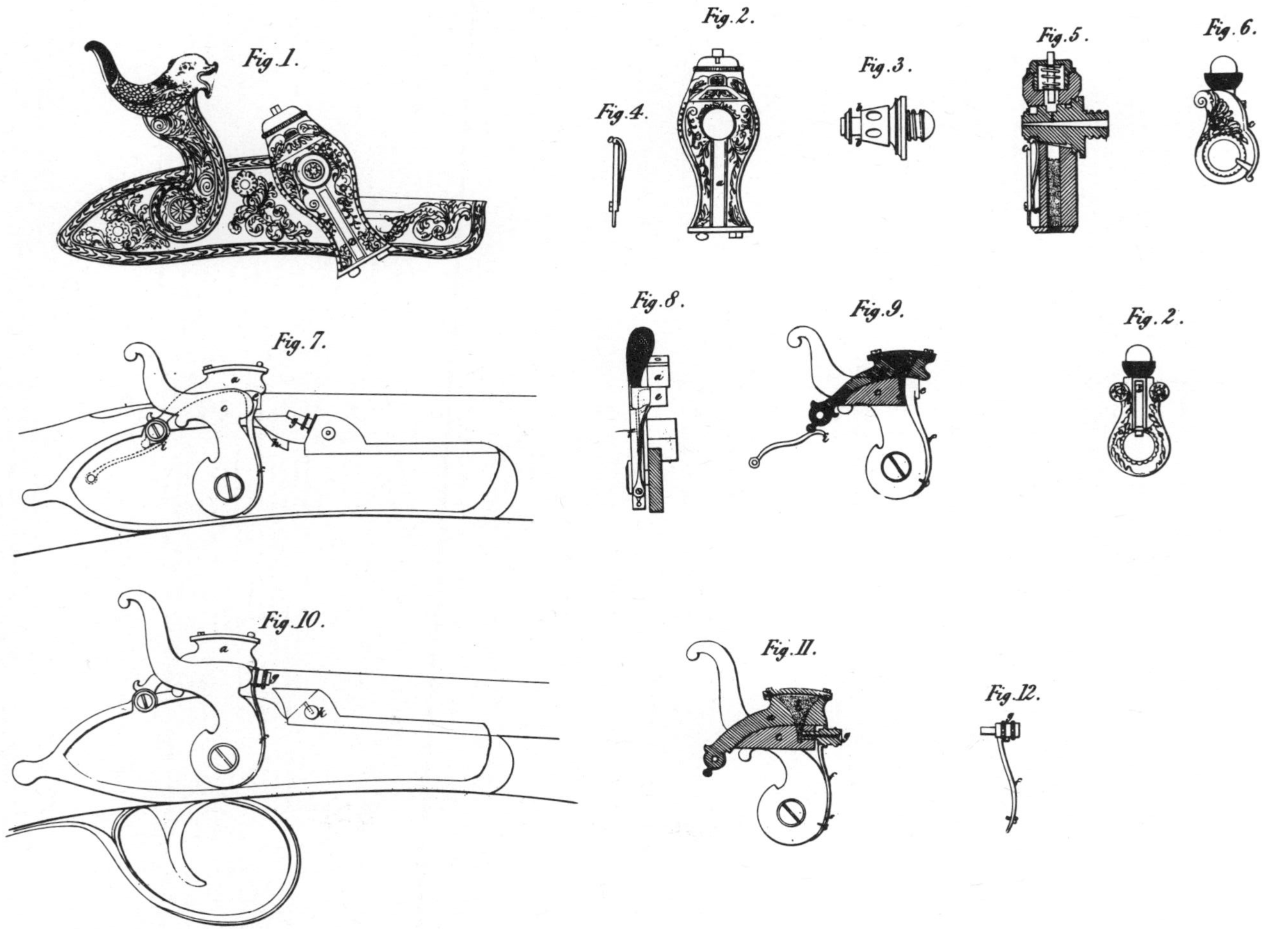

The Enrolled Drawing is partly Colored.

Malby & Sons. Lith.

LONDON: Printed by GEORGE EDWARD EYRE and WILLIAM SPOTTISWOODE. Printers to the Queen's most Excellent Majesty. 1854.

A.D. 1822 . . . . N° 4648.

Locks of Guns and other Fire-arms.

DAVIS' SPECIFICATION.

TO ALL TO WHOM THESE PRESENTS SHALL COME, I Samson Davis, of Upper East Smithfield, in the County of Middlesex, Gun Lock Maker, send greeting.

WHEREAS His most Excellent Majesty King George the Fourth, by His Letters Patent under the Great Seal of Great Britain, bearing date at Westminster, the Twelfth day of February now last past, did, for Himself, His heirs and successors, give and grant unto me, the said Samson Davis, His especial licence that I, the said Samson Davis, my executors, administrators, and assigns, or such others as I, the said Samson Davis, my executors, administrators, or assigns, should at any time agree with, and no others, from time to time and at all times thereafter during the term of years therein expressed, should and lawfully might make, use, exercise, and vend, within England, Wales, and the Town of Berwick upon Tweed, my Invention of "An Improvement upon the Lock for Guns and other Fire-arms," which enables the same lock to be used upon the percussion principle, or with gunpowder, without changing the cock or hammer; in which said Letters Patent is contained a proviso obliging me, the said Samson Davis, by an instrument in writing under my hand and seal, particularly to describe and ascertain the nature of my said Invention, and in what manner the same is to be performed, and to cause the same to be inrolled in His said Majesty's High Court of Chancery within two calendar months next and immediately after the date of the said recited Letters Patent, as in and by the same, reference being thereunto had, will more fully and at large appear.

NOW KNOW YE, that in compliance with the said proviso, I, the said Samson Davis, do hereby declare that the nature of my said Invention, and the manner in which the same is to be performed, is described and ascertained as follows, that is to say:

My said improvement consists in forming and constructing the pan and nipple of the lock so as to be used in discharging guns and other fire arms either with common gunpowder as a flint and steel lock, or with a chemical preparation of powder which ignites by an act of percussion from the cock coming in contact with the nipple attached to the pan, thereby enabling the same lock to be used on the flint and steel or the percussion principle; in either of which cases it is a complete lock without any addition or diminution, the different operations being effected by a mere change in the position of the pan and nipple. A drawing of the lock is hereunto annexed, and is explained as follows:—The part marked A, (Fig. 2,) is a cylinder on which the pan and nipple devolve; B, (Figs. 1, & 3,) is the nipple, having a hole at the top, which communicates to the touch hole in the barrel, and is kept in its proper position by a fly screw C, (Figs. 1, and 3). A cavity under the jaw of the cock marked D, (Fig. 1,) falling on the top of the nipple, which is primed with the percussion powder, ignites the body of powder in the barrel through the channel of the nipple, which communicates to the touch hole, and causes it immediately to explode. E, (Fig. 1,) is a screw which secures the pan and nipple on the cylinder; the fly screw C, (Figs. 1, and 3,) being turned backwards enables the pan to move on the cylinder, and as the nipple rolls down the cavity in the pan marked F, (Figs. 1, and 3,) takes its position exactly opposite to the touch hole in the barrel, and by turning up the fly screw C, to its place, forms a complete lock upon the flint and steel principle, without adding to or taking from it any part whatever.

Davis' Improvements in Locks of Guns and other Fire-arms.

Fig. 4, represents the back view of the pan and nipple, shewing the hole that communicates through the nipple to the touch hole in the barrel, as explained by B, and F, (Figs. 1, & 3).

And I hereby declare that I do not claim the Invention of either of the principles of percussion or flint and steel if taken separately, but that my Invention consists in combining a pan and nipple together that will devolve upon a cylinder, which enables the same lock to be used on either principle, and forms a complete lock upon both without any addition or diminution whatever.

In witness whereof, I, the said Samson Davis, have hereunto set my hand and seal, this Eleventh day of April, in the year of our Lord One thousand eight hundred and twenty-two.

S. DAVIS. (L.S.)

AND BE IT REMEMBERED, that on the same Eleventh day of April, in the year above mentioned, the aforesaid Samson Davis came before our Lord the King, in His Chancery, and acknowledged the Specification aforesaid, and all and every thing therein contained, in form above written. And also the Specification aforesaid was stamped according to the tenor of the Statute in that case made and provided.

J. S. Harvey.

Inrolled the same Eleventh day of April, in the year above written.

LONDON:
Printed by George Edward Eyre and William Spottiswoode,
Printers to the Queen's most Excellent Majesty. 1854.

A.D. 1822. FEBY 12. Nº 4648. (1. SHEET.)

DAVIS' SPECIFICATION.

FIG. 1.

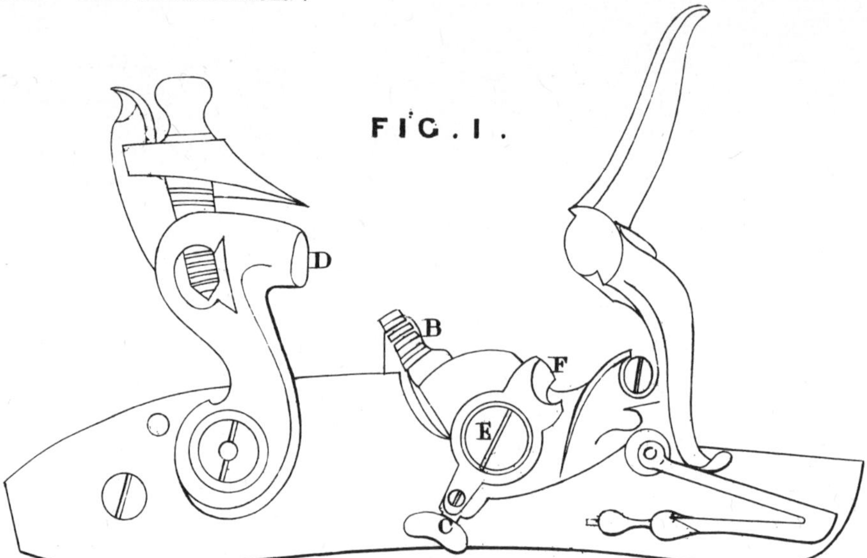

FIG. 2.

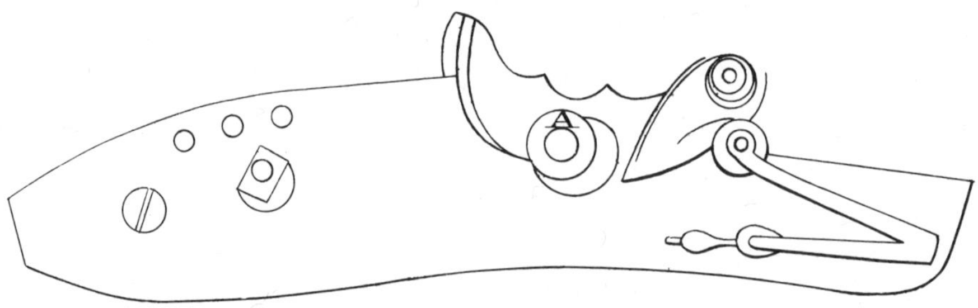

FIG. 4.

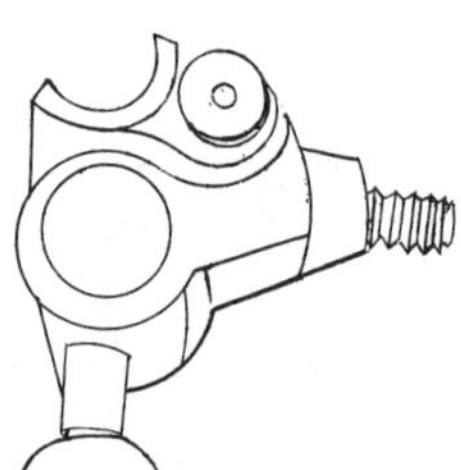

FIG. 3.

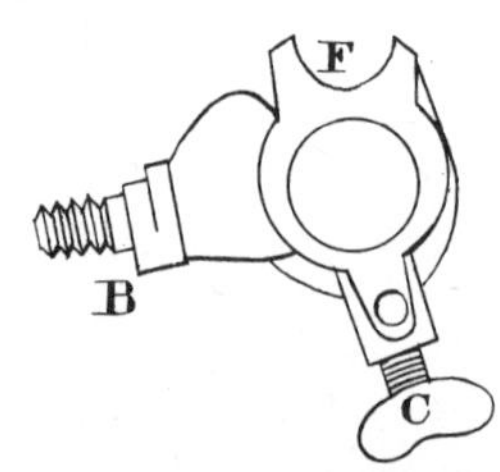

The Enrolled Drawing, is partly Colored.

Malby & Sons, lith.

LONDON: Printed by GEORGE EDWARD EYRE and WILLIAM SPOTTISWOODE,
Printers to the Queen's most Excellent Majesty. 1854.

A.D. 1822 . . . . N° 4727.

Priming Chamber for Fire-arms.

EGG'S SPECIFICATION.

TO ALL TO WHOM THESE PRESENTS SHALL COME, I, Joseph Egg, of Piccadilly, in the Parish of St. James, in the City of Westminster, Gun Maker, do send greeting.

WHEREAS His most Excellent Majesty King George the Fourth, by His Letters Patent under the Great Seal of that part of the United Kingdom of Great Britain and Ireland called England, bearing date at Westminster, the Twenty-sixth day of November, in the third year of His reign, did, for Himself, His heirs and successors, give and grant unto me, the said Joseph Egg, His especial license, sole privilege and authority, that I, the said Joseph Egg, my ex̃ors, adm̃ors, and assigns, or such others as I, the said Joseph Egg, my ex̃ors, adm̃ors, or assigns, should at any time agree with, and no others, from time to time and at all times during the term of years therein expressed, should and lawfully might make, use, exercise, and vend within England, Wales, and the Town of Berwick upon Tweed, my Invention of "**Certain Improvements in the Constructions of Guns and Fire-arms upon the Self-priming and Detonating Principle**;" in which said Letters Patent there is contained a proviso, obliging me, the said Joseph Egg, by an instrument in writing under my hand and seal, particularly to describe and ascertain the nature of my said Invention, and in what manner the same is to be performed, and to cause the same to be enrolled in His Majesty's High Court of Chancery within two calender months next and immediately after the date of the said recited Letters Patent, as in and by the same, reference being thereunto had, may more fully and at large appear.

NOW KNOW YE, that in compliance with the said proviso, I, the said Joseph Egg, do hereby declare that the nature of my said Invention, and the manner in which the same is to be performed, is described and ascertained in and by the Drawings hereunto annexed, and the following description thereof, that is to say:

My Improvements on guns and other fire-arms consist in the constructing of a priming magazine or chamber to be attached on the outside of the barrel, and in the mode of conducting the powder from this chamber to the pan or touch hole, for the purpose of priming the piece, and also in the mode of cutting off the communication between the touch hole and the priming chamber at the time that the piece is discharged. Fig. 1, is a view of the lock and breech part of a gun upon this improved principle, the parts standing as at half cock. Fig. 2, a view of the same after the piece is discharged, or the cock at bearer. Fig. 3, the inside view of the lock detached, the parts in the position of Fig. 1. Fig 4, is also a view of the inside of the lock detached, the parts in the position of Fig 2. Fig. 5, an horizontal view of a double barrel, shewing the priming chamber *a*, formed by a recess between the barrels; the cover of the chamber is in this Figure removed, for the purpose of shewing the groove or recess within. Fig. 6, is a side view of the barrel, the cover being raised. The priming powder (which for this construction of lock must be a detonating composition) is to be deposited in the recess *a*, and there secured by the cover, which fastens by a spring or other suitable contrivance; on raising the muzzle of the piece the powder descends to the broad part of the chamber *b*, and there passes through a small passage or passages *c*, *c*, in a lateral direction, the outer part of this passage is closed by a sliding plate *d*, shewn at Figs 3, 4, and 7, but when the cock is raised as Fig. 1, an aperture *z*, in this plate comes opposite the passage *c*, and permits the priming powder to descend through it to the pan or touch hole *e*; that side of the plate *d*, which slides against the barrel is perfectly flat, as shewn at Figs. 3, 4, and 7, but the outer side of the plate has a small protuberance, as shewn at Figures 1, and 8, in order to form a cap or cover to the pan or touch hole. The channel *z*, above mentioned passes through this protuberance diiect to the touch hole; and hence, when the piece is at half cock, as Fig. 1, the priming deposits itself in the touch hole, but on pulling the trigger the tumbler brings down the lever *f*, which by its connection with the arm of the sliding plate *d*, (seen at Figures 3, 4, and 7,) causes the plate instantly to rise up from its position at Fig. 3, to its position at Fig. 4, and by that means cuts off the communication between the priming chamber and the touchhole. A similar priming chamber may be applied to fire arms intended to be primed with gunpowder, but in that case it will be necessary to enlarge the passage leading from the chamber to the pan or touch hole, and a differently constructed lock must be employed, with a flint cock and hammer. I do not, however, claim any novelty in the lock or any other part of the piece, except in those parts which are marked with letters upon the Drawings and referred to as above. In short my "Improvements in the Construction of Guns and Fire-arms upon the Self-priming and Detonating Principle" are confined to the construction of a priming chamber attached to the barrel or barrels of any description of fire-arms, either as shewn in the Drawing or varied in any other form, place, or position contiguous to the barrel; and in the necessary shield or sliding guard, which also may be varied in shape and position for the purpose of admitting the priming to the touch hole when the piece is loaded, and of cutting of the communication between the chamber and the touch hold when the piece is discharging.

In witness whereof, I, the said Joseph Egg, have hereunto set my hand and seal, this Twenty-fifth day of January, in the year of our Lord One thousand eight hundred and twenty-three.

JOSEPH (L.S.) EGG.

A. D. 1822. Novr. 26. No. 4727

EGG'S SPECIFICATION.

(1 SHEET)

FIG. 1.

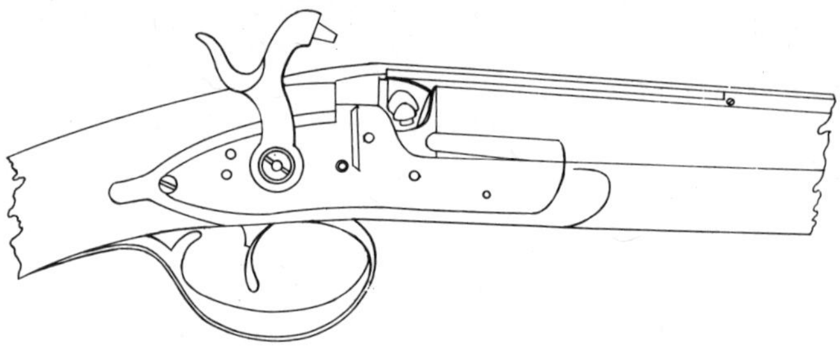

FIG. 2.

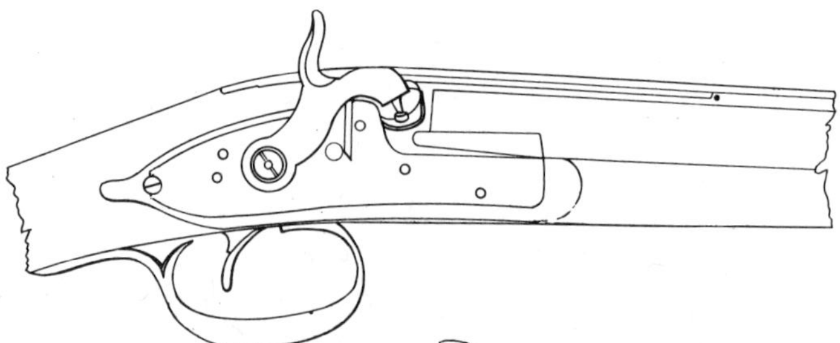

FIG. 3.

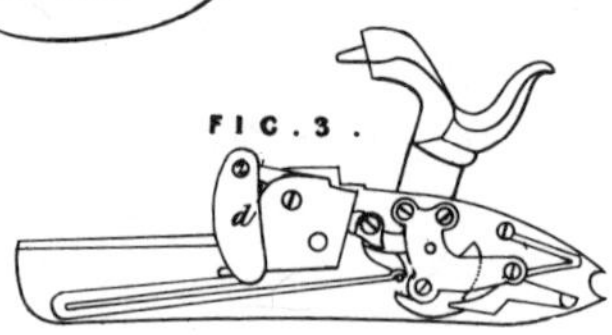

FIG. 7.

FIG. 8.

FIG. 4.

FIG. 5.

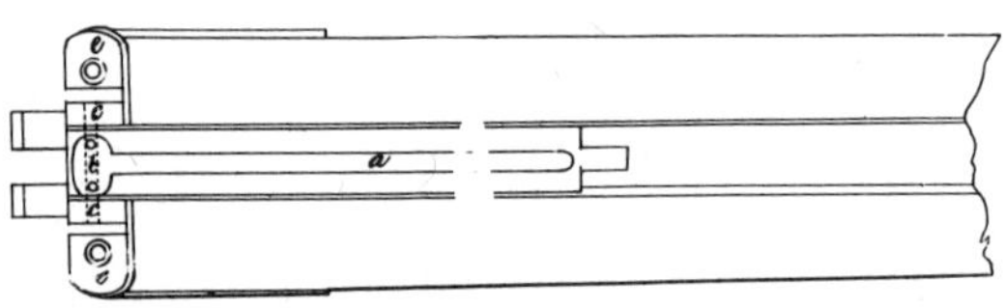

FIG. 6.

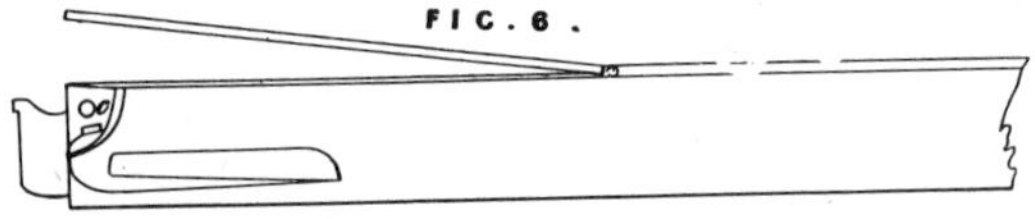

The Enrolled drawing is partly colored.

Malby & Sons lith.

LONDON: Printed by GEORGE EDWARD EYRE and WILLIAM SPOTTISWOODE.
Printers to the Queen's most Excellent Majesty. 1854.

A.D. 1823 . . . . Nº 4861.

Percussion Gun Locks.

DAY'S SPECIFICATION.

TO ALL TO WHOM THESE PRESENTS SHALL COME, I, JOHN DAY, of Barnstaple, in the County of Devon, Esquire, send greeting.

WHEREAS His most Excellent Majesty King George the Fourth did, by His Letters Patent under the Great Seal of His United Kingdom, bearing date at Westminster, the Thirteenth day of November 1823, give and grant unto me, the said John Day, my ex̃ors, adm̃ors, and assigns, His especial licence, full power, sole privilege and authority, that I, the said John Day, my ex̃ors, adm̃ors, and assigns, during the term of years therein m̃entioned, should and lawfully might make, use, exercise, and vend, within England, Wales, and the Town of Berwick-upon-Tweed, my Invention of "**CERTAIN IMPROVEMENTS ON PERCUSSION GUN LOCKS, APPLICABLE TO VARIOUS DESCRIPTION OF FIRE-ARMS**;" in which said Letters Patent there is contained a proviso, that if I, the said John Day, shall not particularly describe and ascertain the nature of my said Invention, and in what manner the same is to be performed, by an instrument in writing under my hand and seal, and cause the same to be inrolled in His Majesty's High Court of Chancery within two calendar months next and immediately after the date of the said Letters Patent, that then the said Letters Patent, and all liberties and advantages whatsoever thereby granted, shall utterly cease, determine, and become void, as in and by the same, relation being thereunto had, will more fully and at large appear.

NOW KNOW YE, that in compliance with the said proviso, I, the said John Day, do hereby declare that the nature of of my said Invention, and the manner in which the same is to be performed, are particularly described and ascertained in and by the Drawing hereunto annexed, and the following description thereof (that is to say):

My said Invention of improvements on percussion gun locks consists in a method or methods of constructing the locks so as to render them more convenient for the purpose of firing by percussion than has hitherto been done. Figure 1, is a section of part of a walking-stick percussion gun, with the lock movements shewn as when the lock is discharged; Fig. 2, an external view of the same, with the lock cocked and the trigger thrown out ready for discharging; and Figure 3, is a view of the under side of part of the gun; the same letters of reference applying to similar parts in all these Figures as well as in all the others. A, A, &c. a tube containing the lock movements, and screwing to the fore part B, of the barrel; it also is fitted up with a patent breech and nipple, as usual for percussion guns. C, the cock or lever, and also the tumbler; D, the trigger, and also the dog or sear; E, the trigger and dog spring; F, the main spring; G, the swivel or bridle; H, H, H, the pins, screws, or centres on which the cock, trigger, and swivel work; I, the nipple or spill to receive the copper percussion caps, and convey the fire into the breech; J, the air hole to allow the excess of the percussion to escape; K, a notch in the cap of the cock for the same purpose; L, L, the screws which secure the main and trigger springs to the tube A, A; M, M, holes to pass a leather thong or silk string, &c. through, as usual in walking sticks, and serving also to ventilate the fumes of the percussion, and in a considerable degree prevent them from passing to the movements of the lock, which purpose is also assisted by the division plate N; O, a female screw or cap, which fits upon the male screw P, of the cock, serving to cock it, and which when taken off prevents the gun from being discharged improperly. Figure 4, is the cock or lever and tumbler C, shewn separately. Fig. 5, the same shewn endways, with the swivel or bridle G, in its place. Fig. 6, shews the swivel or bridle G, separately; and Fig. 7, the trigger and sear. Fig. 8, the end of the cock C, shewing in section the cavity in it which strikes upon the copper caps. Q, in Figs. 1, and 2, a buck-horn handle used in the walking-stick gun, but which may have a wooden butt substituted in its place, as a common gun. Fig. 9, shews another method of carrying my said Invention into effect, and in which the situation of the main spring F is reversed, and the spring is thereby pulled in the act of cocking the lock, in the usual manner, instead of being thrust against, as on the former construction. Fig. 10, shews another method by which the lock movements first described may be applied to a common fowling-piece or other gun, they being affixed in a box or casing R, which is to be let into the under part of the stock, and attached thereto by means of the screw S, which also passes through the break off or the part under which the breech pin of the barrel hooks, as usual. The cock C, is here also made to project outwardly, and has a hook T, for the purpose of cocking it. A guard, shewn by dotted lines at U, may be used or not, as required.

Having thus described the constructions of my said improved percussion gun locks, I shall proceed to mention the advantages which will arise from their adoption; and, first, that the locks consists of fewer parts than the locks in general use, and are consequently less expensive; they are less liable to get out of order, and are more readily cleaned from the effects of the percussion powder, and in this patent improved common fowling piece they are more completely water-proof than when situated as gun locks are usually applied. I do not mean or intend hereby to claim as my improvements any of the parts shewn which are common to gun locks as already in use; but I do hereby claim the use of the main spring acting by its thrusting instead of its pulling power, the prevention to the cocking and discharging of the gun when the female screw O, is taken off, and the enclosure of the lock within the tube A, A, or box R. I may also add that, instead of the bridle G, a roller may be inserted in the end of the main spring F, to act upon the short end of the lever C, or the spring itself may be used without a roller.

In witness whereof, I, the said John Day, have hereunto set my hand and seal, this Eighth day of January, in the year of our Lord One thousand eight hundred and twenty-four.

JOHN (L.S.) DAY.

AND BE IT REMEMBERED, that on the Eighth day of January, in the year of our Lord 1824, the aforesaid John Day came before our said Lord the King, in His Chancery, and acknowledged the Specification aforesaid, and all and every thing therein contained and specified, in form above written. And also the Specification aforesaid was stamped according to the tenor of the Statute made for that purpose.

Inrolled the Tenth day of January, in the year of our Lord One thousand eight hundred and twenty-four.

LONDON:
Printed by GEORGE EDWARD EYRE and WILLIAM SPOTTISWOODE.
Printers to the Queen's most Excellent Majesty. 1854.

A.D. 1823. Nov. 13. Nº 4861.
DAY'S SPECIFICATION.

(1 SHEET)

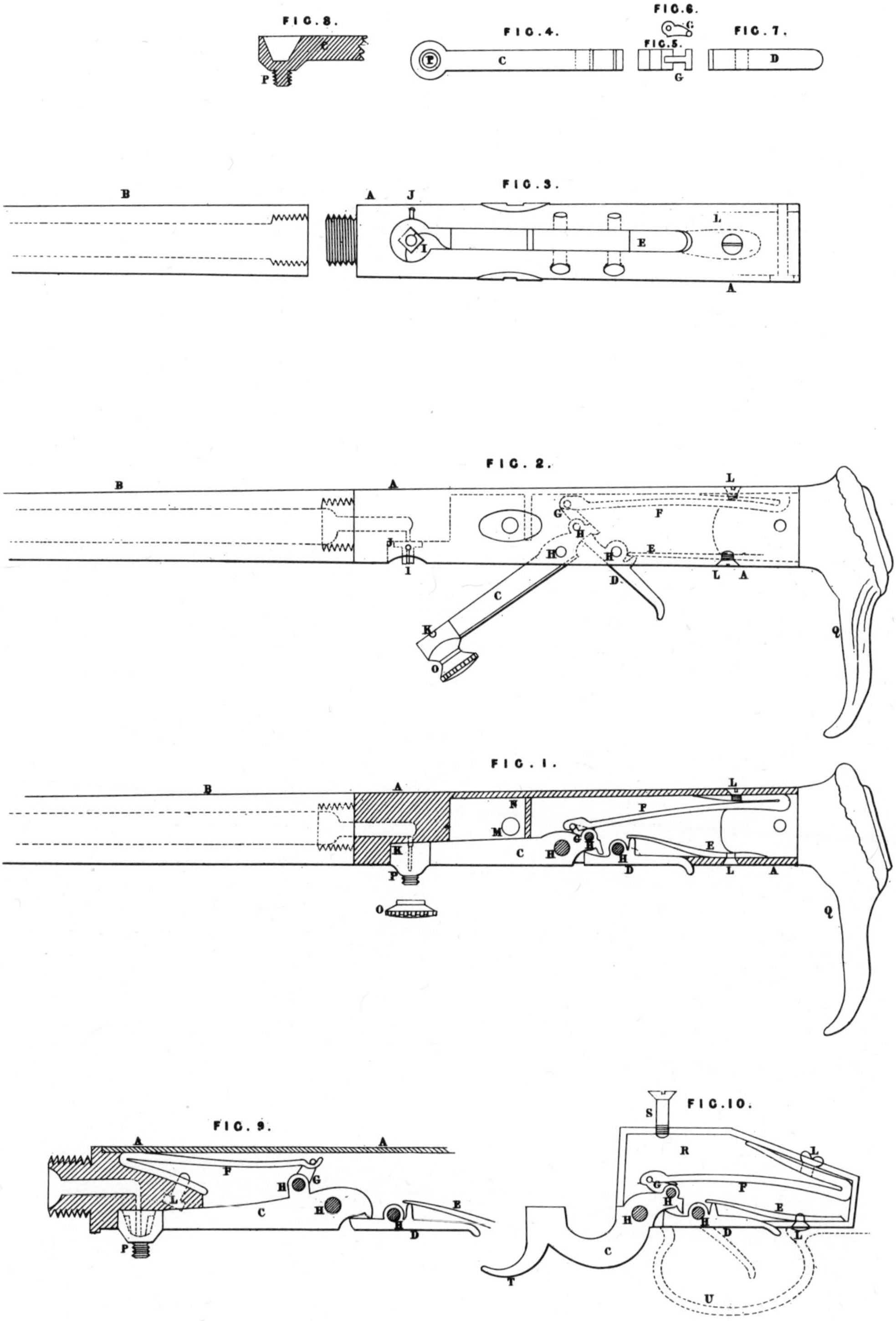

The Enrolled Drawing is Colored.

Malby & Sons lith.

LONDON: Printed by GEORGE EDWARD EYRE and WILLIAM SPOTTISWOODE.
Printers to the QUEEN'S most Excellent Majesty. 1854.

A.D. 1824 . . . . N° 5055.

Lock and Breech of Guns and other Fire-arms.

DAVIS' SPECIFICATION.

TO ALL TO WHOM THESE PRESENTS SHALL COME, I SAMSON DAVIS, of Upper East Smithfield, in the County of Middlesex, Gun Lock Maker, send greeting.

WHEREAS His most Excellent Majesty King George the Fourth, by His Letters Patent under the Great Seal of the United Kingdom of Great Britain and Ireland, bearing date at Westminster, the Eighteenth day of December now last past, did, for Himself, His heirs and successors, give and grant unto me, the said Samson Davis, His especial license, that I, the said Samson Davis, my executors, administrators, and assigns, or such others as I, the said Samson Davis, my executors, administrators, or assigns, should at any time agree with, and no others, from time to time and at all times thereafter, during the term of fourteen years therein expressed, should and lawfully might make, use, exercise, and vend, within England, Wales, and the Town of Berwick upon Tweed, my Invention of an "IMPROVEMENT OR IMPROVEMENTS APPLICABLE TO GUNS AND OTHER FIRE-ARMS;" in which said Letters Patent is contained a proviso, obliging me, the said Samson Davis, by an instrument in writing under my hand and seal, particularly to describe and ascertain the nature of my said Invention, and in what manner the same is to be performed, and to cause the same to be enrolled in His said Majesty's High Court of Chancery within six calendar months next and immediately after the date of the said recited Letters Patent, as in and by the same, reference being thereunto had, will more fully and at large appear.

NOW KNOW YE, that in compliance with the said proviso, I, the said Samson Davis, do hereby declare that the nature of my said Invention, and the manner in which the same is to be performed, is described and ascertained by the Plan or Drawing hereunto annexed and as follows, that is to say:

My said improvement or improvements consist in forming and constructing the parts of a lock and the extra breach of guns and other fire-arms, to prevent accidents arising from the explosion of the percussion cap or other percussion priming, or blowing out the nipple or peg, and to secure the percussion powder from the mischievous effects arising from the introduction of wet. Figure 1, in the annexed Drawing represents the lock and piece in its priming position, with a strong joint at its upper extremity A, by which the lock and but opens from the barrel C, and fore end D, to receive the percussion priming. It is fastened by a bolt E, Figures 1 and 2, which renders it perfectly safe and leaves a cavity to receive the exploded cap or other priming, and effectually prevents any wet finding its way to the prime. The breach not being reduced in one part more than another, makes it particularly strong. F, Figures 1, and 3, the nipple in the centre of the plug to receive the percussion cap; G, Figure 1, the cavity which receives the exploded cap. Figure 2, is the extra breach, shewing two screws which fasten it in its proper place, with a strong wire bolt which passes through the extra breach and shank of the plug; Figure 3, the lock, shewing the works when at half cock; Figure 4, the bridle, which not only secures the tumbler but keeps the sliding striker in its proper place; Figure 5, the seer; Figure 6, the sliding striker; Figure 7, the tumbler, shewing the hole A, in which the sliding striker works. H, Figures 1, and 3, the cock as fastened to the tumbler, which draws back the sliding striker to the extremity of the action, when by drawing the trigger I, Figure 1, the main spring K, Figure 3, moves the tumbler forward, which carries the sliding striker with great velocity against the percussion cap or other priming fastened on the nipple, and discharges the piece.

In witness whereof, I, the said Samson Davis, have hereunto set my hand and seal, this Eighth day of June, in the year of our Lord One thousand eight hundred and twenty-five.

SAMSON DAVIS. (L.S.)

AND BE IT REMEMBERED, that on the same Eighth day of June, in the year above mentioned, the aforesaid Samson Davis came before our Lord the King, in His Chancery, and acknowledged the Specification aforesaid, and all and every thing therein contained, in form above written. And also the Specification aforesaid was stamped according to the tenor of the Statute in that case made and provided.

JA[s] TROWER.

Inrolled the same Eighth day of June, in the year above written.

LONDON:
Printed by GEORGE EDWARD EYRE and WILLIAM SPOTTISWOODE,
Printers to the Queen's most Excellent Majesty. 1854

A.D. 1824, Dec 18. Nº 5055 (1 SHEET)

DAVIS' SPECIFICATION.

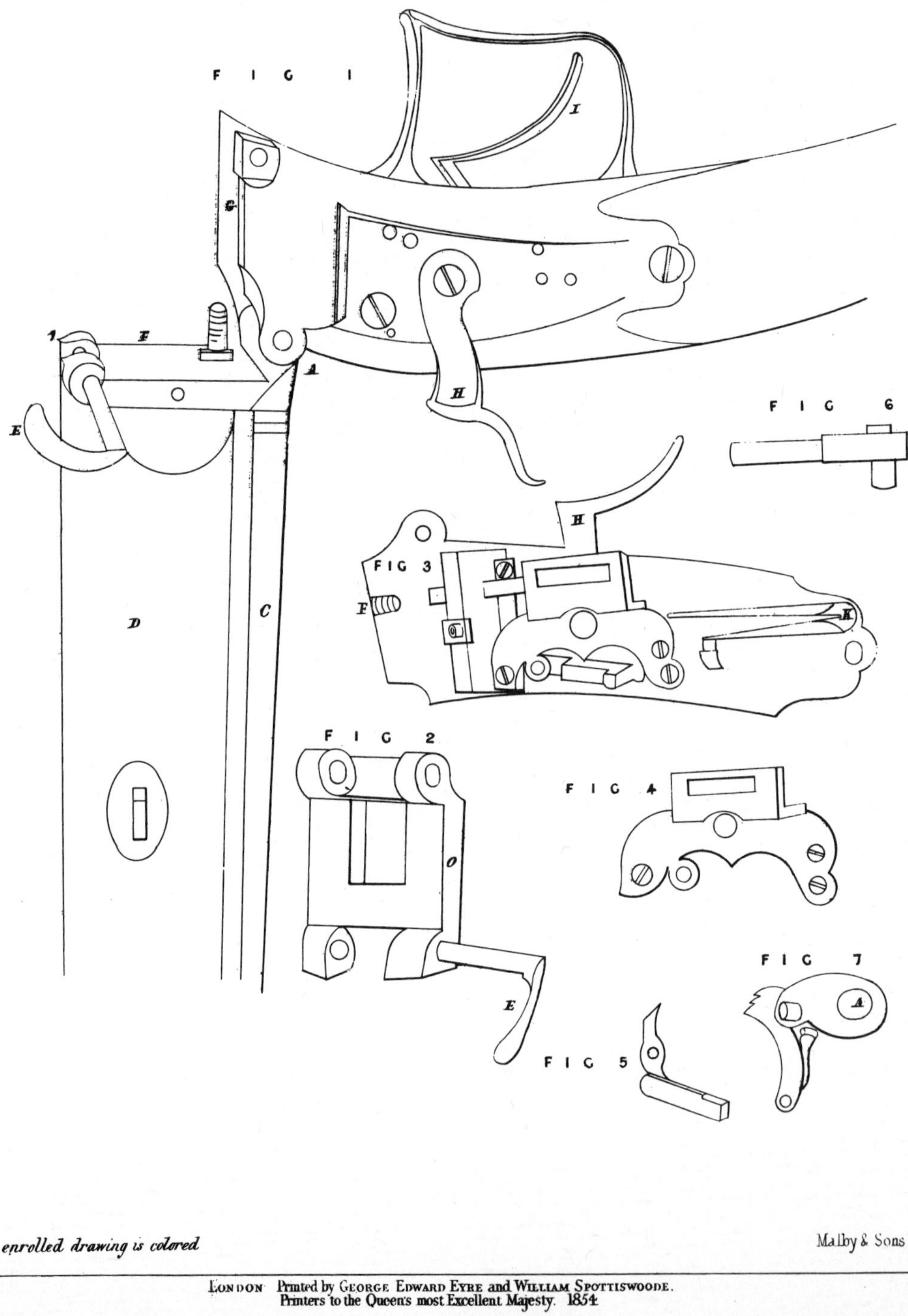

The enrolled drawing is colored

Malby & Sons lith.

LONDON Printed by GEORGE EDWARD EYRE and WILLIAM SPOTTISWOODE.
Printers to the Queen's most Excellent Majesty. 1854.

A.D. 1825 . . . . Nº 5099.

Sliding Lock for Fire-arms.

MOULD'S SPECIFICATION.

TO ALL TO WHOM THESE PRESENTS SHALL COME, I, JACOB MOULD, of Lincoln's Inn Fields, in the County of Middlesex, Gentleman, send greeting.

WHEREAS His present most Excellent Majesty King George the Fourth, by His Letters Patent under the Great Seal of Great Britain, bearing date at Westminster, the Nineteenth day of February, in the sixth year of His reign, did, for Himself, His heirs and successors, give and grant unto me, the said Jacob Mould, His especial licence that I, the said Jacob Mould, my ex̃ors, adm̃ors, and assigns, or such others as I, the said Jacob Mould, my ex̃ors, adm̃ors, or assigns, should at any time agree with, and no others, from time to time and at all times during the term of years therein expressed, should and lawfully might make, use, exercise, and vend, within England, Wales, and the Town of Berwick-upon-Tweed, the Invention of "CERTAIN IMPROVEMENTS IN FIRE-ARMS;" in which said Letters Patent there is contained a proviso, obliging me, the said Jacob Mould, by an instrument in writing under my hand and seal, particularly to describe and ascertain the nature of the said Invention, and in what manner the same is to be performed, and to cause the same to be enrolled in

His said Majesty's High Court of Chancery within six calendar months next and immediately after the date of the said recited Letters Patent, as in and by the same, reference being thereunto had, will more fully and at large appear.

NOW KNOW YE, that in compliance with the said proviso, I, the said Jacob Mould, do hereby declare the nature of the said Invention to consist in the formation of a series of touch-holes along the barrel or breech of any fire-arm, and in the application thereto of a sliding or other adjusting lock, to act by percussion or otherwise, whereby any number of charges not exceeding the number of touch-holes may be inserted in the barrel at one time, and fired successively. And in further compliance with the said proviso, I, the said Jacob Mould, do hereby describe the manner in which I perform the said Invention, by the following description thereof, reference being had to the Drawing annexed, and the Figures marked thereon (that is to say):

Description of the Drawing.

The same letters of reference are placed on corresponding parts in the several Figures. Figure 1, is a complete view of the said Invention, shewing the application of the principle to a common gun. The other Figures are different parts of the said gun in detail. Figures 2, and 3, are two views of the barrel of the gun, with a plate A, A, of iron, steel, or other metal attached to it, and which may be done by screws, brazing, or otherwise. Through this plate is made as many touch-holes into the barrel of the gun as the number of charges it is to contain, appointing a touch-hole for every separate charge; and these touch-holes are made at such distances from each other as the size of the charge requires; and all the touch-holes, except the one nearest to the mouth of the barrel, are furnished with valves *f, f, f, f,* to close the touch-holes when required. These valves serve also as stops to the lock, by which it may be placed in the proper situation for communicating with the several touch-holes. Figure 4, exhibits the stock of the gun with a plate B, of iron, steel, or other metal, screwed or otherwise secured to it. These plates A, A, Figures 2 and 3, and B, in this Figure, are made in such forms that when the barrel is fitted to the stock, an aperture or opening is left between them for the bar of the sliding case which contains the lock to move in, as seen at G, Figure 1. Figure 5, exhibits the stock without the plate B, by which the trigger bar C is seen, furnished with angular projections D, D, on its under side and at each end. These angular projections are so fitted to the stock, that by the usual motion of the trigger E, when pulled to discharge the gun, the trigger bar rises on a level, as is shewn by the dotted lines, against the sear of the lock, and will spring it at its several positions. Figure 6, is another view of the trigger bar C, shewing the opening *g*, through which the trigger pin passes to act upon it. Figure 7, is the trigger pin E, and the fulcrum of its leverage is at the pin hole *h*, near the end. Figure 8, is a back view of the lock case F, having attached to it a bar I, which is made with projections on it, in order that it may be confined in its situation behind the plates A and B, Figure 1, and slide freely along the aperture G, Figure 1. Figure 9, is a front view of the lock case F; *a*, is the stud in it by which the lock is secured to the case; *b*, is the axis on which the priming box K turns, having on it a groove *i*, which is the medium of communication between the priming box and the pan of the lock. *c*, *c*, are the screws which secure the bar I, to the lock case. Figure 10, is the plan of the lock case. Figure 11, is a section of the priming box K. Figure 12, is a full view of the priming box. Figure 13, is a side view of the priming box. Figure 14, is a view of the manner in which the priming box is connected with the cock or hammer of the lock, viz[t], by a bar *l*, which is screwed to the priming box seen at *k*, Figure 1, and the other end of the bar being attached to a hook on the cock or hammer of the lock. After the gun is loaded, the operation of discharging it is thus effected: The lock is to be placed at the touch-hole which communicates with the charge that is nearest the mouth of the barrel; the lock is then to be cocked in the usual manner; and the bar *l*, Figure 1, and Figure 14, which connects the cock or hammer of the lock with the priming box, turns the priming box so far round the axis *b*, Figure 9, as to bring the orifice *d*, Figure 11, of the priming box exactly over the groove *i*, Figures 9, and 10, which is cut on the axis, and thus opens a communication between the priming box and the pan of the lock, and the powder will fall from the priming box along the groove in the axis, into the pan, and thus the gun is primed. As soon as the lock is sprung by the action of the trigger bar, the communication between the pan and the priming box is cut off by the action of the hammer or cock causing the priming pan to revolve on its axis, in consequence of their being connected together. Figure 15, represents a part of a gun on the percussion principle, with cylinders; and the distance or proper situation for the lock to be placed, so that the hammer may act correctly, is by a spring in lieu of the valves mentioned in the description of Figures 2 and 3. Figure 16, is a side view of the spring *m*, detached from the lock, the situation of which is seen at the dotted lines *o*, *o*, Figure 15. Figure 17, is the end view of the spring *m*, shewing the end which is bent to fit into the grooves *n*, *n*, *n*, *n*, is seen in Figure 15.

Although I have described two modes, and certain materials applied to one description of fire-arms, I do not by this confine myself to the modes, or materials, or the fire-arms I have described; but I claim the application of the principle herein-before stated, in whatever form or of whatever materials it may be made, and to whatever kind of fire-arms it can be applicable. In loading the arms, which is done in the usual manner, particular care should be taken not to overcharge them. A little more than one half the quantity of powder used in other arms only is necessary in loading these, in consequence of the compact manner of putting in the charges with tight wads of leather or felt, cut with a machine to fit the bore of the piece. Large grained powder must be used, for this obvious reason: when the top charge is fired the pressure downward is such as to compress fine grained powder in such a manner as to render it slow to ignite. Any number of charges may be loaded and fired from the upper touch-hole, keeping in reserve whatever number of charges the piece may contain below, without affecting them in the least; or, after having disposed of all the reserved charges, it can then be loaded and fired from the lower touch-hole; or, as the lock progresses down the barrel, it may be loaded and fired from either of the intermediate touch-holes, with the advantage always of priming itself. The number of charges to which this principle may be extended seems to be unlimited. Although it has not been extended to more than twelve, there appears no reason why it might not be extended to twenty, thirty, or more, and the charges fired with accuracy, and without the least difficulty. Now whereas fire-arms with a series of touch-holes along the barrel or breech as herein-before described, and with a sliding or other adjusting lock, to act by percussion or otherwise adapted thereto, are, to the best of my knowledge and belief, entirely new, and never before used within that part of His said Majesty's United Kingdom of Great Britain and Ireland called England, His Dominion of Wales, or Town of Berwick-upon-Tweed, I do hereby therefore declare this to be my Specification of the said improvements, and that I do verily believe that this my said Specification doth comply in all respects fully, and without reserve or disguise, with the proviso in the said herein-before in part recited Letters Patent contained: Wherefore I do hereby claim to maintain exclusive right and privilege to the said Invention.

In witness whereof, I, the said Jacob Mould, have hereunto set my hand and seal, the Eighteenth day of August, in the year of our Lord One thousand eight hundred and twenty-five.

JACOB MOULD. (L.S.)

AND BE IT REMEMBERED, that on the Eighteenth day of August, in the year of our Lord 1825, the aforesaid Jacob Mould came before our said Lord the King, in His Chancery, and acknowledged the Specification aforesaid, and all and every thing therein contained and specified, in form above written. And also the Specification aforesaid was stamped according to the tenor of the Statute made for that purpose.

HARVEY.

Inrolled the Nineteenth day of August, in the year of our Lord One thousand eight hundred and twenty-five.

LONDON:
Printed by GEORGE EDWARD EYRE and WILLIAM SPOTTISWOODE, Printers to the Queen's most Excellent Majesty. 1854.

A.D. 1825. February 19. No. 5099.

MOULD'S Specification.

(1 SHEET.)

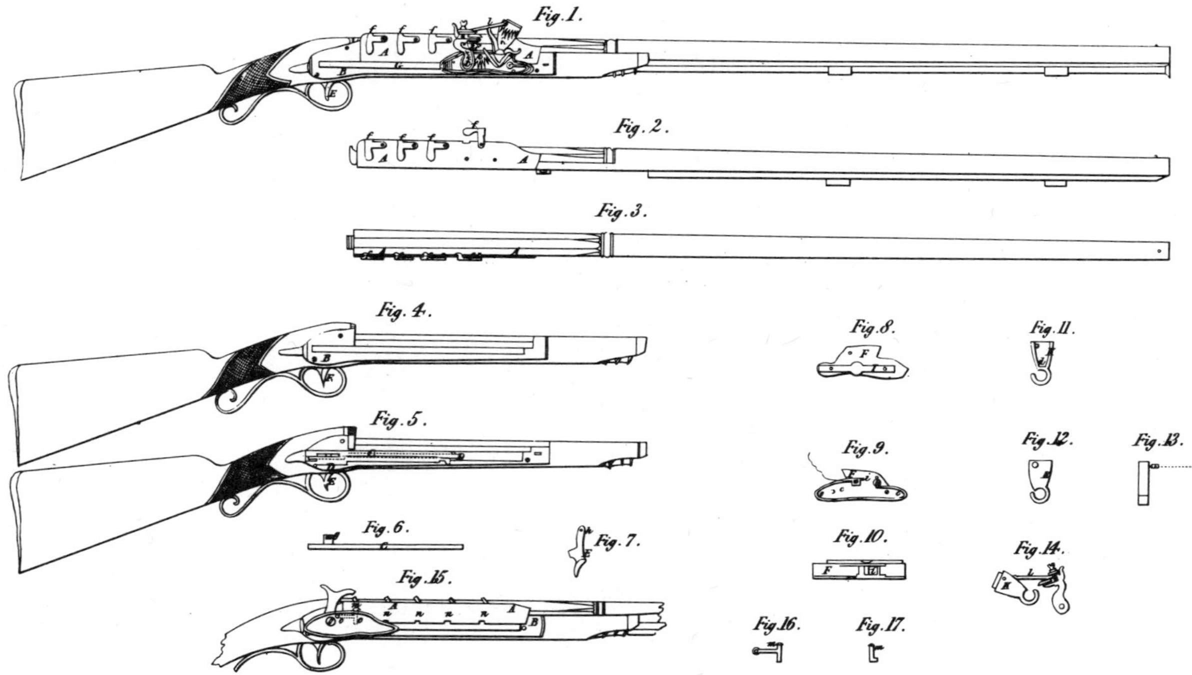

The Enrolled Drawing is Colored.

Malby & Sons Lith.

London: Printed by George Edward Eyre and William Spottiswoode. Printers to the Queen's most Excellent Majesty. 1854.

A.D. 1825 . . . . N° 5106.

Revolving Pellet Carrier, &c. for Percussion Locks of Fire-arms.

MANTON'S SPECIFICATION.

TO ALL TO WHOM THESE PRESENTS SHALL COME, I, Joseph Manton, of Hanover Square, in the Parish of Saint George, Hanover Square, in the County of Middlesex, send greeting.

WHEREAS His present most Excellent Majesty King George the Fourth, by His Letters Patent under the Great Seal of Great Britain, bearing date at Westminster, the Twenty-sixth day of February, One thousand eight hundred and twenty-five, in the sixth year of His reign, did, for Himself, His heirs and successors, give and grant unto me, the said Joseph Manton, His especial licence that I, the said Joseph Manton, my ex̃ors, adm̃iors, and assigns, or such others as I, the said Joseph Manton, my ex̃ors, adm̃iors, and assigns, should at any time agree with, and no others, from time to time and at all times during the term of years therein expressed, should and lawfully might make, use, exercise, and vend, within England, Wales, and the Town of Berwick upon Tweed, my Invention of "Certain Improvements in Fire-arms;" in which said Letters Patent there is contained a proviso, obliging me, the said Joseph Manton, by an instrument in writing under my hand and seal, particularly to describe and ascertain the nature of my said Invention, and in what manner the same is to be performed, and to cause the same to be enrolled in His Majesty's High Court of Chancery within six calendar months next and immediately after the date of the said in part recited Letters Patent, as in and by the same, reference being thereunto had, will more fully and at large appear.

NOW KNOW YE, that in compliance with the said proviso, I, the said Joseph Manton, do hereby declare that the nature of my said Invention, and the manner in which the same is to be performed, are particularly described and ascertained in and by the Drawing hereunto annexed, and the following description thereof (that is to say):

My said improvements in fire-arms chiefly consist in the application of a part which I term a revolving globular pellet carrier to the percussion locks of fire-arms of all descriptions. In order, however, to render this application more fully understood, I shall proceed to refer to the various Figures contained in the Drawing, which, as aforesaid, is annexed to this Specification, and in all of which Figures the same letters of reference indicate the similar parts. Fig. 1, is a plan, and Fig. 2, a side view of part of a double barrelled gun or fowling piece. A, A, the stock; B, B, the lock plates; C, C, the plugs, screwed into the patent breeches D, D; E, E, the pellet carriers, turning upon the plugs; F, F, the magazines for the pellets; G, G, the conducting cylinders of the spring discharging or inflaming punches; H, H, the lock hammers or cocks of the gun. The plugs C, C, have collars at their shoulders near to where they are screwed into the patent breeches upon which the inner ends of the pellet carriers fit their outer ends, turning upon necks provided at the outer end of each plug to receive them, a space being thus left around and at the outer end of each plug for the escape of the flash from the percussion prime, and which finally passes out into the open air through the perforations or openings made in the outer ends of the pellet carriers for that purpose. In Fig. 3, the dark shadow around the plug C, shews this space between it and the pellet carrier E. In Fig. 4, I, I, I, shew three of the openings or perforations in the end of a pellet carrier, and J, the remainder of the circular hole, which turns upon the neck K, of the plug C; L, is the handle or lever of the pellet carrier E, an external view of which is given in Fig. 5. This handle is lodged or caught by the head of the spring catch M, the spring of which is affixed to the inside of the lock plate, and the head of it M, passes through a hole made in that plate. The magazines F, F, for the pellets have dovetailed appendages N, N, one of which is shewn endways in Fig. 6, which can be slided into dovetailed recesses made in the tops of the patent breeches to receive them; they meet in the middle under the elevated rib O, and are secured in their places by the screws P, P. These magazines have each an angular or bent cylindrical cavity in them for the reception of the globular pellets, as shewn in Figs. 7, and 8, and they are closed at their outer ends either by corks or wooden plugs loosely fitted into them by caps made to slide over them, or by plates turning upon pins as centres, or in any other proper manner. The conducting cylinders G, G, of the spring discharging or inflaming punches Q, Q, may be either formed solid with the false breech R, or may be affixed to it, at pleasure. The punches Q, Q, are posited at an angle which fits them, to receive the blows of the cocks or hammers H, H, of the locks to the greatest advantage. They are each provided with a collar S, against which the helical springs lodged within the cylinders G, G, act so as to urge them outwardly, and they are guided at their opposite or inner ends by holes made in the ends of the cylinders G, G, to fit them, the outer ends of the punches passing through other holes made in the screwed nuts T, T, which are fitted into the cylinders G, G. The other, or inflaming ends of the punches, are reduced in diameter, so as to suit them to the size of the globular pellets; a hole is also made in the side of each ring of the revolving pellet carriers of a proper size to receive a pellet within it when brought into the position shewn in the section Fig. 7, and upon bringing it by means of the handle L, into the other position shewn in the section at Fig. 8, the hole with the pellet lodged within it is carried round until it is exactly opposite to the end of the inflaming punch Q, ready to receive the blow from the cock or hammer H, upon the discharge of the lock, the shoulder near the end of the punch being larger than the hole in the revolving pellet carrier E, partly closing it and preventing the flash from entering the conducting cylinders G, G, so as to corrode them or the parts included within them; a groove U, is made partly around the outside of each revolving pellet carrier into which a stud affixed at the end of a spring V, enters and secures it in its place upon the plug C. The two springs are secured to the pellet magazines F, F, by means of screws; one of these springs is shewn in Figure 9, which is an under view of one of the magazines F. W, Figs. 6, and 9, is a raised circular portion of one of the magazines around the hole through which the pellets pass into the carrier. This raised part nicely fits the external cylindrical surface of the carrier so as to accurately close the entrance to the magazine and prevent the explosion of their pellets within it from the flash of the percussion reaching them. Another method of preventing the revolving pellet carriages E, E, from coming off the plugs C, C, is shewn in Fig. 10; X, is a circular recess made in the inner and under side of one of the magazines F. Into this recess a projection Y, from the bottom of the carrier E, shewn in Fig. 4, enters. In this case the carrier E, must be first placed upon the plug C, and then the magazine F, be slided into its place to prevent it from coming off. The magazine being also secured in its place by the screw P, a still different method of effecting the same purpose is shewn in Fig. 11, in which a projecting ledge or border Z, is formed on the outside of the magazine F, which overhangs or projects before a part of the flat end of the revolving carrier E, and thus prevents it from coming off the plug C. The action or motion of the revolving carrier E, is limited by the projection *a*, Fig. 4, coming into contact with the angle *b*, of the magazine F, shewn in Figs. 2, 6, 9, 10, and 11. Fig. 12, is an inside view and Fig. 13, an edge view of one of the cocks or hammers H; *c*, is a recess or cavity made in the cock or hammer to allow a greater length of stroke to be made by it, and yet to confine it in a small compass, the blow upon the head of the percussion or inflaming punch Q, being made by the part *a*, at the back of the recess *c*. Q, Fig. 14 is a section of another percussion punch, having a spare or separate inflaming piece *e*, screwed into its end, and which again is also plugged with a platina screw *f*. In case of applying my said Invention to a single barrelled gun, pistol, &c. it will be necessary to round off the end of the dovetailed appendage F, to the form of the breech; in other respects no difference need be made.

Having thus described the several parts, I shall next proceed to show their action. In loading the fire-arms the percussion globular pellets with which the magazines must be filled will of their own accord descend to the lower part of the bent cylindrical cavities in the magazines, as shewn in Figures 7, and 8. In Figure 7, the pellet carrier is shewn as brought by means of its lever L, (shewn by dotted lines) into a proper position to receive one of the pellets into the hole made in the revolving carrier for that purpose. The carrier is then to be turned into the position shewn in Figure 8, by means of its lever L, (shewn by the dotted lines), and is lodged upon the spring catch M, the pellet being brought thereby opposite to the inflaming punch Q, and the touch hole in the plug C, ready to be inflamed by the blow of the cock or hammer H, on the discharge of the piece. In order to reprime the carrier the spring catch M, is to pressed inwards by the forefinger of either hand, (according to the lock to be primed), and the lever L, will then be disengaged and may be turned by applying the thumb and forefinger to that lever into the position shewn in Figure 7, after which it is instantly to be brought back to its first position and retained therein by the spring catch M, as above-mentioned. I do not mean or intend hereby to claim as my Invention any of the several parts of which my said apparatus is composed that may already be known or in use. I do, however, hereby claim the Invention of the revolving percussion pellet carrier, and the methods of fitting it and securing it upon the plug of the gun or other fire-arm so as to allow of the free escape of the excess of the flash or gases produced by the percussion, and from the exploding of the gunpowder in the charge. I also claim as my Invention the new fixed position of the conducting cylinder G, of the inflaming punch H, and the shoulder at its inflaming end; and, lastly, I also claim the new fixed position of the pellet magazine F, and the manner of preventing the explosion of the pellets in the magazines by means of the raised circular portion W, around the hole through which the pellets pass into the carrier.

In witness whereof, I, the said Joseph Manton, have hereunto set my hand and seal, this Twenty-sixth day of August, in the of our Lord One thousand eight hundred and twenty-five.

JOSEPH (L.S.) MANTON.

A.D. 1825. Feby 26th N° 5106.
MANTON'S SPECIFICATION.

(1 SHEET)

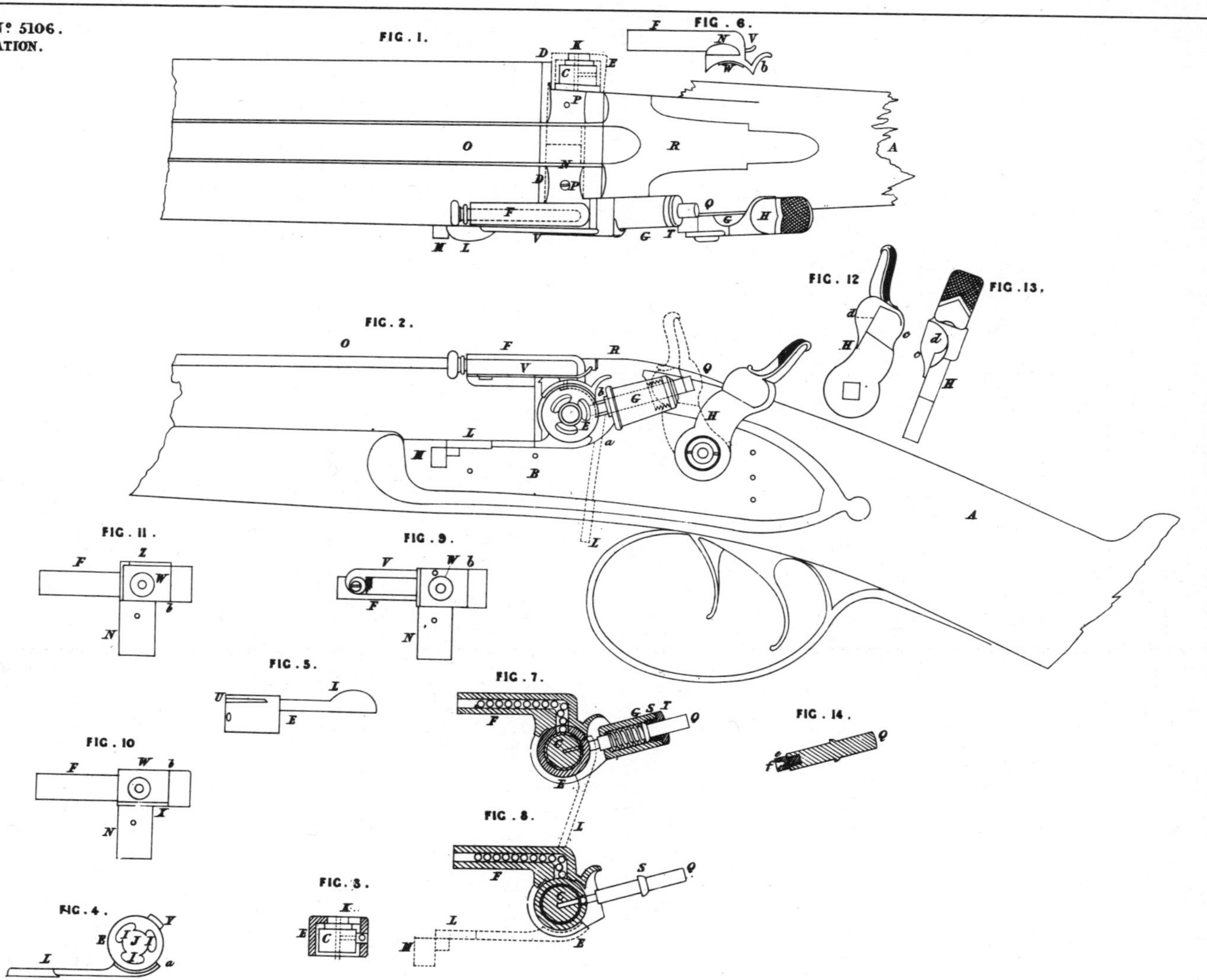

The Enrolled Drawing is colored.

Malby & Sons Litho.

London: Printed by George Edward Eyre and William Spottiswoode.
Printers to the Queen's most Excellent Majesty. 1854.

A.D. 1825 . . . . N° 5175.

Locks of Fire-arms.

RIVIERE'S SPECIFICATION.

TO ALL TO WHOM THESE PRESENTS SHALL COME, I, Isaac Riviere, of Oxford Street, in the Parish of Saint George, Hanover Square, in the County of Middlesex, Gun Maker, send greeting.

WHEREAS His most Excellent Majesty King George the Fourth, by His Letters Patent under the Great Seal of Great Britain, bearing date at Westminster, the Twentieth day of May, in the sixth year of His reign, did give and grant unto me, the said Isaac Riviere, my executors, adm̃ors, and assigns, His especial licence, full power, sole privilege & authority, that I, the said Isaac Riviere, my ex̃ors, adm̃ors, and assigns, during the term of years therein expressed, should and lawfully might make, use, exercise, and vend, within England, Wales, and the Town of Berwick-upon-Tweed, my Invention of "**An Improved Construction, Arrangement, and Simplification of Machinery by which Guns, Pistols, and other Fire-arms are Discharged**; in which said Letters Patent there is contained a proviso obliging me, the said Isaac Riviere, particularly to describe and ascertain the nature of my said Invention, and in what manner the same is to be performed, by an instrument in writing under my hand and seal, and to cause the same to be inrolled in His Majesty's High Court of Chancery within six calendar months next and immediately after the date of the same Letters Patent, as in and by the same Letters Patent, relation being thereunto had, will more fully and at large appear.

NOW KNOW YE, that in compliance with the said proviso I, the said Isaac Riviere, do hereby declare that the nature of my said Invention, and the manner in which the same is to be performed, are particularly described and ascertained in and by these presents, reference being made to the Drawing hereunto annexed, and the following description thereof (that is to say):

My Invention of an improved construction, arrangement, and simplification of the machinery by which guns, pistols, and other fire-arms are descharged consists in dispensing with certain parts of the ordinary gun lock, and substituting other parts for similar purposes, and arranging the mechanism in a different way to that of any machine or lock heretofore made for the discharge of fire-arms, which improvements are exhibited in the Drawing hereto annexed, and will be made fully evident by the following description thereof, that is to say:—Fig. 1, is a section of a gun or fowling piece cut lengthways through the stock and barrel; Fig. 2, is an external representation of the same. This gun is intended to be discharged by means of percussion, but the improved construction of the machinery will apply also to guns discharged by flints. *a*, is the barrel; *b*, the nipple, through which the fire from the percussion passes to the charge; *c*, *c*, is the false breech, through which an aperture is made for the cock *d*, to work in (see Fig. 3). The cock *d*, and the tumbler *e*, are in one piece (see Fig. 4). *f*, is the main spring attached to the under side of the false breech, and connected to the tumbler by a swivel *g*; the the force of this spring acts upwards, and causes the cock, when released from the sear, to fall upon the nipple with great force; *h*, is the sear; and *i*, is the trigger. The sear and trigger are in one piece, as shewn in different positions in the detached Figures 5, 6, and 7. *k*, is the sear and trigger spring which keeps the sear and trigger up to their proper bearing. This improved machinery, though shewn only as applied to a single barrelled gun, may be adapted also to a double barrelled gun, in which case the parts will require to be repeated.

Though I have described the entire construction of the machinery, and the form of all its pieces, yet I wish it to be understood that I do not claim all those parts as new, having used some of them before; but I claim as new the form of the piece constituting the sear and trigger. I also claim as new the attachment of the main spring to the false breech.

In witness whereof, I, the said Isaac Riviere, have hereunto set my hand and seal, this Nineteenth day of November, in the year of our Lord One thousand eight hundred and twenty-five.

ISAAC (L.S.) RIVIERE.

AND BE IT REMEMBERED, that on the Nineteenth day of November, in the year of our Lord 1825, the aforesaid Isaac Riviere came before our said Lord the King in His Chancery, and acknowledged the Specification aforesaid, and all and every thing therein contained and specified, in form above written. And also the Specification aforesaid was stamped according to the tenor of the Statute made for that purpose.

Cox.

Inrolled the Nineteenth day of November, in the year of our Lord One thousand eight hundred and twenty-five.

LONDON:
Printed by GEORGE EDWARD EYRE and WILLIAM SPOTTISWOODE,
Printers to the Queen's most Excellent Majesty. 1854.

A.D. 1825. May 20. N° 5175.

RIVIERE'S Specification.

(1 SHEET)

FIG. 1.

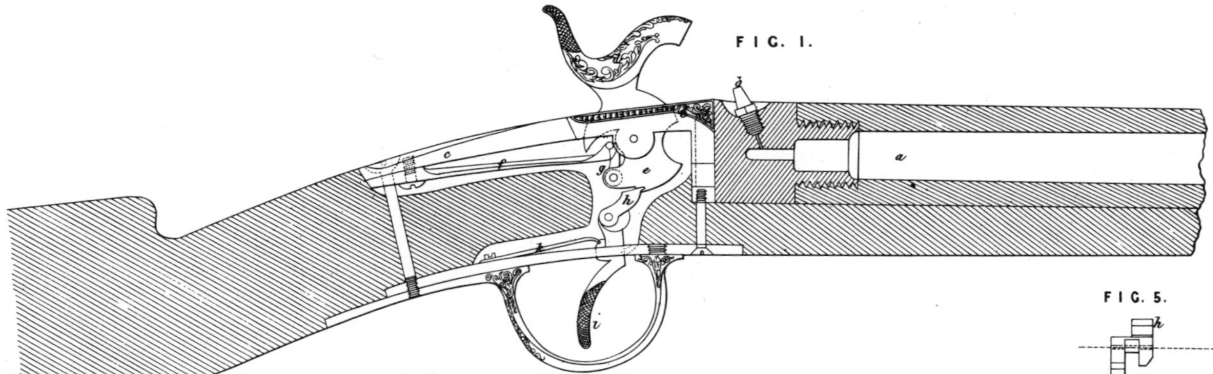

FIG. 4.

FIG. 5.

FIG. 6.

FIG. 7.

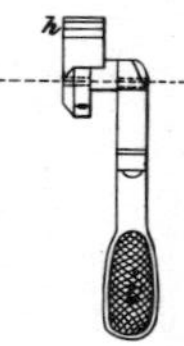

FIG. 2.

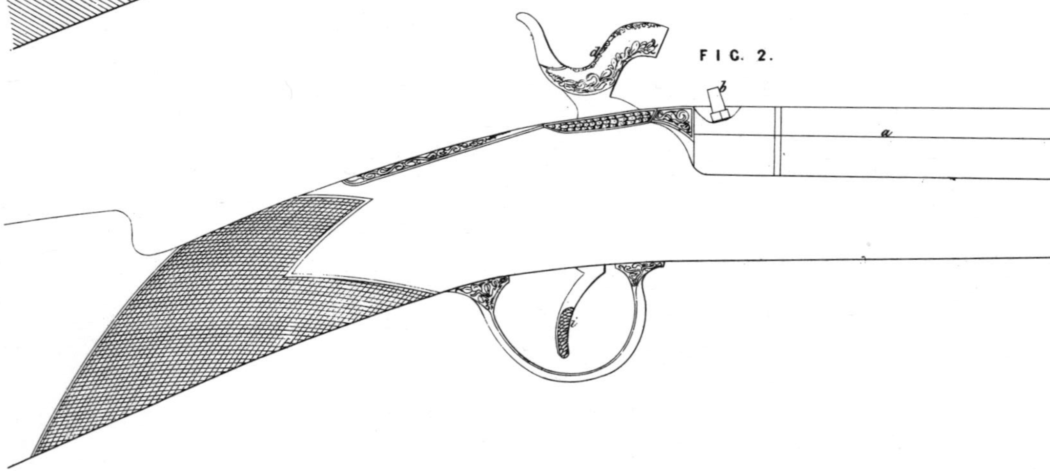

FIG. 3.

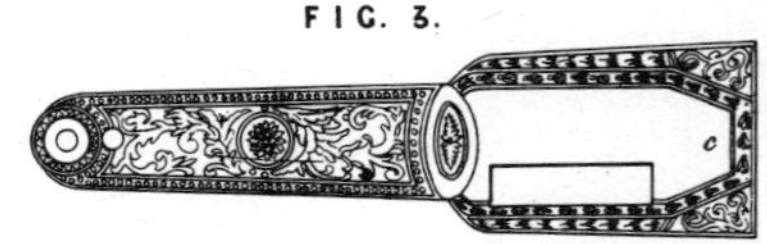

The Enrolled Drawing is colored.

Malby & Sons Lith.

A.D. 1829 . . . . Nº 5838.

Locks of Fire-arms.

MANTON'S SPECIFICATION.

TO ALL TO WHOM THESE PRESENTS SHALL COME, I, George Henry Manton, of Dover Street, Piccadilly, in the County of Middlesex, Gun Maker, send greeting.

WHEREAS His most Excellent Majesty King George the Fourth, by His Letters Patent under the Great Seal of Great Britain, bearing date at Westminster, the Second day of September, in the tenth year of His reign, did give and grant unto me, the said George Henry Manton, my executors, adm̃ors, and assigns, His especial license, full power, sole privilege and authority, that I, the said George Henry Manton, my executors, adm̃ors, and assigns, during the term of years therein expressed, should and lawfully might make, use, exercise, and vend, within England, Wales, and the Town of Berwick upon Tweed, my Invention of "An Improvement in the Construction of Locks in all Kinds of Fowling Pieces and Fire-arms;" in which said Letters Patent there is contained a proviso, obliging me, the said George Henry Manton, by an instrument in writing under my hand and seal, to cause a particular description of the nature of my said Invention, and in what manner the same is to be performed to be enrolled in His

Majesty's High Court of Chancery within two calendar months next and immediately after the date of the said recited Letters Patent, as in and by the same (reference being thereunto had) may more fully and at large appear.

NOW KNOW YE, that in compliance with the said proviso, I, the said George Henry Manton, do hereby declare that the nature of my said Invention, and the manner in which the same is to be performed, are particularly described and ascertained in and by the Drawing hereunto annexed, and the following description thereof (that is to say):

My said improvement in the construction of locks in all kinds of fowling pieces and fire-arms has for its object the lessening of the recoil occasioned by the rapid inflammation of the gunpowder in fowling pieces and other fire-arms fired by means of percussion locks; and I attain this object by giving a greater vent of the gaseous matters generated in the explosion of the gunpowder at the time of firing, although I keep the said vent closed or shut, and protected from wet, and so as to prevent the loss of any part of the charge of gunpowder in the act of loading. And I hereby claim as my Invention, and the object of this my said Patent, any and every mode of attaining these ends; as, however, it is desirable to afford examples of methods of carrying my said improvement into effect, I shall proceed to do so by referring to the Figures which are contained in the Drawing, which, as aforesaid, is annexed to the Specification, and in which said Drawing Figure 1, represents part of a double barreled fowling piece with the left hand lock in its place, the right hand lock being not shewn. *a*, is the lock plate; *b*, the percussion cock, represented as being upon the half cock; *c*, the cover of the vent hole, which is shewn as open in Figure 2, at *d*. This cover is affixed at one end of a bent lever *e*, which turns upon a stud *f*, affixed in the lock plate as shewn in Figure 3; this stud is farther secured by its outer end entering a hole made inside of a bridle *g*, Figure 4, and which bridle is secured to the lock plate *a*, by two screws passing through the plate and fitted into screwed holes made in the bridle to receive them, as likewise shewn in Figure 4.

Figure 5, is a side or edge view of the bridle, and Figure 6, a plan or top view of it. The bridle also served to keep the lever *e* in its place, as shewn in Figures 1, and 2. The lever *e*, is moved or actuated by means of the front edge of the cock *b*, being in contact with the shorter end of it, the lever *e*, being pressed against it by the action of the spring *h*, and its friction roller *i*, shewn separately in Figures 7, and 8, and in contact with the toe of the lever *e*, in Figures 1, 2, and 9. In Figure 1, the cover *c*, is shewn as being pressed up by the action of the spring *h*, and closing the vent *d*, the gun being at half cock. In Figure 2, the cock is shewn in its position when the gun is discharged, and the front part of it has acted upon the shorter end of the lever *e*, and raised it, at the same time lowering its longer end, and has thus removed the cover *c*, from the vent hole *d*, so as to allow the escape of the gaseous matter produced by the explosion, and also at the same time permitting the entrance of air into the barrel of the gun through the vent hole, both of which causes considerably aid in lessening the recoil of the piece. Figure 10, is an underview of the bent lever *e*; and Figure 11, an edge or side view of it. *e*, the part against which the front edge of the cock acts; *j*, the toe of the lever against which the friction roller *i*, of the spring *h*, acts; *k*, Figure 10, the hole in the lever which fits upon the stud *f*, shewn in Figure 3; *c*, the cover of the vent hole. *d*, Figure 12, is another outside view of the bent lever *e*, with the addition of a friction roller *l*, mounted in its shorter end, with a view to lessening the friction occasioned by its sliding along the front edge of the cock *b*, when thought necessary. Figure 7, is an edge view of the spring *h*, and its friction roller *i*; *m*, is a screwed hole intended to receive a screw, which is passed through a hole *n*, Figure 3, made in the lock plate, and it is farther steadied by means of the stud *o*, Figure 8, entering a cap *p*, made in the lock plate, Figure 3, to receive it. I do not mean or intend hereby to claim as my Invention any of the various parts shewn and described herein, which may already be known or in use. But I do hereby claim the opening and closing the vent hole or holes in the barrel or barrels of percussion fowling pieces and other fire-arms which are fired by percussion, and whether effected in the modes herein shewn and described, or in any other method.

In witness whereof, I, the said George Henry Manton, have hereunto set my hand and seal, this Twenty-eighth day of October, in the year of our Lord One thousand eight hundred and twenty-nine.

G. H. (L.S.) MANTON.

FARRER.

AND BE IT REMEMBERED, that on the Twenty-eight day of October, in the year of our Lord 1829, the aforesaid George Henry Manton came before our said Lord the King, in His Chancery, and acknowledged the Specification aforesaid, and all and every thing therein contained and specified, in form above written. And also the Specification aforesaid was stamped according to the tenor of the Statute made for that purpose.

Inrolled the Thirty-first day of October, in the year of our Lord One thousand eight hundred and twenty-nine.

LONDON:
Printed by GEORGE EDWARD EYRE and WILLIAM SPOTTISWOODE,
Printers to the Queen's most Excellent Majesty. 1854.

A.D. 1829. Sept: 2. No. 5838.

MANTON'S SPECIFICATION.

1 SHEET

FIG. 2.

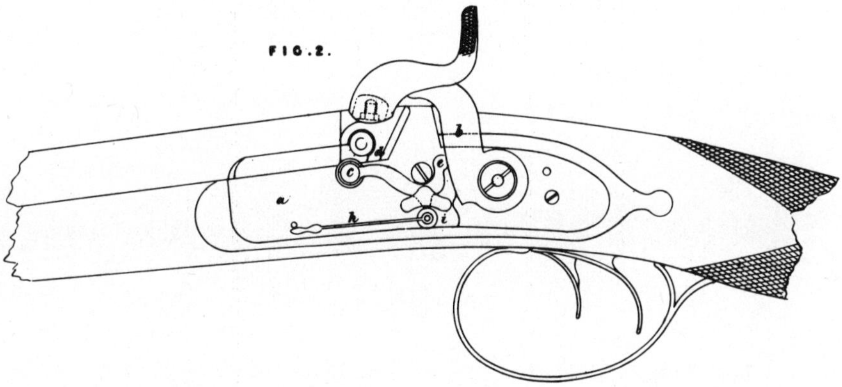

FIG. 1.

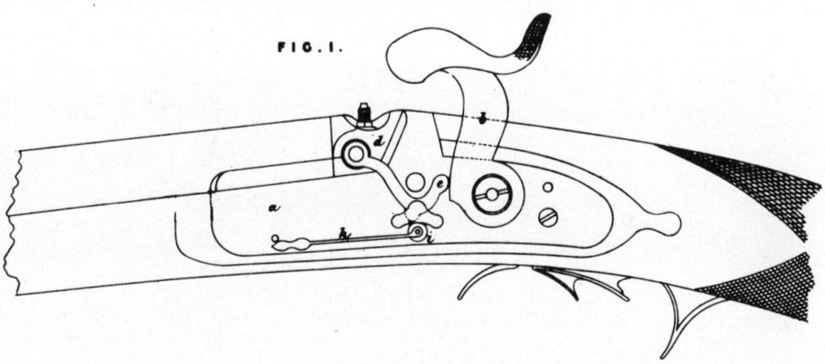

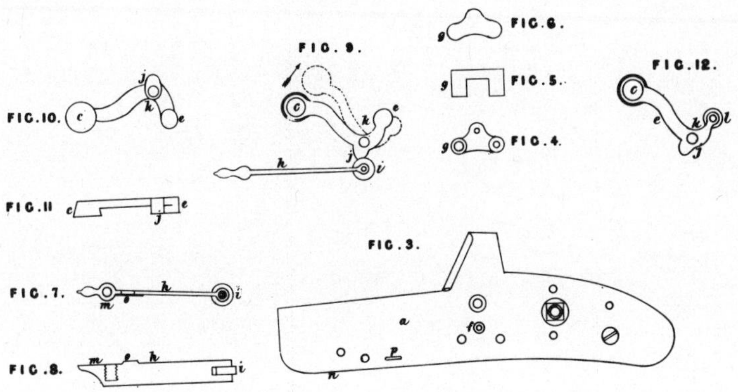

The Enrolled Drawing is colored.

Malby & Sons lith.

LONDON: Printed by GEORGE EDWARD EYRE and WILLIAM SPOTTISWOODE, Printers to the Queen's most Excellent Majesty. 1854.

A.D. 1831 . . . . Nº 6137.

Improvements in Fire-arms and Cartridges, and Machinery for effecting the same.

DEMONDION'S SPECIFICATION.

TO ALL TO WHOM THESE PRESENTS SHALL COME, I, AUGUSTUS DEMONDION, of Old Fish Street, in the City of London, send greeting.

WHEREAS His present most Excellent Majesty King William the Fourth, by His Letters Patent under the Great Seal of Great Britain, bearing date at Westminster, the Thirteenth day of July, in the second year of His reign, did, for Himself, His heirs and successors, give and grant unto me, the said Augustus Demondion, His especial licence, sole privilege and authority, that I, the said Augustus Demondion, my ex̃ors, adm̃ors, and assigns, or such others as I, the said Augustus Demondion, my ex̃ors, adm̃ors, or assigns, should at any time agree with, and no others, from time to time and at all times during the term of years therein mentioned, should and lawfully might make, use, exercise, and vend, within England, Wales, andthe Town of Berwick upon Tweed, the Invention of "**CERTAIN IMPROVEMENTS ON GUNS, MUSKETS, AND OTHER FIRE-ARMS, AND IN CARTRIDGES TO BE USED THEREWITH, AND METHOD OF PRIMING THE SAME, AND IN THE MACHINERY FOR MAKING THE SAID GUNS, MUSKETS, AND FIRE-ARMS, ALSO THE CARTRIDGES AND**

PRIMING, WHICH IMPROVEMENTS ARE ALSO APPLICABLE TO OTHER PURPOSES," communicated to me by a certain foreigner residing abroad; in which said Letters Patent is contained a proviso, that I, the said Augustus Demondion, shall cause a particular description of the nature of the said Invention, and in what manner the same is to be performed, to be inrolled in His Majesty's High Court of Chancery within six calendar months next and immediately after the date of the said in part recited Letters Patent, as in and by the same, reference being thereunto had, will more fully and at large appear.

NOW KNOW YE, that in compliance with the said proviso, I, the said Augustus Demondion, do hereby declare that the nature of the said Invention, and the manner of effecting the same are described and ascertained in manner following, reference being had to the Drawings hereunto annexed, and to the figures and letters marked thereon (that is to say):

I will first describe a gun constructed according to the said improvements, and I will afterwards point out what parts particularly constitute those improvements. Figure 1, of the Drawing is an external view of a soldier's musket constructed according to the said improvements; Figure 2, is a view of the same when it is cocked and ready to be loaded; Figure 3, is a view of the same, shewing the internal structure and arrangement of the parts used to prepare the musket for being loaded, and for firing it off when loaded. Note.—All the Figures on the Drawing are of the full size; N° 1, is the wood stock; M, N, is the barrel lying in a cavity in the stock, bound firmly thereto by iron bands at 2, & 3, in the usual manner, and by screws at 5, Fig. 3, and other screws at 6, and 7, Fig. 28, which screws also pass through the trigger plate N° 8, at 6, and 7, Figure 18. The plate N° 8, carries the trigger H, Z, and the hammer spring and hammer G, F, which operates in discharging the gun, as will be herein-after described, and it is screwed fast to the underside of the stock in the usual manner; a hole is cut through the wood, of which the shape is seen in Fig. 28, and also in Figure 3, to allow room for the parts which operate to discharge the gun. The breech end of the barrel, shewn in section in Fig. 3, is of the shape shewn in Figures 7, and 8, vizt, the top and bottom are flat, the hind or breech end is convex, and the sides are flat, with pivots A, A, projecting out one from each side in the same central line, see Figure 8. The breech P, is not screwed in fast to the barrel in the usual manner, but is moveable about the pivots A, A, to which it is connected. The breech is a block of iron with a breech piece of gun metal or other similar metal let into it, as will be herein-after more particularly described, and the surface of the breech P, is concave, to apply against and fit the convex end of the barrel, see Figures 3, and 7. The shape of the breech is shewn in Figures 3, 4, 5, and 6, and it is connected to the pivots A, A, by two strong plates B, C; Figure 9, 10, and 11, applied one against each side of the breech P; the parts marked B, B, in Figures 10, and 11, projecting inwards into corresponding cavities formed in the sides of the breech P, (shewn at F, E, Figure 6, and by dotted lines at F, E, in Figure 5,) while the circular holes A, (Figures 9, 10, and 11,) of the plates B, C, fit on to the pivots A, A, that project out of the fixed barrel. The breech P, and the side plate B, C, are screwed fast together by two screws at x, and y, Figures 3, and 9, and form thus one piece, which can be turned round on the pivots A, A, as a centre of motion. The end of the plates B, C, are circular, and fit into a circular cavity sunk in the wood stock (see Figures 3, 27, and 28), and lined with a metal plate 9, 9, screwed fast to the wood.

The breech P, is made with a long handle or lever N° 10, projecting back from it towards the butt end of the musket (see Figures 1, 2, and 3), the end of it being in a convenient position for the person who is using the musket to lift it up in order to raise the breech into the position Figure 3.

At the end of the lever N° 10, a curved or bent piece K, L, is jointed to it by a pin at K; at the other end J, of the piece K, L, a tail J, T, S, is fixed to it of the shape shewn in Figures 2, and 3, vizt, the upper part of the tail J, T, S, is an inclined plane, and the lower part curved surface inclined in an opposite direction. The use of this piece K, L, with its tail J, T, S, is as follows:—When the breech P, is in the position Figure 1, with its concave surface applying closely against the convex surface of the breech end of the barrel, the tail J, T, is depressed, as shewn in Figure 1, and then the thick part T, of the wedge J, T, (which is shewn raised up in Figs 2, and 3, and projects out as far as the end *j*, of the lever N° 10,) is depressed so as to bring the thin part J, of the wedge on a level with the upper surface of the lever handle N° 10, and it thus allows a tooth or catch R, at the upper end of a spring R, V, to press forwards into the cavity 12, formed in the end of the piece K, L, Figure 2, and to catch over the end *j*, of the lever handle N° 10, as shown by dotted lines in Figure 1. While the spring R, V, is in that position the end of the lever N° 10, cannot rise, but in order to release the lever N° 10, from the spring R, V, the piece K, L, is lifted up by the thumb; in raising it up the inclined plane J, T, acts against the inclined face of the catch R, and forces it back in opposition to the elasticity of the spring R, V, and when the tail J, T, S, is raised up as far as the pin 6, (put through the tail J, T, S, under the lever handle N° 10,) will allow it to rise, the thick end T, of the wedge projects a little beyond the end of the lever handle, and consequently pushes the spring R, V, back far enough to allow the end *j*, of the lever handle to escape by the catch R, and to be lifted up clear out of its way, and then the person using the gun can lift up the lever handle and turn the breech P, about its axis of motion into the position shewn in Figures 2, and 3. The breech end of the barrel is then quite uncovered, and the cartridge can be put into its place at the breech, as represented at A, Figure 3. The same motion of the lever N° 10, which uncovers the breech end of the barrel for loading it also cocks the gun by the following means:—In each of the side plates B, and C, that are screwed as aforesaid to the breech P, there is a hole D, Figs 3, and 9, which receives and holds fast one of the broad flat ends D, of a small cross bar D, D, (see Figure 12,) placed horizontally under the barrel and parallel to the pivots A, A, of the breech; a short arm 13, with a roller E, at its end, projects out from the cross bar D, D, away from the centre of motion A, in an opposite direction to the lever N° 10, and consequently the arm 13, (being fastened to the plates B, C, as aforesaid) is depressed by them, when the lever N° 10, is raised, and vice versâ; and the arm 13, is of such a length, that while the lever N° 10, is being lifted up towards the position, Figures 2, and 3, the roller E, will be caused to depress a strong bent spring F, G, on which it bears, forcing down the end 14, of that spring against another bent spring Z, which forms the trigger, so that when the lever N° 10, is raised up, as in Figure 3, and has uncovered the opening at the breech end of the barrel, the spring F, G, will have been forced down into the position in which it is shewn in Figure 3, with its end 14, caught and held down by the edge 15, of the trigger spring Z. The gun is thus cocked by the same operation which prepares it for being loaded. A, Figure 3, is the cartridge, which is put into its place in the barrel when the breech P, has been lifted up into the position shewn in that Figure. The cartridge A, contained the priming within itself, viz[t], a small copper tube *a*, filled with detonating or percussion powder, and closed at both ends, is stuck into the cartridge about half an inch in length, being contained within the cartridge. When cartridges are to be made up ready for the ammunition boxes of soldiers, they are fitted with priming tubes *a*, made as shewn by Figures 23, and 24, viz. a piece of tube filled with the detonating powder is bent back upon itself, and the end is again bent up at right *a*, angle to the tube. The double end is put into a cartridge, the short end *y*, bearing against a piece of pasteboard *z*, glued on the end of the cartridge, and then another piece of pasteboard is slipped over the single piece of tube and glued down over the short end *y*. The piece of tube that projects out of the cartridge thus forms a handle to draw the cartridge out by, if it is desired to unload the piece without firing it off; or otherwise a plain straight tube *a*, may be fixed into the cartridge sufficiently tight by gluing it in. But other tubes for containing priming are also made, which can be fixed into any common cartridge that has not been made for guns constructed according to the aforesaid improvements; one of these priming tubes *a*, is shewn by Figure 22, viz., it is a plain straight tube, closed at both ends, but one is made with a sharp prong 7, to stick into a cartridge. The cartridges A, are made up with paper or parchment with ball and common gunpowder, but as they are to be ignited by means of detonating powder, they do not require so great a charge as is usual for common ball cartridge; a touch pan C, Figure 3, is formed on the lower part of the breech P, the form of which touch pan is shewn in Figures 3, 4, and 6, viz., there is a triangular opening in front of the touch pan, (see Fig. 4,) the inclined sides B, B, of that opening Fig. 4, rising up from the side edges of the touch pan to a small semicircular opening A, Figure 4, which when the breech P, is shut down into the position Figure 1, fits over a corresponding semicircular opening or cavity seen at D, Figures 3, and 8, in the lower part of the bore of the barrel. The cartridge A, ought to be put in in such position that its priming tube *a*, will drop into the cavity D; but if from hurry or neglect it has been put in with the tube *a*, in an improper position, the inclined edges B, B, Fig. 4, catch hold of the said tube and force it down till it is enclosed in the circular cavity between the bottom of the breech and the bottom of the bore of the barrel; when the cartridge has been so forced into its right position, the piece of tube *a*, that projects out of it bears against the touch pan *c*, at its upper side, and bears partly upon its under side upon a small piece of steel *r*, screwed on for that purpose under the breech end of the barrel, see Figures 3, and 8. The operation of the gun is as follows:—Suppose it to be in the position Figure 2, loaded with the cartridge and its priming tube, and cocked, the lever N° 10, is to be pulled downwards by its end K, L, into the position Figure 1. In descending the curved surface T, S, of the tail, J, T, S, comes in contact with the inclined face of the catch R, and forces that catch gradually back, until the thick part T, of the wedge has passed the edge *s*, of the catch R, which allows the catch R, to press forward into the cavity 12, and to catch over the end of the lever N° 10, so that when the lever N° 10, and the breech P, are shut down into the position shewn in Figure 1, they are prevented from rising up by the spring and catch R, V, as herein-before stated; the roller E, at the end of the short arm 13, has in the same time been raised up by the motion of the plates B, C, about the pivots A, A, off from the spring G, F, into a nearly horizontal position under the barrel. The gun is then ready to be fired off by pulling back the trigger H, Z, so as to draw back the edge 15, of the spring Z, and release the edge 14, of the hammer spring F, G, which flies up instantly and causes the hammer F, formed on its end to strike a sharp quick blow on the copper priming tube *a*, the bottom of the touch pan *c*, having the effect of an anvil, as if the copper tube *a*, were laid on a hard steel anvil and struck a smart blow with a hammer. The blow of the hammer F, on the copper tube *a*, causes the detonating powder contained therein to explode, and to ignite the powder contained in the cartridge A. Note.—The gun cannot go off by the trigger being accidentally pulled while the breech P, is raised up, as in Fig. 2, because the roller E, at the end of the arm 13, then bears on the spring G, F, and keeps it down, and the part *m*, *l*, Fig. 3, of the hammer spring must rise or swell out sufficiently for the roller E, always to intercept the spring and prevent it flying up to its full extent, except when the end of the lever handle N° 10, is brought quite down into its place under the catch R, in order to prevent any possibility of the priming tube being struck so as to discharge the piece, except when the breech is fairly shut down in its place, as in Figure 1. The flame and gasses which are liberated by the explosion of the detonating powder in the pan C, escape through the opening 20, in the trigger plate, see Fig. 3, and 18, and thence escape out under the stock between the arms U, V, of the guard O, S, V, these arms U, V, being made strong and broad to protect the hand of the person who fires off the gun from being injured by the flash or by the gasses that are liberated. Note.—To uncock the gun without discharging it, the trigger must be pulled back with one hand, while with the other the lever N° 10, is gently pulled down from the position Fig. 2, into the position Fig. 1; when the gun is cocked, a small prong 21, on the under side of the hammer spring projects down below the wood, as in Fig. 3, but when it is not cocked, that prong is contained within the wood, and is not visible. Figures 39, to 50, represent the mode of fixing the bayonet on the muzzle of the musket; M, Q, is a tube which fits over the muzzle end of the barrel and of the stock, and can be slided up and down thereon, being guided by a tail N, V, screwed to the tube M, Q, at *x*, which is bedded in a groove A, C, Fig. 47, formed in the under side of the stock, see Figures 39, 40, and 43. A spring S, T, is let into the stock, see Figures 40, 39, and 43, and 47; and a rounded nob T, is formed at the end of that spring, which is forced out by it, and projects through a slit R, in the tube M, Q, whenever the tube is slided away from the end of the muzzle into the position Figure 43, so that it cannot fall or be pulled off. But by pressing that nob inwards the tube M, Q, can be slided up over it into the position Figures 40 and 39, and then the nob T, being released projects out against the lower edge of the tube M, Q, and prevents its sliding back, see Fig[s] 40 and 39. The bayonet is formed with only a half tube or semicircular socket B, C, at Fig[s] 45, and 48, at its lower end, in which an oblong hole K, is cut to fit on to a corresponding stud L, Figure 43, or 46. To fix the bayonet the tube M, Q, is drawn down into the position Figure 43, and the bayonet put on the stud L; the tube M, Q, is then slided up over the socket B, C, of the bayonet into the position Figure 40, and the bayonet is then firmly held, for the nob T, prevents the tube M, Q, from being pushed back, and the bayonet cannot be turned round to pull it off from the barrel so long as the tube M, Q, covers the socket B, C. To unfix the bayonet the nob T, is pressed inwards, and the tube M, Q, slided down into the position Figure 43. The bayonet can then be taken off the stud L. Note.—The end V, of the guide N, V, of the holding tube when it is removed from the barrel serves as a screw driver to unscrew the parts of the gun. The bayonet F, H, G, H, Figure 40, is fixed on the under side of the barrel instead of at the side, and because the gun is not loaded at the muzzle, and the use of a ram rod is not therefore required, the elbow 22, 23, of the bayonet may be shorter than usual, and the bayonet consequently somewhat stronger. The trigger shewn in Figure 3, is in one piece, viz., it consists of a single spring to which the finger piece H, is fixed fast. Figure 13, shews another form of the trigger, viz., the trigger is a bent lever H, Z, *i*, mounted on a centre pin at *n*, supported in cocks fixed fast to the trigger plate N° 8. The arm *i*, of the trigger is pressed towards the hammer by a spring P. The operation of this Figure is apparent from the Drawing. also the bent lifting piece K, L, at the end of the lever N° 10, may be constructed as shewn in Figure 14, viz., the piece K, L, Figure 14, is moveable about a joint pin K, and is bent down at *l*. The tail K, *l*, passes through a slit in the lever N° 10, and is received in a hole in a sliding bolt T, S, held to the under side of the lever

N° 10 by a screw at 24, on which it can slide backwards and forwards. The catch R, Fig. 14, is made as already before described for Figures 2, and 3. By lifting up the piece K, L, the tail *l*, pushes the sliding bolt T, S, against a tongue *s*, projecting out of the spring R, V, and forces the catch R, back until it releases the end *i*, of the lever N° 10. In pulling down the lever N° 10, to prepare the piece to be discharged, the lower part of the tail J, T, O, acts in a similar manner to that already described for Figure 2, and the tail K, *l*, draws away the sliding T, S, from the tongue *s*, and allows the catch R, to drop into the cavity in the end of the bent piece K, L; also the hammer spring may be made as shewn by Figure 26, instead of being of the form Fig. 3. Figures 4, 5, 6, 7, 8, 9, 10, 11, 12, 15, 16, 17, 18, 19, 20, 21, 22, 23, 24, 25, 27, 28, 29, are detached views of the several parts above described, drawn of the real size; and their object and operation having been already pointed out they will need no further explanation than that which the Drawing affords, and their forms and strength may be varied, provided they effect the purposes herein-before described. The breech [illegible], is made as herein-before stated, of a block of iron and a breech piece of gun metal, or other hard and durable metal, of the shape of a cylindrical bolt with a head, is let into that block of iron, as shewn in Figures 3, and 4. The breech piece when worn out or damaged by the action of the powder can be taken out and replaced without making an entire new breech; also a small strip of gun metal or other similar alloy is let into the side plates B, C, at *p*, see Figure 9, at that part which works against the sharp edge of the breech end of the barrel when the breech is lifted up and down.

The cartridge may be made with a ball a little larger than the bore of the barrel, and a chamber of corresponding size is bored out at the breech end of the barrel to receive the cartridge, see Fig. 3, whereby the ball is projected with greater force, and the cartridge cannot slip down in the barrel towards the muzzle in loading; but whether the ball is larger than the bore of the barrel or not the cartridge must be so, and must be received in a chamber, as in Fig. 3, to prevent its slipping down too far for the hammer to strike the priming tube *a*. Cartridges A, made as aforesaid, with ball larger than the bore of the barrel, will not fit muskets of the ordinary construction of the same calibre, but cartridges made for common muskets may be used in muskets made according to the said improvements by fixing into such common cartridges the priming tubes shewn in Figure 22.

I have shewn in the Drawing only a soldier's musket constructed according to the said improvements, but the same may be applied in the construction of single or double barrelled fowling pieces, rifles, pistols, and such other fire arms as are usually constructed with flint locks or percussion locks, with only such variations in the proportions and workmanship as are required by the nature and object of the particular kind of fire arms for which they may be applied, and as will be obvious to a competent gunsmith.

In making a double barrelled fowling piece there will be only one moveable breech, but two arms 13, two triggers Z, H, and two hammer springs G, F, and two touch pans C; also a thin piece of metal must be fixed to the wood stock, and stand up therefrom so as to be received in a slit between the two touch pans whenever the breech is shut down, to prevent communication between the said touch pans in order to prevent the possibility of the explosion of the priming tube of one of the cartridges igniting the charge of the other barrel.

Figures 32, 34, 35, 36, 37, and 38, represent the machinery which may be used in manufacturing certain parts of the musket or gun or other fire arm constructed according to the said improvements. Figure 38, represents part of a machine for forming the concave surface of the breech. P, L, in Figure 38, is a cylinder, on the surface of which very fine teeth are cut obliquely to the axis like the teeth of a file. The cylinder L, is fixed on a shaft M, M, mounted in bearings not shewn in the Drawings, and the shaft L, may be turned round by means of toothed wheels or by straps to communicate motion to it from a water wheel or steam engine or other first moving machine. The circumference of the cylinder L, is exactly of the curvature intended to be given to the concave surface of the breech P, which (having been previously forged something like the required shape) is held and pressed up against the cutting surface of the cylinder while the cylinder is turned round with suitable velocity to cut or grind the surface of the breech P, very truly to the proper concave curvature. The breech end of the barrel is first forged to something near its intended shape with flat sides and a flat end, and the pivots A, A, roughly formed on it, and it is afterwards finished by machinery shewn in Figures 32, 35, and 37. The end of the barrel is first ground to a truly flat surface in a machine, Fig. 32, by slipping it on to a stem or mandril A, projecting out from the middle of a circular boss J, formed on the end of a shaft M, that lies in bearings not shewn in the Drawing. The face of the boss J, is cut with teeth like the teeth of a file; and to grind the end of the barrel it is pressed up against the cutting face of the boss J, while the shaft M, is turned round with suitable force and velocity, until the end of the barrel is gradually ground to a truly flat surface at right angles to the length of the barrel. The shaft M, is then stopped, the barrel taken off the stem A, and removed to a machine for grinding the sides of the breech end of the barrel to truly flat surfaces all round the pivots A, and reducing the pivots A, A, to their proper diameter. The operating parts of this machine are shewn in Figures 34, and 35. M, M, are two shafts supported in bearings on the same central line, not shewn in the Drawings. On the end of each shaft M, a circular boss A is formed, with a circular hole D, in the centre thereof, of such diameter as exactly to fit the pivots A, of the barrel when the same are finished, and the face of each boss A, all around the circular hole D, is cut with sharp fine teeth, see *a*, *a*, Fig. 35. An adjusting screw, not seen in the Drawings, is supported in suitable framing, and bears against the end of each of the shafts M, in order to move the shafts lengthwise towards each other, or they may be pressed towards each other by a weight acting to force one end of a lever against the end of each shaft M, or by many other contrivances known to mechanicians. A spindle or mandril C, of the diameter of the bore of the barrel, projects out between the faces of the shafts M, from a large flat plate E, which is supported in suitable fixed framing not shewn in the Drawings; the mandril C, being exactly at right angles to the length of the shafts M, M. The operation of grinding the sides of the breech end of the barrel and the pivots A, is a follows:—The shafts

M, are drawn apart in their drawings till their cutting faces *a*, *a*, are far enough apart to allow the barrel to be slipped on to the mandril C; the shafts M, are then moved towards each other by their setting screws, or other means of adjustment, until the holes D, come up to the pivots A. The shaft M, is then turned round with a suitable velocity (by wheel work or other connection with a water wheel, steam engine, or other power), and the cutting edges of the holes D, take off the metal of the pivots A, gradually as the shafts M are turned round, the said shafts being constantly pressed up to their work (by their setting screws or other means of forcing them towards each other) until the pivots A, are reduced to their proper diameter all their length. By that time the cutting faces *a*, *a*, will have come up to touch the sides of the barrel, and the shafts M, are still turned round and pressed up to their work until the cutting surfaces *a*, *a*, have ground the sides of the barrel to truly flat surfaces, leaving the pivots A, projecting out therefrom of the proper diameter. Note.—The distance from the centre line of the shafts M, to the face of the plate E, is a very little more than the intended radius of the convex end of the barrel. The next operation is to grind the breech end of the barrel to a correct circular surface. This is done, first in a rough manner with the file, or by other tools, and then the barrel is put into a machine, the operating parts of which are shewn in Figures 37, and 36. A plate J, J, Fig. 36, with a circular cavity A, formed into it, is fitted on each of the pivots A; from the back of each of the plates J, J, a neck or gudgeon B, projects, and the plates J, J, with the barrel enclosed between them, are laid in bearings C, B^2, Fig. 37, adapted to receive the gudgeons B; the bearing C, B^2, are formed at the end of a block or bed of iron A, Fig. 37, fixed solidly upon a frame, not shewn in the Drawing. Note.—The circular sockets A, which receive the pivots A, A, are not exactly concentric with the gudgeons B. But when the plates J, J, are placed in their bearings at C, Fig. 37, the centre of the gudgeons B, is a very little above and on one side of the centre of the sockets A; the bearings C, B^2, are so far apart as to receive the end of the barrel between them; and the part of the bed A^2, contained between the bearings, is hollowed out to a concave circular surface *i, i,* of exactly the curvature of the breech P, that surface being cut with sharp fine teeth like a file, and hardened. To grind the convex end of the barrel the caps D^2, are screwed down upon the necks or gudgeons B, the end of the barrel bearing against the cutting surface *i, i,* The barrel is then laid hold of by the muzzle, and worked backwards and forwards in its bearings until its end is ground to the proper convexity, that is, until it fits correctly to the cutting surface *i, i.* The excentricity of the sockets A, Fig 36, to the gudgeons B, will cause the convex circular end of the barrel to be a little excentric in respect to its pivots A, A, the upper edge of that convex end being a very little nearer to the centre of the pivots A, than the lower edge is; but the arch of the breech P, is concentric with the pivots A, A. By this arrangement when the breech and barrel are in their places on the stock, and are shut down as in Figure 1, the concave end of the breech, and the convex end of the barrel fit against each other with sufficient accuracy for the breech to answer its purpose. But when the breech is raised up, the lower edge of the breech gradually departs from perfect contact with the convex end of the barrel, and clears it by the time it has risen up to the position Figure 3, whereby the breech moves more freely, and is less liable to bind against the barrel than it would be if the two surfaces were made perfecly concentric, so as to fit to each other throughout the motion of the breech.

Or, instead of grinding the end of the barrel excentric to its pivots A, as above directed, it may be ground to fit exactly to the breech, and then filed away a little at the upper part or a little at both the upper and lower parts, the main object being to ensure a sufficiently tight fitting of the breech when the same is shut down, but yet to allow it to move freely up and down. Therefore when the end of the barrel is ground eccentrix to its pivots A, the excentricity must be very trifling, and it will depend upon the finish of the workmanship how much excentricity it is advisable to give. If the end of the barrel is not ground excentric, the plated Figure 36, may be dispensed with, and the end of the barrel ground by laying its pivots A, in the bearings C, Figure 37, and working the barrel backwards and forwards. Figures 30 and 31, represent the operating parts of the machinery used in making the copper priming tube *a*, and filling them with detonating powder; a small tube of copper, Figure 31, is drawn through a conical or tapering hole at A, in a steel plate E, and thence through a round hole at *i*, in another steel plate C, of the exact diameter of the priming tube. The upper part of the hole A, is square, but the lower part semicircular, and it is a little larger than the priming tube. A small piece of steel *f*, with a semicircular prong *h*, at its lower end, see Figures 30, and 31, is screwed fast on the inside of the plate E, and descends below the top of the opening A. Before putting the copper tube into the machine it is cut open a little way at the end that is pushed forwards into the hole A, and then as fast as the tube is drawn through the holes A, and *i*, it is opened by the prong *h*, into the form of a trough with a semicircular bottom, see Figures 30, and 70. A hopper or reservoir D, containing detonating powder is fixed between the plates E, and C, see Figures 30, and 31, and the open copper trough passes through that reservoir (see Figure 31) in being drawn through the draw plates E, and C, and gets filled with detonating powder; as each part of the trough comes successively to the hole *i*, and is drawn through it, the edges of the trough are forced together, and the copper formed into a tube filled with detonating powder. B, is a sponge, fastened to the plate C, filled with glue or some other glutinous matter which covers over and closes the joint where the two edges of the copper meet, in order to prevent any powder from escaping. Note.—Instead of a copper tube or plain strip of copper may be drawn thro' the holes A, and *i*, and will be formed into a trough and become filled with detonating powder in passing thro' the copper D, and then be closed and formed into a tube in being drawn through the hole *i*. The filled tube is afterwards cut into proper lengths for the priming tubes with nippers which at the same time close the ends of the tubes, and they are then fitted into the cartridge A, or else prepared with a sharp end for fitting into common cartridges as herein-before described. Note.—I have only shewn the machinery, Figures 30, and 31, adapted for

A.D. 1831. July 13. Nº 6137.
DEMONDION'S Specification.

(In three parts) Part 1.

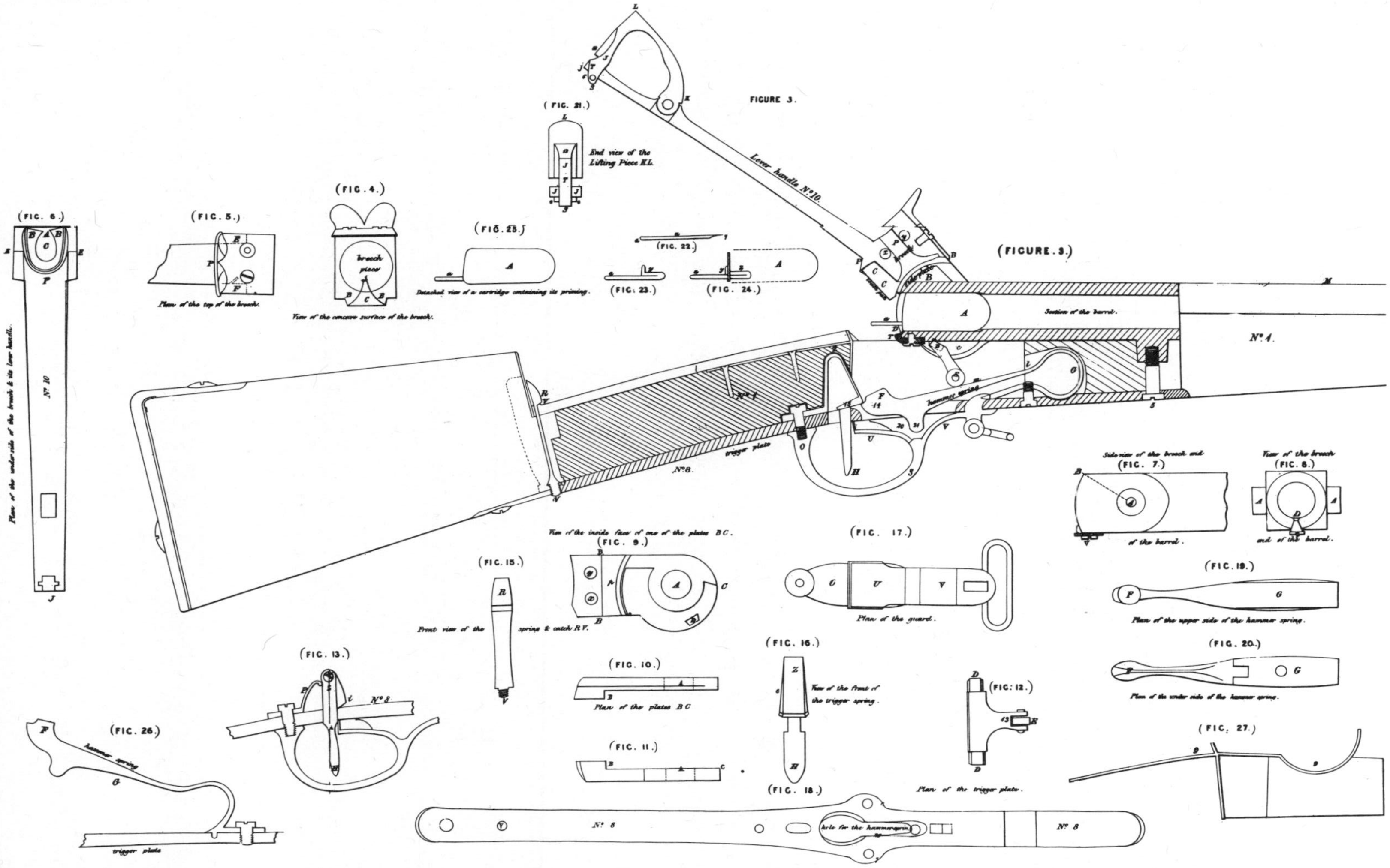

The enrolled drawing is colored.

London: Printed by George Edward Eyre and William Spottiswoode, Printers to the Queen's most Excellent Majesty. 1859.

Malby & Sons, lith.

drawing one tube and filling it, but the same may be adapted for drawing several tubes at a time according to the number of holes A, and C, and plates *f*, *h*. In describing all the parts of a gun constructed with the improvements communicated to me as aforesaid, I have described many things which form no part of the same, and it remains for me to state in what those improvements consist; and I do hereby declare that I claim as part of the said improvements, the breech P, moveable about pivots A, by means of a lever handle N° 10, furnished with a lifting piece K, L, and spring R, V, to prevent the lever handle N° 10, from rising, or else to allow it to be released and lifted up in the manner herein-before described. The said breech being combined with an arm 13, a hammer spring G, F, and a trigger Z, H, so that the same movement which raises the moveable breech to prepare the piece for being loaded, also cocks the same, and the said breech being also provided with a touch pan C, on its under side to receive the blow of the hammer F. But I do not claim the use of a breech moveable about a centre, nor of any of the other parts separately, but only when they are combined in the manner and for the purposes herein-before described; also I claim, as another part of the said improvements, that particular arrangement and construction of the parts of the bayonet used in fixing and unfixing it which I have described in reference to Figures 39, to 50; also I claim, as another part of the said improvements, the method of priming and loading at one operation by means of a cartridge in which is placed a copper tube filled with detonating powder, a part of which tube is inclosed within the cartridge, and the other part remains outside of the cartridge at one of the ends of it, and which latter part of the tube is received in the touch pan to receive the blow of the hammer in the manner herein-before described. And I claim further, as part of the said improvements, the machinery which I have described in reference to Figures 34, 35, 36, and 37, for grinding the flat sides and the pivots at the breech end of the barrel, and for grinding the convex end of the barrel, and also the machinery described in reference to Figures 30, and 31, for manufacturing the priming tubes and filling the same with detonating powder. But I do not claim any of the separate parts of the said machinery which I have described in reference to Figures 30, 31, 34, 35, 36, and 37, but only when they are combined as I have described them.

In witness whereof, I, the said Augustus Demondion, have hereunto set my hand and seal, this Fourth day of January, One thousand eight hundred and thirty-two.

AUGUSTE (L.S.) DEMONDION.

MARTIN.

AND BE IT REMEMBERED, that on the Fourth day of January, in the year of our Lord 1832, the aforesaid Augustus Demondion came before our said Lord the King, in His Chancery, and acknowledged the Specification aforesaid, and all and every thing therein contained and specified, in form above written. And also the Specification aforesaid was stamped according to the tenor of the Statute made for that purpose.

Inrolled the Thirteenth day of January, in the year of our Lord One thousand eight hundred and thirty-two.

LONDON:
Printed by GEORGE EDWARD EYRE and WILLIAM SPOTTISWOODE,
Printers to the Queen's most Excellent Majesty. 1854.

A.D. 1831 . . . . N° 6166.

Locks and Breeches of Fire-arms.

BINGHAM AND DUPE'S SPECIFICATION.

TO ALL TO WHOM THESE PRESENTS SHALL COME, we, William Bingham, of Saint Mary Hall, Esquire, and William Dupe, Gun Maker, both of Oxford, send greeting.

WHEREAS His present most Excellent Majesty King William the Fourth, by His Letters Patent under the Great Seal of Great Britain, bearing date at Westminster, the Twenty-fourth day of September, in the second year of His reign, did, for Himself, His heirs and successors, give and grant unto us, the said William Bingham and William Dupe, His especial licence, sole privilege and authority, that we, the said William Bingham and William Dupe, our ex̃ors, adm̃ors, and assigns, or such others as we, the said William Bingham and William Dupe. our ex̃ors, adm̃ors, or assigns, should at any time agree with, and no others, from time to time and at all times during the term of years therein mentioned, should and lawfully might make, use, exercise, and vend, within England, Wales, and the Town of Berwick upon Tweed, our Invention of "**Certain Improvements on Fire-arms of different Descriptions**;" in which said Letters Patent is contained a proviso that we, the said William Bingham and William Dupe, or one of us, shall cause a particular description of the nature of our said Invention, and in what manner the same is to be performed, to be inrolled in His said Majesty's High Court of Chancery within six calendar months next and immediately after the date of the said in part recited Letters Patent, as in and by the same, reference being thereunto had, will more fully and at large appear.

NOW KNOW YE, that in compliance with the said proviso, we, the said William Bingham and William Dupe, or one of us. do hereby declare that the nature of our said Invention, and in what manner the same is to be performed, is particularly described and ascertained by the following description thereof (reference being had to the Drawing hereunto annexed, and to the figures and letters marked thereon, that is to say):

The improvements for which a Patent has been obtained consists in an alteration of the present mode of firing percussion fire-arms now in use upon a new principle; it is adapted in the present instances to an iron case or handle similar to that used for air guns, to which a wooden fore arm or handle before and a wooden butt behind may be attached, or the iron case continued into an iron stock throughout. The principle may be used with a wooden stock throughout, and with the common percussion locks, with a slight alteration in the setting of them in the butt, together with an alteration of the nipple in the breech.

Description of the Drawing.

Letter A, is perspective view of the lock at full cock; N° 1, is the lock plate; 2, the main spring; 3, the bridle; 4, hammer, one piece with tumbler; 5, trigger, one piece with the seear; 6, the seear spring. Letter B, is a perspective view of same, lock quite down. Letter C, is a geometrical view of same, lock uncocked and quite down, with the main spring more at ease. Letter D, is the lock plate, made of best iron, and rather thicker than usual, and with bevilled edges to fit water-tight into the case or stock. N°s 1, 2, 3, are the screw holes of the bridle. N°s 1, 2, are countersunk a little to receive corresponding male countersunk of the bridle legs or pillars; 4, is the axis hole of the tumbler, rather larger than common in order to keep the tumbler firmly in its place; 5, is the screw hole seear spring, also countersunk; 6, is the stay hole of the seear spring; 7, is the main spring screw hole, also countersunk; 8, the stay hole of the main spring. Letters E, F, G, are different views of main spring; E, side of the spring nearest the lock plate; F, outside of spring nearest the eye when fitted to lock; G, is side view of the same spring. In G, N° 1, is the main spring hook; 2, is the stay of the spring; 3, is the male countersunk to fit into lock plate; 4, the screw to connect it with ditto. Letters H, and I, are different views of the bridle, which is made stronger than usual to hold the tumbler in place. Nos 1, 2, are the holes through which the screws pass to attach the bridle to the plate; 3, has no leg or pillar, but rests upon the seear through which the screw passes; 4, the hole to receive the bridle axis of the tumbler, which is made rather larger than usual to keep the tumbler from becoming loose. Letter I, is a side view of the same bridle. N° 1, is one of the pillars or legs of bridle; 2, is a male countersink to fit into corresponding ditto of lock plate; 3, is the screw to attach it to ditto; 4, is the hole N° 3, in letter H, which has no leg, but rests upon the seear itself. Letters J, K, are representations of hammers, one piece with the tumblers. Letter J, is a side or geometrical view of a tumbler. N° 1, is the hammer head; 2, is the bridle axis; 3, is the swivel to connect the hammer with the main spring; 4, are the ratchet nicks or bents. Letter K, is a perspective view of tumbler. N° 1, is hammer head in an unfinished state, left large enough to reduce into any shape the maker may fancy best adapted for striking the cap. N.B. For a double gun it is the reverse way. 2, the bridle axis of tumbler, made very strong; 3, plate axis of tumbler, made also very strong to keep the tumbler axis firm in lock plate; 4, the bent holes half cock; 5, swivel to attach it to main spring hook. Letter L, is a view of trigger, one piece with steear, with the seear spring bearing upon it. N° 1, is the seear or catch to fit into bent holes of tumbler; 2, is the screw hole, by means of which a screw passing through hole N° 4, in letter I, attaches it to the lock plate; 3, is the seear spring, bearing upon the rest of the sears; 4, is the trigger tail or handle, in an unfinished state to shew the forging; 5, screw hole to attach it to lock plate. Letter M, is a side view of seear spring; 1, is the countersink to fit into female ditto of the lock plate; 2, is the screw to attach it to ditto; 3, is the stay to hold it firm in ditto. Letters

N, O, P, Q, are views of breeches of gun; N, is a geometrical view of ditto; O, is a perspective view of ditto. N° 1, is the breech screw; 2, part joining to barrel; 3, hut to hook into false breech or break off, made rather wide to allow of nipple being screwed into it firmly; 4, nipple itself. P, is a perspective view of inside of breech. N° 1, is the cavity of breech in form of a cone; 2, is the bottom of the cone immediately above the end of nipple; 3, the screw. Letter Q, is a section of a new breech capable of being used in the present gun. 1, is the breech itself; 2, is the narrow passage leading from the cone to magazine; 3, is the priming passage from top of breech, down which the priming passes; 4, is a concavity filled by the priming; 5, the cavity enclosing the spiral spring; 6, the termination of the cavity and the stop to the spiral spring to prevent it touching the priming; 7, the screw through which the striking pin passes to strike the priming. Letter R, is a Drawing of spiral spring and striking pin. 1, is the rest or stop of the spiral spring next the priming to prevent it touching it; 2, the catch of the striking pin, which bears upon the spring, and when struck, compresses it down to Figure 1; 3, is the striking pin head. Letter S, is a Drawing of the screw which holds the striking pin in its place. 1, is the countersink, which allows the head of the striking pin to descend to the priming when struck by the hammer; 2, the screw into the hut; 3, the projection resting on the hut when screwed down. Letter T, is a view of the slide and the cap hole open, through which the instrument passes when the nipple is capped. N° 1, is the flap or slide drawn back from hole; 2, the point by which the slide is pulled back; 3, the hole through which the instrument is passed to cap the nipple. N.B. The capping instrument is the same as the capping instrument of Dixon's patent walking stick gun; 4, the nipple appearing through; 5, the breech and barrel; 6, cock outside for cocking gun.

Letter U, is a Drawing of the inside iron case N° 1. The outside of block inside of which is the but screw, as in Figure 20, letter *b*, upon which the narrow end of lock plate is screwed; 2, hole through which trigger passes; 3, hole through which waste caps fall; 4, flaps of false breech or break off, upon which broad end of lock plate is screwed; 5, division of guard to cover finger. Letter *b*, is a section of the whole lock enclosed in an iron case, which can either be continued in an iron stock throughout, or have a wooden butt screwed on behind, and a wooden handle or fore part attached to it, as in letter to Figure 8. N° 1, 2, 3, 4, 5, 6, are the same as before described in letter *a*; 7, outside hammer to cock and uncock the gun; 8, the nipple; 9, tumbler hammer head; 10, hole through which an instrument passes to cap the nipple; 11, the slide to cover ditto; 12, a spring with small roller, upon which the slide 11, passes to keep slide firm; 13, handle to pull back slide. The nipple is capped at half cock by the introduction of the instrument described in letter T, N° 3, through holes N° 10, from which the slide 11, is withdrawn by pulling the point 13, back over the roller spring 12, then return the slide 11, so as to cover hole 10, cock the lock, as in letter A. The trigger N° 5 being pulled will release the tumbler hammer, which will strike the cap upon the nipple 8, ignite the powder contained in it, and communicate to the load in breech. Then half cock the gun, and the pieces of the cap will fall through the hole at bottom of iron case, always open at 14, and through the division of the guard at 15; should the remnants of the cap adhere to the nipple the capping instrument will remove them. 16, is the side of false breeching or break off; 17, is the screw hole upon the flap or false breech of the lock; 18, is the hole in iron case through which trigger passed; 19, guard itself; 20, the other screw hole of lock plate by side of butt screw; 21, the screw to which a wooden or iron butt may be attached; 22, hut hooked into false breeching; 23, break off; 24 barrel of gun; 25, section of breech; 26, cone or chamber in ditto; 27, tube between cone and breech; 28, spring to keep the guard in its place; 29, pin attached to spring and passing through the end of guard to keep it in its place; 30, pin, which being pressed down forces down the spring 28, and liberates the guard 19; 31, screw to hold spring. Letter W, is a view of fresh method of capping nipple through hole at bottom, as in letter U, Figure 3, through which waste caps pass. N° 1, is the guard turned half round; 2, is the finger guard made broad; 3, the pin on the spring as in letter U, Figures 28, 29, 30, 31, to keep the guard in its place, with side bevilled to allow guard to slip over it easily; 4, pin, which, when pressed, being same as 30; letter *b*, releases the guard for capping; 5, the hole through which the instrument is introduced to cap the nipple; 6, stay to prevent guard going past the pin at N° 3; 7, continuation of iron case to hold the wooden handle firm; 8, wooden handle continued, same as in other guns; 9, 10, pins which fasten the iron to the wood. Letter X, is a view of single gun in case, with cocking hammer at half cock; N° 1, is cock, with head formed according to fancy; 2, slide which covers the capping hole at top; 3, thumb piece to moove slide with; 4, end of iron case containing female screw, as N° 20, letter V, into which butt screw enters; 5, lock plate fitted very clean; 6, screw of lock plate into female screw of butt, as in letter U, Fig, 1; 7, screw of lock plate into flap of break off, as in letter U, Figure 4; 8, 9, 10, screws of bridle; 11, breech; 12, barrell; 13, continuation of iron case which supports the wood of the handle or fore part of gun; 14, dotted line to mark hole between the division of the guard.

Description of Plate II.

Letter A, is a view of the plate for making a stock, cut into its first form, and ready for bending out of a piece of best Bristol plate from 20 to 30 guage, or any thickness most agreeable to the maker, or most portable and convenient.

Letter B, is a Drawing of the second form of plate bent when rather warm; the dotted lines are to shew what is cut out of the plate to adapt it to the shape of a stock when bent upon the block C. They may be cut out either before or after.

Letter C, is a Drawing of the metal block, upon which the bended plate B, is placed when hot, and struck into the required curve by a stamp or driver for that purpose. The dotted lines are the same as those in B; 1, 2, 3, are dotted lines showing that part of the butt which does not require so much bending as the other part on the block; 4, is the wooden stand upon which the block rests; 5, is the metal stamp or driver.

Letter D, is another view of stamp half size; 1, 2, are the ends of the fork, curved outwards to prevent them bruising or nipping the plate in driving; 3, is a section of metal block; the curve may be either more or less round or oval, according to the fancy of the maker.

Letter E, is the view of the stock formed upon block C, with the dotted lines in B, and C, cut out, and ready for the reception of F, just above it. N°s 1, 2, 3, 6, 7, 8, 9, mark the places of the corresponding Figures in F, on the top or upper side N° 4; 10, ditto in the middle N°s 5; 12, mark the places of the corresponding Figures in F, in the bottom or lower side.

Letter F, is a view of the butt plate, stop handle plate, and false breech or break off, in the form they are shewn in F. N°s 1, 2, is the butt plate, which is rather thinner than the stock, and rather curved to fit the hollow of the butt, and tapering from 1 to 2. N° 3, 4, 5, is the stop, which is made $^{1}/_{4}$ or more thick, and cuts off all communication between the lock and butt, and strengthens the stock into which it is rivetted; 3, is the small tongue which lies between the lips of the bend of the stock, and when brazed, rivetted, and filed down, forms N° 3, in section I, and finished stock K; 4, is the handle flap to which lock plate is attached; 6, 7, is the handle plate, made the same as 1, 2, and tapering from 7 to 6. N°s 8, 9, 10, 11, 12, is the false breech; 8, is the point; immediately above 7, (the point of junction between the handle, flap, & breech,) is a rabbit continuing to 10 and 12, round which, over the handle flap, the handle of the stock fits; 9, is the outside rim, which, when put together, is on a level with the top of the handle and barrels; N° 10 and 11, are the flaps or rests for the lock plates to screw to; N° 12, is the bottom of the breech, resting upon corresponding bottom in E, into which a rivet or screw passes; N.B. The breech is $^{1}/_{2}$ or $^{3}/_{8}$ thick, or whatever thickness the maker may find best. N.B. All the Figures in F, fall into places corresponding with the Figures in E, and then the stock is ready to close up.

Letter G, is a section of the false breech; N° 9, is the outside rim, level with the barrels; N° 8, the rabbit, upon which the stock is rivetted or screwed; N° 7, the point where the handle flap joins it; 10, the screw hole of lock plate; 13, the hut hole; 12, the screw at the bottom. N.B. These Figures correspond with those in E, and F.

Letter H, is an elevation of false breech; N°s 13 and 14, are the hut holes; N°s 15, and 16, the outside flaps or guards, which the outside cocks fall. The other Figures correspond with those in E, F, G.

Letter I, is a section of the stock E, when closed up and rivetted with F, placed inside it.

The same coincidence takes place in the Figures as in E, and F, and G, and H; the heads of the screws or rivetts appear and show their relative positions; they are afterwards cut off and smoothed. At N° 3, the stop intervenes between lips of the projecting point at 3; here it is filed into shape. At N° 7, the handle flap joins the false breech 8; the handle of the stock covers and is rivetted to the rabbit N° 9; the handle of the stock rises to a level with the outside rim (vide Drawing G), 0, 0, 0, 0. The inside is filed up and cleaned and got ready for putting in the heel plate; if the stock is not hardened, the inside should either be painted, varnished, or japanned. 14, is the place for the lock plate cut out. N° 15, is the trigger hole.

Letter I, is a perspective view of the proceeding stocks screwed up, with F, fixed in. N°s 1, and 2, show the screws over the butt plate; N° 3, 4, and 5, over the stop; 6, and 7, over the handle plate; 8, 9, and 12, over the false breech; X, is the ramrod pipe, which is to be beat out of the rim when hot. The butt is then hammered out into shape after being filed and cleaned inside; the heel plate is brazed on without a tail; the outside is cleared of rivets, filed, and polished.

Letter K, is a Drawing of a stock when finished, polished, and engraved, so as to resemble any other gun, and hardened with beautiful colours. N° 1, is the heel plate, with the dotted lines to shew where it is brazed on; 2, is the termination of the engraving, to shew a false tail, as in a wood stock. N.B. The heel plate may be made to open with a spring, and the inside lined or cased so as to admit of gun furniture, spare ramrods, &c. 3, is the same as in letters I, J, and F, and shews what is the appearance of the end of the stop 3, in letter F, when finished; 4, is the cocking hammer, which may be ornamented ad libitum; 6, is the fastening of barrels. N.B. The barrel should rest upon two or three solid pieces of leather, so as not, or if so, but just to touch the sides of the stock at 6; the leather will make the barrel springy as well as preserve it.

Letter L, is a soldier's musquet of last regulation, Tower mark upon butt, 1805; upon the barrel, by the side of lock, 12; sent to Patentee by Board of Ordnance. 1, shows the heel plate, which if made to open as far as the stop at 2, would contain either a soldier's canteen or amunition for a long battle; 2, is the same sort of stop as in F, and I, but without any flap for lock plate to rest on. The dotted line shows its place. 3, is the end of the lock plate to rest on a tongue of iron; 4, is the end of the breech, which may be let into the stock at the handle, the same as the musquets are now; 5, is the other end of the lock plate resting in the same manner as 3; 6, 8, 10, and 12, are the ramrod pipes, which may be stamped or bent out of solid in a mould into the required form; 5, 7, 9, and 11, are places for solid leather rests, upon which barrels should lie, which are also rivetted in, and will considerably lessen the recoil; 13, is the trigger guard; 14, ramrod. The lock plate is left bright. 1, 13, 8, 10, 12, and 14, are left bright; the rest of the stock is dark, being hardened or covered with leather if not. N.B. The advantage of this stock is, it will never rust; it is lighter than the wooden ones; it will endure almost for ever; the stroke of a sabre will only require brazing to repair it; it will carry a soldier's canteen or amunition in the butt; and no climate or violence can affect; lastly, it is considerably cheaper, and takes infinitely less time to make, than the present wooden ones.

The improvement claimed consists in the principle of firing the gun behind, according to the method and arrangement described or set forth in the annexed Specification; also in the particular parts described or contained in that arrangement, videlicet:—

1st the countersinking of the screw holes of the lock plate, and the corresponding male countersinks.

2nd, the straight main spring.

3rd, trigger one piece with seear.

4th, hammer one piece with tumbler.

5th, nipple placed behind breech, or in a hut behind breech, and principle of firing gun by that means behind.

6th, also principle of firing gun behind by means of breech described in letters Q, R, S, of plan I, as also for the principle or invention of aforesaid breech itself.

7th, also for firing gun by means of a lock contained in a metal cover, stock, or case.

8th, also for the metal cover, stock, or case described in the annexed Specification.

In witness whereof, I, the said William Bingham, for myself and the said William Dupe, have hereunto set my hand and seal, the Twenty-third day of March, in the year of our Lord One thousand eight hundred and thirty-two.

WILLIAM (L.S.) BINGHAM.

MORRELL, Extra.

AND BE IT REMEMBERED, that on the Twenty-third day of March, in the year of our Lord 1832, the aforesaid William Bingham came before our said Lord the King, in His Chancery, and acknowledged the Specification aforesaid, and all and every thing therein contained and specified, in form above written. And also the Specification aforesaid was stamped according to the tenor of the Statute made for that purpose.

Inrolled the Twenty-fourth day of March, in the year of our Lord One thousand eight hundred and thirty-two.

LONDON:
Printed by GEORGE EDWARD EYRE and WILLIAM SPOTTISWOODE,
Printers to the Queen's most Excellent Majesty. 1854.

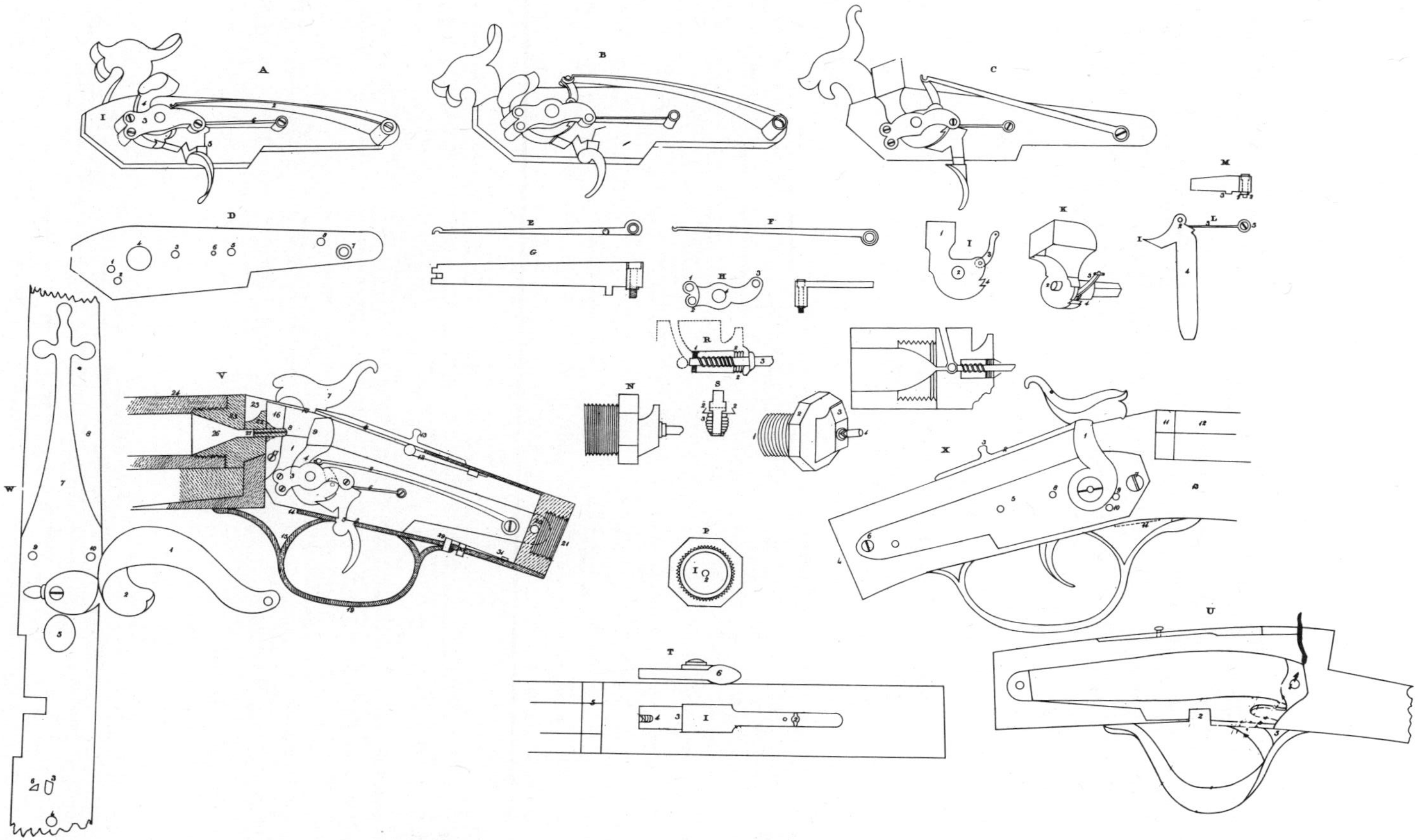

The Enrolled Drawing is Coloured.

Malby & Sons, Lith.

London: Printed by George Edward Eyre and William Spottiswoode.
Printers to the Queen's most Excellent Majesty. 1854.

A.D. 1834 . . . . N° 6572

Application of Conical Detonating Primers directly to Breeches of Fire-arms; and Mechanism for Exploding the same.

MANTON'S SPECIFICATION.

TO ALL TO WHOM THESE PRESENTS SHALL COME, I, John Augustus Manton, formerly of Calcutta, in the East Indies, lately residing with my brother at the Small Gun Office, in the Tower of London, but now in Holles Street, Cavendish Square, in the County of Middlesex, Gun Maker, send greeting.

WHEREAS His most Excellent Majesty King William the Fourth, by His Letters Patent under the Great Seal of Great Britain, bearing date at Westminster, the Thirteenth day of March, in the fourth year of His reign, did give and grant unto me, the said John Augustus Manton, my executors, administrators, and assigns, His especial licence, full power, sole privilege and authority, that I, the said John Augustus Manton, my executors, administrators, and assigns, and such others as I, the said John Augustus Manton, my executors, administrators, and assigns, should at any time agree with, and no others, from time to time and at all times hereafter during the term of years therein mentioned, should and lawfully might make, use, exercise, and vend, within England, Wales, and the Town of Berwick upon Tweed, my Invention of "Certain Improvements in Fire-arms;" in which said Letters Patent there is contained a proviso, that I, the said John Augustus Manton, shall cause a particular description of the nature of my said Invention, and in what manner the same is to be performed, by an instrument in writing under my hand and seal, to be enrolled in His said Majesty's High Court of Chancery within six calendar months next and immediately after the date of the said in part recited Letters Patent, as, reference being thereunto had, may more fully and at large appear.

NOW KNOW YE, that I, the said John Augustus Manton, do hereby declare that the nature of my said Invention, and the manner in which the same is to be performed, are particularly described and ascertained in and by the Drawing hereunto annexed, and the following description thereof (that is to say):

Figure 1, represents a patent breech of a fowling piece or other fire-arm, into which is fitted by screwing a hardened steel cup A, having a conical cavity within it; this cup A, is shewn separately in Figure 2, and in section in Figure 3. Into the steel cup A, wooden cones, or cones formed out of any other fit and proper material or materials, are fitted, one of which is shewn at B, in Figure 4, and in section at Figure 5. These wooden or other cones B, are made with a cylindrical hole in each, as shewn in the section Figure 5, in order to receive an iron cylindrical pin C, shewn separately in Figure 6. The upper end of the cylindrical hole in the cone B, is widened a little, as shewn in Figure 5; at the lower end of this cylindrical hole a detonating pellet D, (shewn separately in section at Figure 7, and in plan at Figure 8,) is securely fastened, with a waterproof varnish or other proper cement. Figure 9, represents the cock or hammer of the gun lock, into the front of which is screwed a hardened steel striking pin E, shewn edgeways in Figure 10, and in plan at Figure 11, in the latter of which Figures are also shewn the two gaps or notches F, F, for the wrench or turns-screw to enter, by means of which the steel pin E, is screwed into the cock. The effect of this arrangement will be, that the iron pin C, in the centre of the wooden or other cone B, upon being struck by the steel pin E, on the pulling off the lock, will be instantaneously driven down upon the detonating pellet D, and explode it, and thereby discharge the fire-arm. It should be observed that the lower part of the conical cavity within the steel cup A, is contracted into a smaller cone and terminated by a small cylindrical hole, and as shewn in the section, Figure 3, and around the edges of this small cylindrical hole the blow or effect of the iron pin C, in striking upon these edges is received and explodes the detonating pellet. In order to apply this improvement to a plain gun barrel, without a patent breech, it will be necessary to have a piece of iron firmly affixed, by brazing or otherways to the side of the barrel, in order to receive the hardened steel cup A, and other parts, as shewn in Figure 1, and also to make the iron cylinders C, C, longer, or to project out above the wooden or other cones B, B, for about a quarter of an inch, in which case the cock or hammer of the gun lock need only be made flat, similar to those employed in exploding the common copper caps, and not to be provided with the steel pin E, as above mentioned. Figure 12, represents another of my said improvements in fire-arms, whereby a fowling-piece or other fire-arm may be rendered capable of priming itself. G, is a small steel nipple, similar, but shorter, to those commonly used in copper-cap guns, and which is fitted by screwing into the patent breech of the gun. This nipple G, is shewn in section in Figure 13. H, in Figure 12, is the lock plate; the lock being constructed on the back-action principle, and well known to gun makers, need not be farther described here. I, is a lever, turning upon pivots underneath the bridle J; K, is a bent spring acting against the shorter end of the lever I; a friction roller L, is mounted on pivots into the longer arm of the lever I, and which acting against a projecting part M, of the vertical slider N, always tends to keep the slider up; O, is a horizontal slider or striking pin, which is generally held back by the action of the spring P. The vertical slider N, serves to raise up or convey a metallic primer Q, received from the primer magazine R, beneath the barrel of the gun to its station above, as shewn in Figure 14, or opposite to the nipple G, shewn in Figure 12, and where it remains ready to receive the blow from the cock of the gun, communicated to it through the agency of the striking pin O, and thus, on pulling the trigger, the primer Q, is rapidly driven or forced against the nipple G, and the detonating prime contained within a cavity formed to hold it at one end of the said metallic primer is exploded by the percussion and fires the gun. This vertical slider N, is shewn separately in Figure 15. Figure 16, is a section through the lock plate H, taken at the dotted line S, S, in Figure 12, shewing the vertical slider M, N, as having received one of the metallic primes, Q, and raised it to its station above. There is a mortice T, T, made in the inner edge of this vertical slider M, N, as shewn in Figure 16, to receive the point of a stop-screw U, and thus to limit its motion upwards and downwards. Figure 17, represents another section of the lock plate H, taken at the dotted line V, V, in Figure 12, and shewing the back end of the horizontal striking pin O, and its spring P, and also the lever I, and friction-roller L, and bridle J, the friction-roller being seen in contact with the projecting part M, of the slider N. Figure 18, is a separate view of the horizontal slider O, which has a solid ring W, formed upon it to act against the screwed socket X, shewn in Figure 14, which is screwed into the lock plate H, after the slider is inserted, and thus to prevent the slider from being withdrawn. The spring P, has a cap or notch made at its upper end, which receives the hook Y, within it, and thus preserves the slider in a proper position. Figure 19, is a metal tube R, or magazine, to contain the metallic primers and likewise a spiral spring; these are shewn in the section of the tube at Figure 14. The tube R, has a slit Z, Z, made along it part of its length, as shewn in Figure 19, and a notch or gap *a*, at one end of the said slit. Into this gap a stop screw *b*, which is screwed in a metal cylinder *c*, can be introduced, and which serves to confine the spiral spring from acting whilst the magazine is being filled with the metallic primers, and as shewn in Figure 14; when the magazine is filled the stop screw must be removed from the gap *a*, but not until the magazine is put into its place underneath the gun barrel, and as shewn by dotted lines in Figure 12. *d*, in Figure 20, is an external view of one of my metallic primings, and a section of which is shewn in Figure 21; *e*, in Figure 21, exhibits a small cavity, formed in one end of the primer, to contain the detonating prime. The magazine R, has its open end placed close to the vertical slide N, and when the latter is depressed or slidden down by the action of the thumb of the sportsman, upon the projecting part of it M, it will receive a primer into a semicircular gap formed upon the end of the slider, and which, upon suffering the slider to ascend, will be raised with it and lodged between it and another semicircular gap made in the lock plate H, and will be retained there by the action of the lever I, and its spring K, until it is driven rapidly and forcibly against the nipple G, by the action of the main spring of the gun lock, when, as aforesaid, the detonating prime will be exploded and the gun discharged. The metallic priming thus exploded will drop out through a side opening *f*, when the lock is at the half cock. Another priming will, of course, be received upon the upper end of the vertical slider N, every time it is depressed, until the whole contents of the magazine are expended. The metal cylinder *c*, will then project beyond the tube R, and prevent the vertical slider N, from rising. I do not mean or intend hereby to claim as my Invention any of the individual parts herein shewn and described, but only their combinations and applications to the above purposes, and particularly the application of conical detonating primers directly to the breeches of fire-arms, either in conical metal cups screwed into the breeches or formed in the breeches themselves, and also the arrangement for discharging fire-arms by means of small metallic detonating primers contained in a magazine, and supplied one at a time, as required.

In witness whereof, I, the said John Augustus Manton, have hereunto set my hand and seal, this Thirteenth day of September, One thousand eight hundred and thirty-four.

JOHN AUGUSTUS (L.S.) MANTON.

AND BE IT REMEMBERED, that on the Thirteenth day of September, in the Fifth year of the reign of His Majesty King William the Fourth, the said John Augustus Manton came before our said Lord the King, in His Chancery, and acknowledged the instrument aforesaid, and all and every thing therein contained and specified, in form above written. And also the instrument aforesaid was stamped according to the tenor of the Statute made in the fifty-fifth year of the reign of His late Majesty King George the Third.

Inrolled the Thirteenth day of September, One thousand eight hundred and thirty-four.

SOUTHGATE, Extra.

LONDON:
Printed by GEORGE EDWARD EYRE and WILLIAM SPOTTISWOODE,
Printers to the Queen's most Excellent Majesty. 1854.

A. D. 1834. March 13. Nº 6572.
MANTON'S SPECIFICATION.

(1 SHEET)

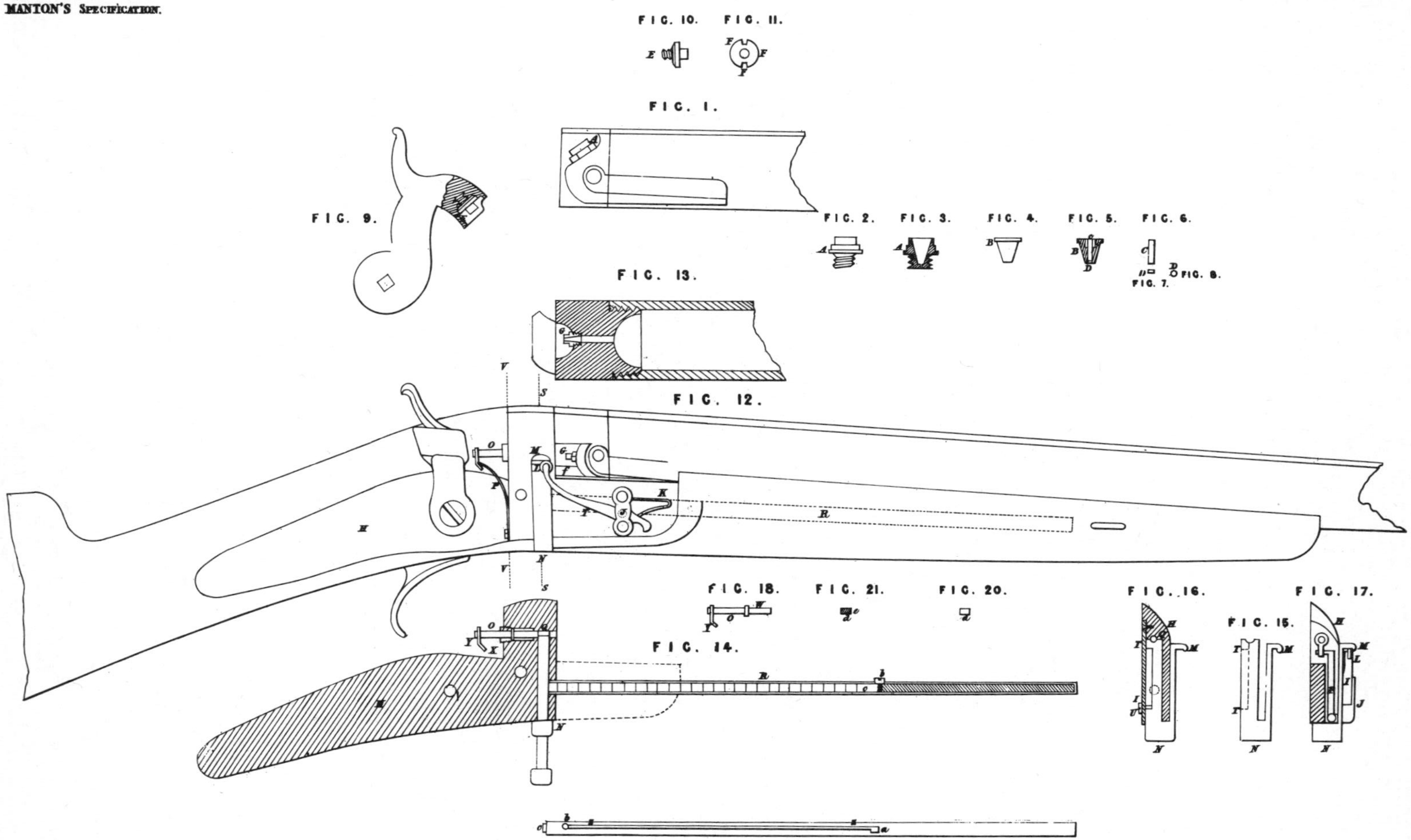

The enrolled drawing is partly colored.

Malby & Sons, Lith.

LONDON: Printed by GEORGE EDWARD EYRE and WILLIAM SPOTTISWOODE.
Printers to the Queen's most Excellent Majesty. 1854.

A.D. 1834 . . . . N° 6611.

Fire-arms.

HEURTELOUP'S SPECIFICATION.

TO ALL TO WHOM THESE PRESENTS SHALL COME, I, Charles Louis Stanislas, Baron Heurteloup, of Holles Street, Cavendish Square, in the County of Middlesex, send greeting.

WHEREAS His present most Excellent Majesty King William the Fourth, by His Letters Patent under the Great Seal of Great Britain, bearing date at Westminster, the Twenty-Second day of May, in the Fourth year of His reign, did, for himself, His heirs and successors, give and grant unto me the said Charles Louis Stanislas Baron Heurteloup, His especial license, that I, the said Charles Lewis Stanislas, Baron Heurteloup, my ex̃ors, adm̃ors, and assigns, or such others as I, the said Charles Louis Stanislas, Baron Heurteloup, my ex̃ors, adm̃ors, or assigns, should at any time agree with and no others, from time to time and at all times during the term of years therein expressed, should and lawfully might make, use, exercise, and vend, within England, Wales, and the Town of Berwick upon Tweed, my Invention of "Improvements in certain Parts of certain Descriptions of Fire-arms," in which said Letters Patent is contained a proviso, that I, the said Charles Louis Stanislas, Baron Heurteloup, shall cause a particular description of the nature of my said Invention, and in what manner the same is to be performed, to be inrolled in His said Majesty's High Court of Chancery, within six calender months next and immediately after the date of the said in part recited Letters Patent, as in and by the same, reference being hereunto had, will more fully and at large appear.

NOW KNOW YE, that in compliance with the said proviso, I, the said Charles Louis Stanislas, Baron Heurteloup, do hereby declare the nature of my said Invention to consist,—

First, in the adaptation or fitting up of the stock of a musket or other the like fire-arm to receive, and of the lock plate and hammer to bring forward to the touch hole, a small pipe or tube made of soft metal or other substance, which may be easily divided, containing the priming, which priming is detonating powder, and the end of which said pipe is cut off or otherwise divided or separated from the remainder, and the powder in the end so cut off or separated, detonated or exploded by the falling or blow of the said hammer when the fire-arm is discharged, in manner hereinafter more particularly explained.

And, secondly, in so arranging and forming the hut plug and breech of a double or single barrelled gun or other such fire-arm, that the barrel may be detached from the break off, and the hut released from its hold by the action of a lever, in such manner as to enable the cartridge to be put in at the breech end of the barrel, after which it is rammed tightly to its final position, by the mere act of replacing or re-adjusting the barrel to the breech by means of the same lever by which it was detached.

And in further compliance with the said proviso I, the said Charles Louis Stanislas, Baron Heurteloup, do hereby describe the manner in which my said Invention is to be performed by the following statement thereof, reference being had to the Drawings annexed, and to the Figures and letters marked thereon, that is to say:

Description of the Drawings.

In the Drawing marked A, Figure 1, is a side view of the but end of the stock of the lock, and of part of the barrel of an ordinary fowling piece. The but end of the stock is bored in a line from A, to B, at which latter point the cavity for the lock begins so as to receive what I call the priming tube, which is a long small tube or pipe made of soft metal, and filled with detonating powder; the part colored pink here represents the priming tube in the bore extending through its whole length into and along the cavity made, for the lock, till it reaches the guide nipple C, through which it passes, and it is thus directed to the touch hole D. It will be seen that just below the touch hole E, is a ledge or shelf at E, on which the end of the priming tube rests; and I will now proceed to describe, first the hammer itself, and then its action on the priming tube. First, then, as to the hammer itself, it is made forked at the end, so as to bestride the guide nipple C, when it falls. The dark blue line F, is a steel blade or knife extending over the fork the full width of the head of the hammer, and G, is a piece of steel which I call the striking piece, and is screwed on to the head of the hammer by the screw H, which keeps itself and the blade F, firmly to its place. When the hammer is let down by pulling the trigger, it describes in falling the arc shown by the dotted lines, the fork at the end bestrides the guide nipple while the steel blade F, cuts a small piece off the end of the priming tube, which piece remains on the ledge or shelf E, which shelf I call the anvil, where the striking piece G, comes in contact with it, and by the blow causes the powder in it to detonate, thus communicating fire to the touch hole, while the steel blade F, covers the hole at the end of the guide nipple, and thus, interposing itself between the end cut off of the priming tube and what remains in the guide nipple, effectually prevents the fire communicating to the remainder of the priming. In order more accurately to describe the construction of the head or upper part of the hammer, I have shown its several parts separately at Figures 2, 3, 4, 5, and 6. Figure 2, is a plan of the hammer head, shewing the end 1, forked so as to enable it to bestride the guide nipple; Figure 3, is a side view of the steel blade or knife which cuts the priming tube; Figure 4, is a plan of the same, shewing an open slot J, at its hinder end, for the purpose of adjustment on the hammer head; Figure 5, is a plan of the striking piece, the hole for the screw being oblong also for the purpose of adjustment, and Figure 6, is a side view of the same; and it may be as well here to state that similar letters and colors are used to denote similar parts in all the Figures of this Drawing. Now, having described that the proper portion of priming for each discharge is obtained by cutting off a piece from the end of the priming tube, it follows of necessity that some means must be provided of advancing a fresh portion of the priming tube under the knife after each discharge of the fire arm, and this I effect on the drawing back of the hammer to half cock, by means of a small contrivance connected with that movement, which I shall now proceed to describe:—Figure 7, is a view of the inside of the hammer or cock; K, is a small spring catch let into a recess cut for it, but projecting at the end K a little way out from it, as shown in Figure 8, which is a front view of the cock or hammer; Figure 9, is a plan; and Figure 10, a side view of the said spring K, shown separately. It is this spring K, projecting from the inside of the cock, which, when the cock is drawn back to half cock, acts upon the mechanism; which causes the priming tube to advance the required distance to bring it under the knife after each discharge, and which mechanism is clearly shown in Figure 11, which is a plan or top view of the lock; L, is the square, or, as in this case, the hexagon arm, on which the cock is fitted, the cock being removed here in order not to confuse the Figure. M, is a ratchet wheel of six teeth, having a milled boss beneath, as shown at Figure 12; this ratchet wheel turns on a centre N, firmly fixed to the solid piece O, of the lock plate, the milled boss being let into the solid piece; P, is a spring paul or catch to secure the ratchet wheel from moving, except in the direction of the arrow, and that only when required; Q, is another spring, with a friction roller R, at one end of it; the piece O, of the lock plate it will be seen is bored to receive the priming tube, which it will be observed passes through it, between the milled boss of the ratchet wheel and the friction roller R; the object of the spring Q, and roller R, being to press the soft metal priming tube against the milled boss of the ratchet wheel sufficiently to indent it slightly, and thus give the milled boss sufficient hold of the tube to force it forward when the ratchet wheel is turned round. In order to describe the action of this mechanism, the hammer must be supposed to be down after a discharge, when the spring catch K, which is on the inside of the cock, will take into one of the teeth of the ratchet wheel M. On drawing the cock back again the spring K, will force the ratchet wheel round one tooth or notch, and the priming tube will be advanced in proportion; by this time the hammer will have arrived at half cock, and the spring K, will have disengaged itself from the tooth of the ratchet wheel, so that the cock may afterwards be drawn back to full cock without any further interference with the priming mechanism; Figure 13, is a view of the interior of the lock showing the aperture at *r*, into the bore made for the priming tube, through which aperture the friction roller is admitted to press upon the priming tube; Figure 14, is an end view of Figure 13; Figure 15, is a separate view or plan of the spring Q, and friction roller R; Figure,16, is a side view of the same; Figure 17, is a separate view or plan of the spring paul or catch P, and Figure 18, is a side view of the same; Figure 19, is a side view of the breech, touch hole, anvil, and hut of the fire-arm, and Figure 20, is an end view of the same; Figure 21, is a side view of the break off of the fire-arm shewing the guide nipple, and Figure 22, is an end view of the same; Figure 23, is a separate and side view of the priming tube, and Figure 24, is an end view of the same; Figure 25, is a plan of that part of Figure 1, which is immediately over it.

Having now described the first part of my said Invention, which relates to the adaptation of the stock and lock to receive, and the application thereto of the priming tube aforesaid, I will proceed to the description of the second part of my said Invention, which is more particularly confined to improved means of loading at the breech, and which, for the more easy understanding of the Figures, I have described without the priming mechanism, though

I would wish it to be understood, that the said priming mechanism is equally capable of being adapted to fire-arms made on the principle of those I am now about to describe, as of those I have already described, but for the purpose of this part of the specification, I have only availed myself of so much of the former part as relates to the soft metal priming tube filled with detonating powder, which I describe as the priming used in each case.

The second part of my said Invention is particularly described by the Figures on the Drawing marked B. Figure 1, is a side view of the barrel of a double barrelled gun, which is the fire-arm I have selected in this case to illustrate my Invention. Instead of having a hut formed on the plug at the but end of the barrel, as is usual, I have two huts A, and B, formed on to the solid central piece between the two barrels; C, is a round stud or pin, the use of which will be explained hereafter; and D, is a mere steadying or guide pin, the position of which in the stock will be shown by other Figures; the hut B, it will be seen has an oblong hole E, in it, and it may be as well here to state, that I accomplish the loading at the breech by tilting up the but end of the barrel, and depressing the muzzle end, while the barrel turns on a fulcrum, introduced through the oblong hole E. Figure 2, is a but end view of Figure 1; F, F, being the nipples against which the hammer strikes to detonate or explode the priming, and G, G, are two grooves to admit a short piece of my aforesaid priming tube. Figure 3, is a side view, partly in section, of the barrels placed in the stock; H, is the fulcrum pin. The steadying pin D, is in a mortice cut for it, and it will be observed that the mortice and the two cavities cut for the huts A, and B, are all longer than the respective parts which are to occupy them, the reason of which is, that in order to tilt up the breech, for the purpose of loading the fire-arm, I am obliged to separate the barrels from the break off and plugs by advancing them along the stock, and the extra size of the mortice and cavities, before-mentioned, is to allow for that action, which is obtained by means of the mechanism colored green, and which will be hereafter explained. Figure 4, represents a side view of the barrels when in the position required for loading; it will be seen that the barrels have turned on the pin or fulcrum H, the guide or steadying pin D, has been raised partly out of its mortice, and the hut A, has been completely raised out of the cavity cut in the stock for its reception, and the barrels have been advanced along the stock, for it will be seen that if the but end, as shown in this Figure, were to be depressed, it would form the arc described by the dotted lines, and would not come in contact with the break off till acted upon by the mechanism colored green, by which alone the barrels can be drawn back to the break off when required, as will be hereafter explained. It will be seen that there is a projecting piece I, forming part of the break off, and it is scarcely necessary to observe that there is one to each barrel. These projecting pieces are, in fact, the plugs being formed in one piece with the break off, instead of being in one piece with the huts as usual. These plugs enter the barrels when the barrels are drawn back to the break off, and their object is two fold. First, to completely stop the breech end of the barrels, and secondly, to ram the cartridge tight; this latter object is generally attained by the ramrod, applied from the muzzle end of the gun, but as the chamber which holds the charge or cartridge is somewhat longer than the barrel, it stands to reason, if the cartridge J, is put into this chamber at the breech end, and is of the full length of the chamber, and is then drawn back against the plug I, the forcing of the plug into the chamber must ram the cartridge, the advance of which is resisted by the narrowing of the barrel at K. Figure 5, is a view of the barrels with their but ends brought down again upon the stock, but still separated from the break off, the huts A, and B, and the guide pin D, occupying the muzzle ends of their respective cavities. Before I proceed to describe the mechanism colored green, for advancing the barrels from and drawing them back to the break off, I may as well describe the sort of cartridge which is required for this improved fire-arm (that is to say):—The cartridge should be made with a cup of wood or lead at the end next the plug I, bored to receive a short piece of my said priming tube as shown by Figures 6, 7, 8, and 9. Figure 6, is a side view of a cartridge complete, with the fuse or priming tube. Figure 7, is a section of the cup and tube shown separately. Figure 8, is a front view of the cup and tube looking into the cavity of the cup and tube, and Figure 9, is a side view of the cup and tube.

I will now proceed to describe the mechanism colored green, for advancing and drawing back the barrels on the stock. It is a lever movement, and I have here shown the trigger guard used for that lever, though it may perhaps be thought preferable that the guard should be fixed as usual, and a separate lever applied for that purpose; as this, however, would be adding to the mechanism, I prefer to use the guard itself as the lever. Figure 10, is a plan of that part of the stock in which the breech ends of the barrels rest, and slide forward and backward. L, is the cavity for what I call the fulcrum hut B. M, is the mortice for the guide pin D, and N, is the cavity of the hinder hut A. P, is a plug shown separately at Figures 11, 12, 13, 14, 15, and 16; the top part of this plug is shown in plan in Figure 10, which I am now describing, and is an excentric, having a cavity Q, cut in it to receive the pin C. When the plug is in this position, the excentric projection, which is at *e*, enters a cavity cut for it in the hut A, as shown in Figures 1, and 3. One end of the guard R, is fastened to a square on the lower extremity of the plug P, and forms a lever, by which it is turned.

When the guard is pushed out from the stock, as shown by dotted lines at S, the plug P, turns and draws forward the barrels by means of the pin C, being in the cavity Q. When the guard arrives at right angles with the stock, as shown by red lines at T, the cavity Q, has arrived at its greatest point of action, and the barrels will be sufficiently advanced from the break off to allow the muzzles being depressed, as shown in Figure 4, (that is to say) they will be in the position shown at Figure 5. When the muzzles have been depressed, the cartridges put in, and the barrels again brought to the position shown in Figure 5, and the guard gradually moved back into its proper place, the excentric boss at *e*, acts upon the hut A, and drives the barrels back until the cavity Q, again begins to act upon the pin C, and draws the barrels close up to the break off, as before and as shown in Figure 3. The metal piece containing the cavities and mortice L, M, N, may either be made in one piece with the break-off, as here shown, or separately. In order to give a more perfect idea of the plug P, I have given separate views of its various parts. Figures 11, and 12, are side elevations, and Figure 13, is a plan of the said plug. Figure 14, is a side view, and Figure 15, a plan of a screw nut, which keeps the plug from rising in the hole bored for it through the stock, and Figure 16, is a side elevation of the plug and nut and the end of the guard, all screwed up to their respective places. Figure 17, is a plan, and Figure 18, a transverse section of a fulcrum pin, which passes through the oblong hole in the fulcrum hut B, and on which the barrels turn. Figure 19, is a plan of the breech end of the barrels, shewing the nipples and part of the break off. Now, whereas I claim as my Invention, First, The improvements in that part of a fire-arm commonly called the lock, hereinbefore described, whereby I am enabled to present a succession of priming in a tube or case to the touch hole, and cut or break off or otherwise separate as much as is necessary for each discharge from the remainder of the same, and explode the part so separated, for the purpose of the discharge, on an anvil or shelf contrived for the purpose, without igniting any other portion of it, and in the Drawing annexed I have described a manner of accomplishing all these objects without any other action being required of the operator, but the ordinary one of bringing the cock, after it has been let down for the discharge, back to half cock.

And secondly, The Improvements or new arrangements of those parts of a musket, pistol, carbine, fowling piece, or other the like fire-arm, commonly called the break off, hut plug breech, and barrel, hereinbefore described, whereby I am enabled to advance the barrel from the plug, tilt it up from the stock, load at the breech, bring the barrel down upon the stock again, and ram the cartridge, by bringing the barrel back again to the plug with the aid only of a single lever, which may be the trigger guard itself, if required, as hereinbefore shown.

And such my Invention being, to the best of my knowledge and belief, entirely new and never before used within that part of His said Majesty's United Kingdom of Great Britain and Ireland called England, His said dominion of Wales, or town of Berwick upon Tweed, I do hereby declare this to be my Specification of the same, and that I do verily believe this my said Specification doth comply in all respects, fully and without reserve or disguise, with the proviso in the said hereinbefore in part recited Letters Patent contained, wherefore I do hereby claim to maintain exclusive right and privilege to my said Invention.

In witness whereof, I, the the said Charles Louis Stanislas, Baron Heurteloup, have hereunto set my hand and seal this Twenty-first day of November, in the year of our Lord, One thousand eight hundred and thirty-four.

CHARLES LOUIS STANISLAS, (L.S.) BON. HEURTELOUP.

FARBER.

AND BE IT REMEMBERED, that on the Twenty-second day of November, in the year of our Lord 1834, the aforesaid Charles Louis Stanislas, Baron Heurteloup came before our said Lord the King in his Chancery, and acknowledged the specification aforesaid, and all and every thing therein contained and specified in form above written. And also the specification aforesaid was stamped, according to the tenor of the statute made for that purpose.

Inrolled the Twenty-second day of November in the year of our Lord, One thousand eight hundred and thirty-four.

LONDON:
Printed by GEORGE EDWARD EYRE and WILLIAM SPOTTISWOODE, Printers to the Queen's most Excellent Majesty. 1854.

A.D. 1834. May 22. No. 6611.

HEURTELOUP'S Specification.

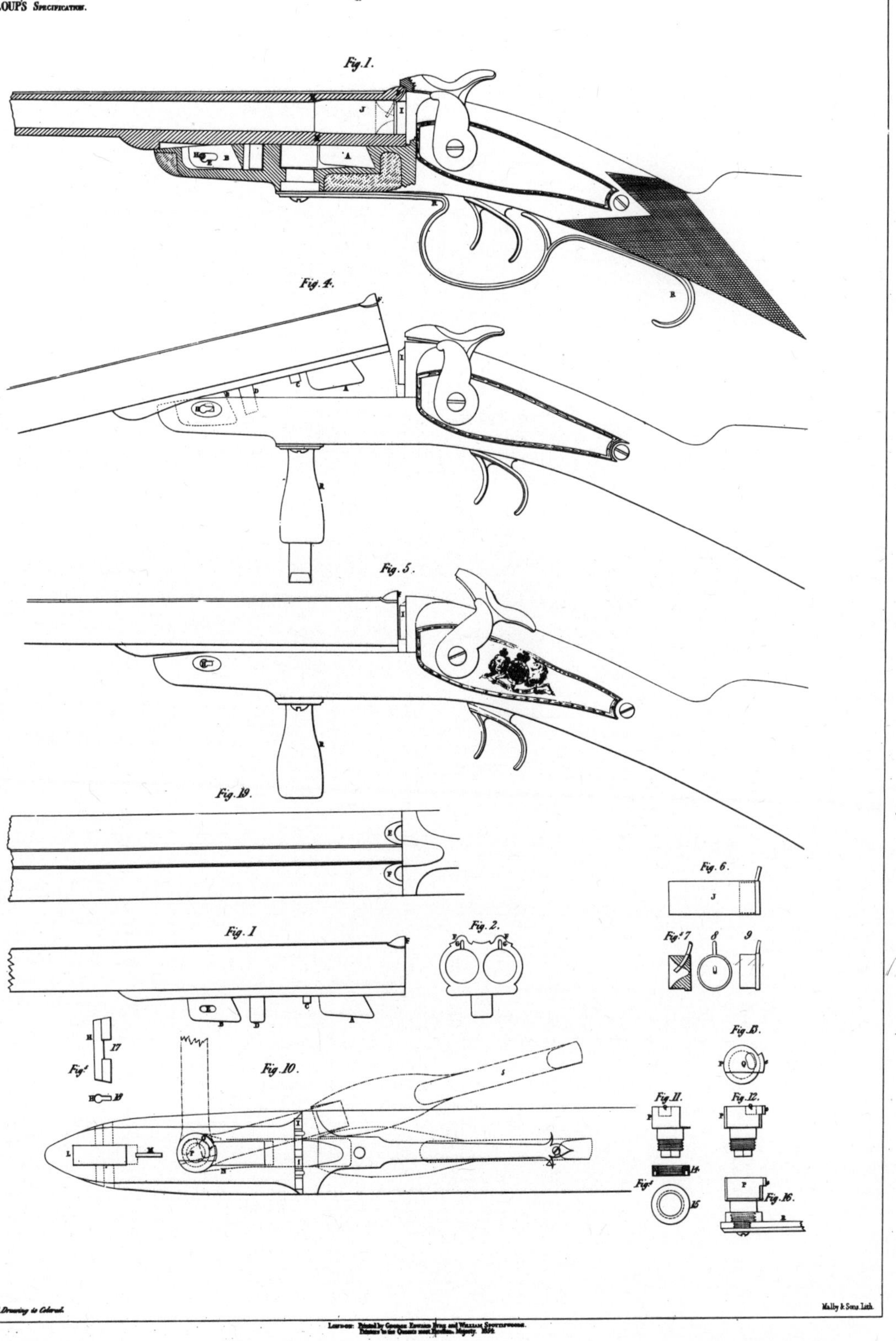

The Enrolled Drawing is Colored.

Malby & Sons, Lith.

London: Printed by George Edward Eyre and William Spottiswoode, Printers to the Queen's most Excellent Majesty. [illegible]

A.D. 1834 . . . . N° 6675.

Fire-arms and Projectiles,

SHRAPNEL'S SPECIFICATION.

TO ALL TO WHOM THESE PRESENTS SHALL COME, I, Henry Shrapnel, of Salisbury, in the County of Wilts, Major General in the Army and Colonel in the Royal Artillery, send greeting.

WHEREAS His present most Excellent Majesty King William the Fourth, by His Letters Patent under the Great Seal of Great Britain, bearing date at Westminster, the Sixth day of September, in the fifth year of His reign, did, for Himself, His heirs and successors, give and grant unto me, the said Henry Shrapnel, His especial license, full power, sole privilege and authority, that I, the said Henry Shrapnel, my exors, admors, and assigns, or such others as I, the said Henry Shrapnel, my exors, admors, or assigns, should at any time agree with, and no others, from time to time and at all times during the term of years therein expressed, should and lawfully might make, use, exercise, and vend, within England, Wales, and the Town of Berwick upon Tweed, my Invention of "IMPROVEMENTS IN FIRE-ARMS OF VARIOUS DESCRIPTIONS, AND IN AMMUNITION FOR THE PURPOSES OF FIRE-ARMS;" in which said Letters Patent is contained a proviso, that I, the said Henry Shrapnel, shall cause a particular description of the nature of my said Invention, and in what manner the same is to be performed, to be inrolled in His said Majesty's High Court of Chancery within six calendar months next and immediately after the date of the said in part recited Letters Patent, as in and by the same, reference being thereunto had, will more fully and at large appear.

NOW KNOW YE, that in compliance with the said proviso, I, the said Henry Shrapnel, do hereby declare the nature of my Invention, and the manner in which the same is to be performed, are fully described and ascertained in and by the following description thereof, reference being had to the Drawings hereunto annexed, and to the figures and letters marked thereon (that is to say):

Figure 1, represents the first part of my Invention, which consists in certain sights or points, whereby a musket may be pointed with more certainty. It is well known that in discharging muskets, particularly where the object fired at is at a distance, that it requires the barrels of the musket to be elevated considerably above the objects fired at, and to obtain a knowledge of the quantity of elevation requires considerable practice, which is seldom sufficiently pursued by the privates of a regiment; at the same time the commanding officer has no positive means of so directing a number of men to give the proper elevation as to ensure an effectual fire. *x*, in Figure 1, is the fixed sight, and *o*, *o*, a moveable sight, which is caused to move up and down by the screw *Z*. It will therefore be evident that if from practice the plate at *o*, be graduated in such manner that it be found that by pointing the gun in such manner that a line drawn from the moveable sight when affixed at any point of the graduated scale will cross the fixed sight *x*, and also the object fired at, and that the elevation of the barrel will be such as to hit the object or mark at a certain distance, then the officer in command, judging the distance of an enemy, will be able to direct the men to set the moveable sight at a particular graduation, the men will then take aim in such manner as to see the enemy, in a line with the moveable sight *o*, and the fixed sight *x*, the barrel being consequently at a proper angle; and it is the application of such moveable sight *o*, in conjunction with the sight *x*, which constitutes the first point of my Invention.

Figures 2, 3, 4, and 5, shew the second point of my Invention, and consists of the application of the prop *a*. Figure 2, shows the prop as used in a horizontal position; Figure 3, shows the prop in a vertical position. This prop turns on a hinge at one end, and has a spring catch at the other, by means of which the person using such gun or fowling piece may either use it in a horizontal or vertical position, the object being to keep the hand which supports the piece as far from the lock and barrel as possible, to avoid accident in case of bursting of the barrel or blowing off the lock. Figures 4, 5, merely show variations from the horizontal prop, Figure 2, those at Figures 4 and 5, not being moveable, but they may be hollow for the sake of lightness.

Figure 6, shows the third point of my Invention, and consists of a moveable hinge or joint *a*, having a rack or screw plate *s*, which is to be operated on by the screw Z, by which any required degree of bend may be obtained to a wooden stock, and thus a gun may readily be made to suit a variety of persons.

Figure 7, shows the fourth point of my Invention, and consists in contracting the bore of the barrel just above that portion thereof as receives the powder, as is indicated from *a*, to *b*. It will therefore be evident that the ball or shot cannot approach nearer to the powder than the distance from *a*, to *b*, and this space will be occupied by air; by this means it will be found that a much greater degree of effect will be produced by guns so constructed.

The fifth point of my Invention consists in an improvement of that part of a lock called the steel or hammer, that is to say, the part against which the flint strikes; and the improvements consists in making the same with projecting sides *a*, Figure 8, which stand at right angles to the face against which the flint strikes; by this means the sparks will not be allowed to spread laterally, but will be guided downwards into the pan and thus more completely ensure the discharge of the piece.

Figures 9, and 10, show the sixth point of my Invention, and consists of an improvement of locks for discharging by detonating balls; *a*, being a hollow tube or magazine, with a cover at the end. This magazine is to contain the detonating balls. *b*, is a sliding plate, which moves between two plates *c*, *c*. The plate *b*, is connected to the cock by a hinge joint, and consequently moves with it, as will be evident on inspecting the Drawing; and it will be seen that the only difference between Figures 9, and 10, is, that one of the locks is made with a straight sliding plate *b*, and the other is made with a circular plate *b*. In cocking either of these locks the sliding plate will be drawn over the recess near the touch hole, and there being a hole in the sliding plate *b*, which coincides with the opening at the end of the magazine, such hole will, when the cock is down, receive one of the detonating balls, and in cocking the lock such ball will be drawn forward and drop through the hole in the plate *b*, and fall into a recess near the touch hole. On pulling the trigger the plate *b*, will be drawn back, and allow of the blow being given to the detonating ball.

The seventh point of my Invention consists in the application of a similar prop to pistols as those shown applied to fowling pieces and other fire-arms; but in this instance the object of such prop is, that the pistol may be held by the hand grasping the prop and thereby enabling the person to ward off the cut of a sword. See Figure 11.

The eighth point of my Invention consists in constructing the barrels of rifle guns with parallel grooves, see Figures 12, and 13, by which means more correct fireing will be obtained; and Figure 12, also shows the ninth point of my Invention, which consists in having two or more touch holes, in order more effectually to insure the ignition of the charge.

The tenth point of my Invention consists in constructing hollow shot to be filled with small shot, as shewn in Figure 13, the interstices of the small shot being filled up with tin or other metal which is equally brittle, and the outer shell of the hollow shot should be so cast as to give way immediately the shot comes in contact with any object; the small shot will then be distributed, and do considerably more execution than when fireing with solid shot. This part of the Invention consisting in filling up the interstices with tin or other metal, as above described.

And, lastly, I would remark, that in fireing balls with muskets I prefer that they should be formed of a compound of tin and lead or iron; and when iron is used it is desirable to have them washed with tin in order to prevent oxydation.

In witness whereof, I, the said Henry Shrapnel, have hereunto set my hand and seal, this Sixth day of March, in the year of our Lord One thousand eight hundred and thirty-five.

HENRY (L.S.) SHRAPNEL.

A. D. 1834. SEPTEMBER 6. Nº 6675.
SHRAPNEL'S SPECIFICATION.

(1 SHEET)

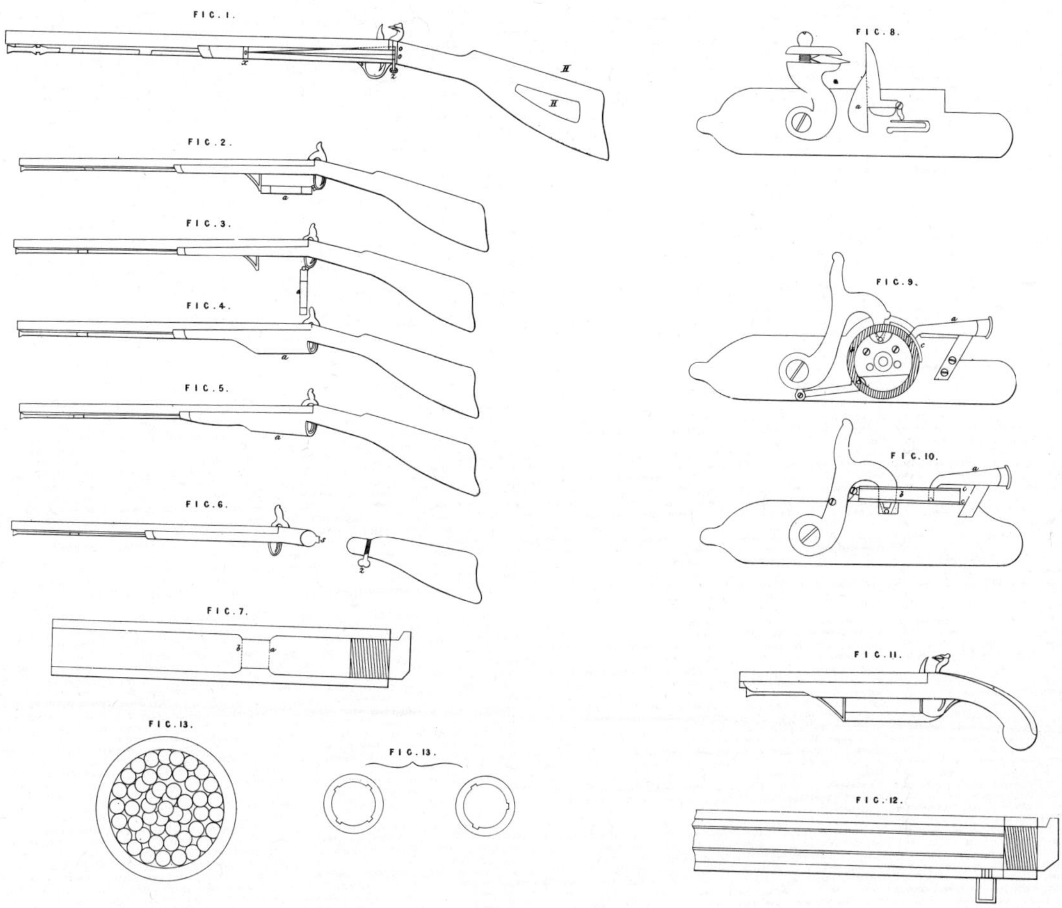

The Enrolled Drawing is partly Colored.

Malby & Sons lith.

LONDON: Printed by GEORGE EDWARD EYRE and WILLIAM SPOTTISWOODE, Printers to the Queen's most Excellent Majesty. 1854.

A.D. 1835 . . . . N° 6755.

Machinery for Cutting out and Shaping Gun Stocks, &c.

DODDS' SPECIFICATION.

TO ALL TO WHOM THESE PRESENTS SHALL COME, I, Isaac Dodds, of Horsley Iron Works, in the Parish of Tipton and County of Stafford, Engineer, send greeting.

WHEREAS His present most Excellent Majesty King William the Fourth, by His Royal Letters Patent, bearing date at Westminster, the Twenty-ninth day of January, in the fifth year of His reign, did, for Himself, His heirs and successors, give and grant to me, the said Isaac Dodds, His special licence, full power, sole privilege and authority, that I, the said Isaac Dodds, my executors, administrators, and assigns, or such others as I, the said Isaac Dodds, might at any time agree with, and no others, from time to time and at all times during the term of years therein expressed, should and lawfully might make, use, exercise, and vend, within England and Wales, and the Towns of Berwick-upon-Tweed, my Invention of "**Improvements in Machinery for Cutting and Shaping Wood and other Materials, Part or Parts of which Machinery are applicable to other Useful Purposes**; in which said Letters Patent is contained a proviso, obliging me, the said Isaac

Dodds, by an instrument in writing under my hand and seal, particularly to discribe and ascertain the nature of my said Invention, and in what manner the same is to be performed, and to cause the same to be inrolled in His Majesty's High Court of Chancery within six calendar months next and immediately after the date of the said in part recited Letters Patent, as in and by the same, reference thereunto had, will more fully and at large appear.

NOW KNOW YE, that in compliance with the said proviso, I, the said Isaac Dodds, do hereby declare that the nature of my said Invention, and the manner in which the same is to be carried into effect, will be fully understood by the following discription thereof, reference being had to the Drawings hereto appended (that is to say):

My improvements in machinery for cutting and shaping wood and other materials, part or parts of which machinery are applicable to other useful purposes, consists in certain mechanical arrangements designed for the purposes of grinding saws or other cutting instruments in machinery employed for cutting and shaping wood or other materials into various forms, or in shifting or sliding the bed or carriage upon which the wood or other article to be cut is fixed, in order that such saws or other cutting instruments may be so brought into operation upon the wood or other material intended to be cut, or that the wood or other material may be so brought under the operation of the saws or other cutting instruments, as to cause the described forms to be given to the material by the action of the saws or other cutting instruments when put into operation.

In order to render these improvements perfectly evident, I have, in the accompanying Drawings, represented my novel arrangements of machinery in several Figures and under several modifications, suited to different purposes; that is, to the cutting of different shapes in wood and other materials. Fig. 1, is a side elevation of a machine for cutting irregular forms out of planks, boards, or blocks of wood or other material; Fig. 2, is a plan or horizontal view of the same; and Fig. 3, is a vertrical section taken transversely through the machine, nearly in the plane of the saws; the respective letters of reference pointing out the same parts in these three Figures.

This machine is designed to cut out and produce the first rough forms in wood for gun stocks, staves for casks, curved pieces for the felloes of wheels, backs, arms, and other erregular shaped parts of chairs and sofas, ships timbers of various descriptions, and for a great variety of other erregularly shaped pieces required in carpentering, cabinet making, coach and boat building, pattern making, and many other purposes. A, A, denotes the side standards of iron which support the machinery, mounted upon foundations of stone, iron, timber, or other substantial base. B, B, are the side bars, railways, or ribs of the framing on which the carriage C moves longitudinally. This framing B, supporting carriage C, is mounted upon transverse bars, railways, or ribs D, D, which allow of its moving laterally, the rails D being fixed upon pillars or walls resting upon the foundation. E, E, E, represent three pieces of timber intended to be cut into erregularly curved shapes. These pieces are fixed upon the carriage C, by clamps or other fastnings, and are supported by transverse blocks or bearers beneath, resting upon the rails of the carriage. These three pieces of timber are so fixed upon the carriage as to correspond with the the situations of three vertical saws F, F, F, mounted in a vibrating frame G, G, which saws are intended to act simultainously in cutting the three pieces of timber into similar curved shapes by a parollel action. The vibrating saw frame G, slides up and down vertically upon perpendicular guide rods H, H, fixed to the side standards, and is actuated by means of a rod I, connected to it and to a crank K below. The crank shaft L, which is the main driving shaft of the machine, is driven by a pully and band leading from any first mover. On this crank shaft there is an excentric M, carrying a click O; this click falls into the teeth of ratchet wheel P, fixed on the end of the axle or shaft Q. Aong pinion R on the shaft Q takes into the teeth of a rack in the under side of the horrizontal bar S, which bar is affixed to the carriage C; hence it will be perceived that by giving rotary motion to the ratchet P, and its axle Q, the carriage C, with the planks, boards, or blocks of wood or other material placed thereon, will be made to advance slowly against the edges of the saws. On the side of the carriage is to be affixed the pattern or model T, which shall direct the cutting or shape to be given to the planks, boards, or blocks by the action of the saws. The manner of effecting this will be best explained by the horrizontal view of the machine, Fig. 2. A friction roller or any other smooth surface intended to operate as a wiper is mounted in one of the side standards of the machine at V, and two weighted levers W, W, are mounted on the opposite side, carrying anti-friction rollers X, X, which press against the side of the frame B, B, for the purpose of forcing up the frame B, and the carriage C, so that the model T shall, as the carriage advances, continually bear against the wiper V.

Having now explained the construction of the several parts of this machine a very slight description of its mode of operating will be sufficient to render the improvement evident. The rotary motion of the crank shaft L driven by the band and rigger causes the vibrating frame G, G, carrying the saws F, F, F, to move rapidly up and down upon its perpendicular guide rods H, H; at the same time the rotation of the excentric M, gives such reciprocating movements to the rod N, and clik O, as to cause the ratchet wheel P to be gradually driven one tooth every rotation of the crank. This rotary movement of the ratchet wheel P, with its axle Q and pinion R, gradually advances the carriage C, upon the longitudinal railways B, B, bringing forward the planks, boards, or blocks E, E, E, against the cutting edges of the vibrating saws F, F, F. This simple progressive movement of the carriage would cause the saws to slit the planks, boards, or blocks, in longitudinal right lines; it is, however, to be remembered that the frame B, B, slides upon the transverse railway D, D, and that as the carriage C advances its lateral position is determined by the inequalities on the edge of the model T, which is pressed against the wiper V, by the weighted levers W., W; it follows, therefore, that as the carriage C, with the planks, boards, or blocks advances, it will, with the frame D, D, be made to slide laterally to and fro, in obedience to the form of the guide or model T, and consequently that the saws as they vibrate will cut the material under operation is certain curved lines in perfectly parallel coincidence with the fixed pattern or model T.

It remains now to explain the manner in which the blades of the saws are at all times made to assume positions coincident with the direction of the cutting lines, that is, parallel to the face or figure of the model. The clips or hilts *a*, *a*, into which the ends of the saw blades F, F, are inserted, are attached to the top and bottom rails of the vibrating frame G, G, by means of pivets or centers, as shewn in the detached view of one of the saws at Fig. 4, which mode of attachment allows the blades of the saws to turn horrizontally, for the purposes of assuming positions which shall at all times be parallel with the operating part of the surface of the model T. In order to preserve this parallelism of the saw blades a small tail lever *b*, *b*, extends from the hilts of each saw, and these tail levers are all connected by a rod *c* to one end of a parallel lever *d* turning on pivets fixed in the rails of the saw frame. The reverse end of these parallel levers are connected to arms *e*, *e*, extending from a sliding piece *f*, inserted into a long slot in the side frame of the carriage. It will now be perceived, by thus connecting the saws to the sliding piece *f*, that as the carriage C is made to slide to and fro laterally by the erregular surface of the model T, the saw blades will be turned to different angles, at all times preserving their correct cutting positions and perfect parallelism with the model.

Having explained the construction and operation of the several parts of my improved machine for cutting of different shapes in wood and other materials into various forms for useful purposes, I desire it to be understood that I do not intend to claim as new all the parts of the said machine, as many of those parts are common to other sawing machinery; neither do I intend to confine myself to the particular arrangement shewn in the foregoing Figures; but I claim as the particular feature of novelty invented by me in connecting with a sawing machine for cutting and shaping wood and other materials, first, the adaptation of the transverse railways or ribs D, D, upon which the frame with the material to be cut may be slidden laterally; secondly, the longitudinal railways or ribs B, B, upon which the carriage for conducting the articles to be cut is gradually advanced to the saws; thirdly, the attachment of a model or pattern T, for the purpose of determining the forms of the pieces to be cut by the action of the saws; and, fourthly, the manner of mounting the saw blades so as to enable them to turn into the correct cutting lines.

My machine for cutting grooves, mortices, tenans, and which may be employed for plaining small surfaces, is shewn in the Drawings at Fig. 5, a side elivation, and Fig. 6, a plan or horizontal view. A, is rigger affixed to a rotary shaft, carrying a crank B; to this crank is connected a rod C, which at the reverse end is attached to the sliding carriage E, holding the cutter. D, D, D, D, are four cranks or levers, on the ends of which the frame H is mounted; this frame H is formed with grooves or ribs within, for the carriage E to move to and fro upon. G, is the frame supporting the machinery, which rests upon the foundation. I, is a vertical table to which the article intended to be cut must be made fast. First this table is intended to slide in two directions, for the purpose of adjustment, to suit the dimensions of the hole intended to be cut, which adjustable movement is affected by screws. The table I, may be brought to any desired angle, by means of a quadrant rack, as shewn by dots, for the purpose of cutting the mortices or groves at any desired angle of obliquity. The rotary action of the shaft and crank B, gives the reciprocating movement to the cutter fixed in the carriage E, by sliding the carriage to and fro on the groves or ribs of the frame H. In order to increase the depth of the grove or recess to be cut in the article placed upon the table I, the carriage H, mounted on the cranks or levers D, D, D, D, must be lowered. This is done by a click connected to an excentric on the crank shaft B, the end of which click takes into a ratchet wheel L. The axle of this ratchet wheel carries two pinions M, M, which respectively take into two segment racks or toothed wheels F, F, on the shalf or axle of two of the cranks or parallel levers D, D. It will now be perceived that by the gradual rotary movement of the ratchet wheel L, the frame H upon the cranks or parallel levers D, D, will be gradually lowered to such an extent as shall cause the sliding cutter to cut the mortice or groove to any required depth. K, is a balance weight attached to an arm extending from one of the parallel levers D, by means of which the frame and

Parallel motion for Morticing or Bevilling, applied with the Machine.

FIG. 7.

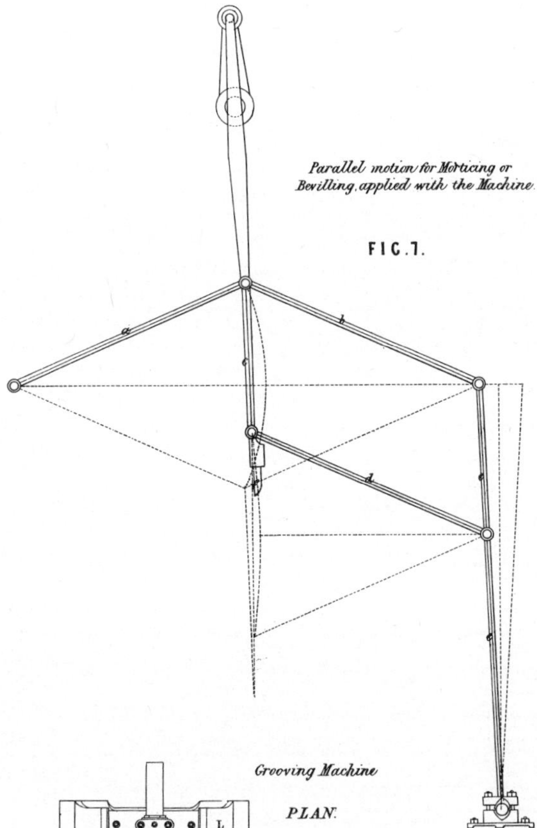

Grooving Machine

PLAN.

FIG. 6.

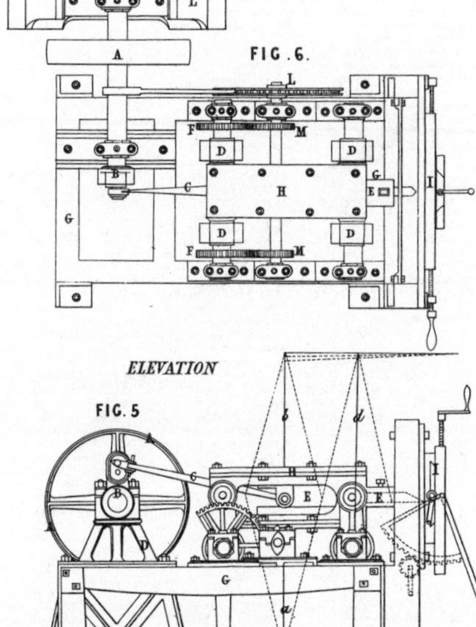

ELEVATION

FIG. 5

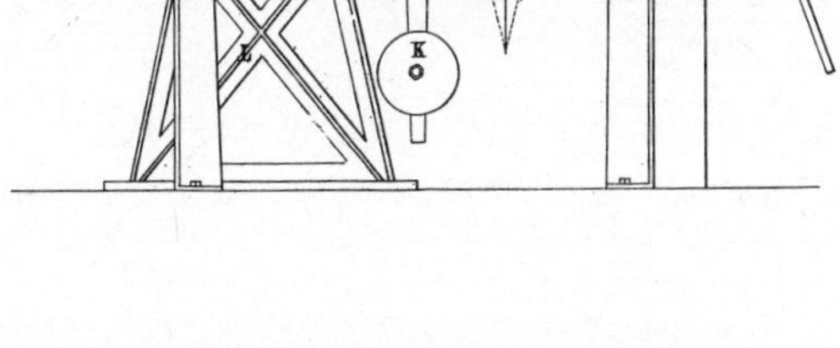

The Enrolled drawing is Colored.

Malby & Sons, Lith.

London: Printed by George Edward Eyre and William Spottiswoode Printers to the Queen's most Excellent Majesty. 1854.

carriage is balanced in whatever position it is placed. I sometimes apply, in place of the guide frame H, and parallel levers D, D, a parallel motion, shewn detached at Fig. 7, which is peculiarly adapted to the machine last described for cutting at right angles to the face plate or table. This parallel motion carries the cutting point at all times perfectly parallel to itself, by means of the peculiar arrangements and exact proportions of the levers *a*, *b*, *c*, *d*, *e*, shewn in Fig. 7, and also by outlines in the situation in which it would be applied at Fig. 5. Instead of the chissel-pointed cutter last described, I sometimes insert into the socket of the carriage E, Fig. 5, or into the socket *f*, Fig. 7, a key-hole saw for cutting wood, and then give to the chuck placed in front of the table on which the wood to be cut is fixed, such a rotary or reciprocating motion, or any other erregular movement by hand or by machinery, as shall enable me to cut out shapes or forms of various descriptions. The paralell motion last described may be applied to other useful purposes, for instance, to steam engines commonly denominated "vibrating pillar engines," as it is capable of communicating an extra paralell motion, by attaching the bars, links, or levers *c* and *d* to engines of that description for the purpose of working the rod of an air pump or other pump attached at *f*. Fig. 8, represents an elivation, an apparatus for cutting the eyes of spade, shovel, or screw handles. Fig. 9, is a plan or horrizontal view of the same; *a*, is the base or bench; *b*, a standard supporting an axle driven by a rigger or pulley; at the end of this axle a crank *c* is fixed, to which is attached a rod *d*, connecting the crank to a sliding bar *e*, *e*. This bar moves in slots in standards *f*, *f*, (one of which is shewn in the front view, Fig. 10.) To this sliding bar *e*, the cross heads *g*, *g*, of a saw *h*, are affixed; these cross heads are also connected to the square guide rods *i*, *i*, which slide through square apertures in the pullies *k*, *k*, turning on hollow bushes in the standards *f*, *f*. A block or poppet head *l*, is affixed on the bench *a*, and a large bush passes through it, one end of the bush carrying a pulley *m*, and the other end receiving the chuck plate *n*, on which the chuck, with the piece of wood *o* to be cut, is fixed in the ordinary way. The saw *p* is attached to the two guide rods *i*, *i*, and is streched by right and left handed screws *q*. The saw *p* having been passed through a hole previously perforated in the shovel head or piece *o* intended to be cut, rotary motion is given to the crank, which causes the saw to move reciprocally in a horizontal direction. The piece to be cut may, during the action of this saw, be turned round by hand, so as to bring the saw to act upon it in any desired way, or it may be driven round by giving to the pullies slow motions by racks, screws, or any other convenient means which are well understood by mechanics.

In witness whereof, I, the said Isaac Dodds, have hereunto set my hand and seal, this Twenty-eighth day of July, in the year of our Lord One thousand eight hundred and thirty-five.

ISAAC (L.S.) DODDS.

Bond, Extra.

AND BE IT REMEMBERED, that on the Twenty-eighth day of July, in the sixth year of the reign of His Majesty King William the Fourth, the said Isaac Dodds came before our said Lord the King, in His Chancery, and acknowledged the instrument aforesaid, and all and every thing therein contained and specified, in form above written. And also the Instrument aforesaid was stamped according to the tenor of the Statute made in the fifty-fifth year of the reign of His late Majesty King George the Third.

Inrolled the Twenty-ninth day of July, One thousand eight hundred and thirty-five.

LONDON:
Printed by GEORGE EDWARD EYRE and WILLIAM SPOTTISWOODE, Printers to the Queen's most Excellent Majesty. 1854.

A.D. 1835 . . . . N° 6829.

Priming Magazine, &c. for Fire-arms.

EGG'S SPECIFICATION.

TO ALL TO WHOM THESE PRESENTS SHALL COME, I, Joseph Egg, of Piccadilly, in the County of Middlesex, Gun Maker, send greeting.

WHEREAS His present most Excellent Majesty King William the Fourth, by His Letters Patent under the Great Seal of Great Britain, bearing date at Westminster, the Ninth day of May, in the fifth year of His reign, did, for Himself, His heirs and successors, give and grant unto me, the said Joseph Egg, His especial licence, full power, sole privilege and authority, that I, the said Joseph Egg, my ex̃ors, adm̃ors, and assigns, or such others as I, the said Joseph Egg, my ex̃ors, adm̃ors, or assigns, should at any time agree with, and no others, from time to time and at all times during the term of years therein expressed, should and lawfully might make, use, exercise, and vend, within England, Wales, and the Town of Berwick-upon-Tweed, my Invention of "Improvements on certain Description of Fire-arms;' in which said Letters Patent is contained a proviso, that I, the said Joseph Egg, shall cause a particular description of the nature of my said Invention, and in what manner the same is to be performed, to be inrolled in His said Majesty's High Court of Chancery within six calendar months next and immediately after the date of the said in part recited Letters Patent, as in and by the same, reference being thereunto had, will more fully and at large appear.

NOW KNOW YE, that in compliance with the said proviso, I, the said Joseph Egg, do hereby declare that the nature of my said Invention, and in what manner the same is to be performed, is particularly described and ascertained in and by the following description thereof, reference being had to the Drawing hereunto annexed, and to the figures and letters marked thereon, that is to say:

My improvements on fire-arms consist in applying peculiarly constructed priming magazines to be attached on the barrel of fowling pieces, pistols, and such like fire-arms; and in the mode of conducting the detonating tube or primer from the magazines to the pan, cup, or touch-hole, for the purpose of priming the piece; and also in a mode of cutting off the communication between the touch hole and priming chamber or magazine at the time the piece is discharged.

Description of the Drawing.

Figure 1, is a view of the lock and breech part of a gun upon this improved principle, the parts standing as at half cock. Fig. 2, a view of the same after the piece is discharged or the cock at bearing. Fig 3, the inside view of the lock detached, the parts in the position of Fig. 1. Fig. 4, is also a view of the inside of the lock detached, the parts in the position of Fig. 2. Fig. 5, an horizontal view of a double barrel, showing the recess between the barrels before the magazine is attached. Fig. 7, the magazine detached, and which consists of two tubes, when for a double barrel of any length preferred, fastened to a top piece, and which for this construction of lock must be filled with short tubes containing detonating composition, and then deposited in the recess *a*, and there secured by a screw or other suitable contrivance. On raising the muzzle of the piece the priming descends to the broad part of the chamber *b*, and there passes through a small passage or passages *c, c*, in a lateral direction. The outer part of this passage is closed by a slide or conductor, shewn at Figures

3 and 4; but when the cock is raised, as at Fig. 1, an aperture *z*, comes opposite this passage *c*, and permits the priming to descend through it to the slide or conductor, and which empty's itself into the pan or touch-hole *e*. The head of the slide or conductor has a small protuberance, as shewn at Figures 1 and 8, which serves as a cap or cover to the pan or touch-hole, and which passes the magazine as shewn at Fig. 1; and the pulling the trigger, the tumbler brings down the lever *f*, which, by its connection with the arm of the slide or conductor *d*, seen at Figures 3 and 4, covers the communication, as at Figure 3, to the position of Fig. 4, and by that means cuts off the communication between the priming chamber and the cup or touch-hole.

It will be evident from the above description, that a similar priming magazine may be applied to fire-arms intended to be primed with gunpowder, but in that case it will be necessary to enlarge the passage leading from the chamber to the pan or touch-hole, and a different constructed lock must be employed, with a flint cock and hammer. I do not, however, claim any novelty in the lock or any other part of the piece, except in those parts which are marked with letters upon the Drawing referred to as above. In short, my improvements in the construction of guns and fire-arms upon the self-priming and detonating principle are confined to the construction of a priming chamber or magazine attached to the barrel or barrels of fire-arms as above described, though the same may be varied to suit other forms, place, or position contiguous to the breech, false breech, or barrel, and in the necessary shield, slide, or guide or conductor, which also may be varied in shape or position for the purpose of admitting the priming to the touch-hole when the piece is loaded, and of cutting off the communication between the chamber and the touch-hole when the piece is discharging.

In witness whereof, I, the said Joseph Egg, have hereunto set my hand and seal, this Ninth day of November, in the year of our Lord One thousand eight hundred and thirty-five.

JOSEPH (L.S.) EGG.

A.D. 1835. May 9. No. 6829.

EGG'S Specification.

(1 SHEET.)

FIG. 1.

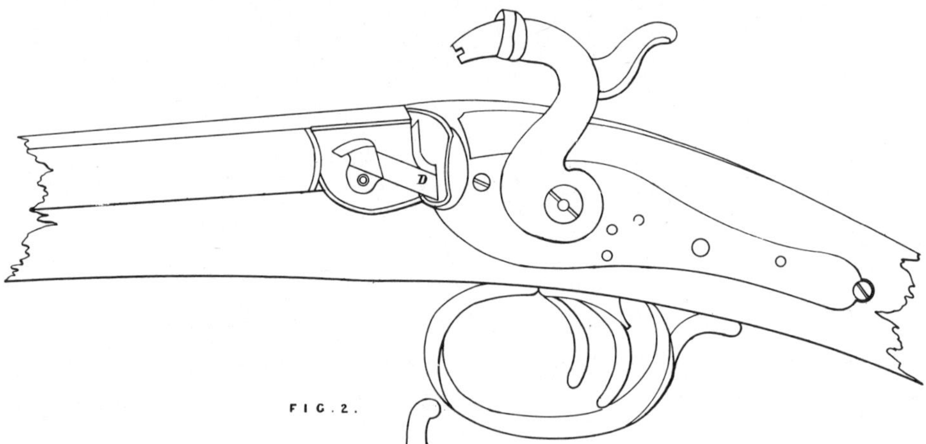

FIG. 2.

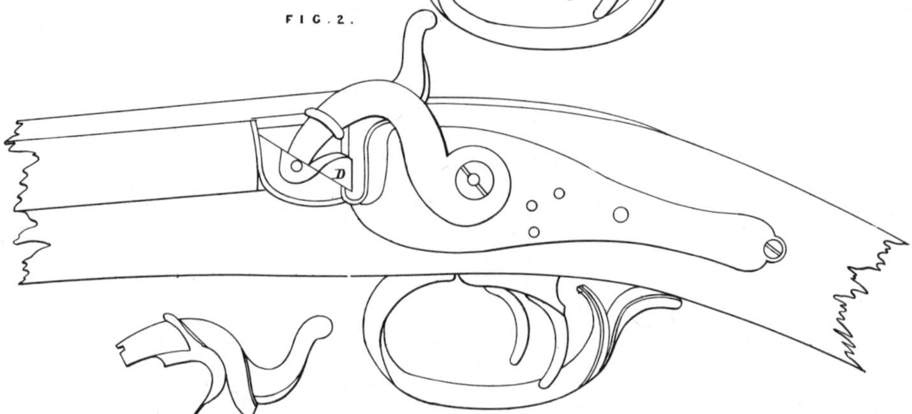

FIG. 3.

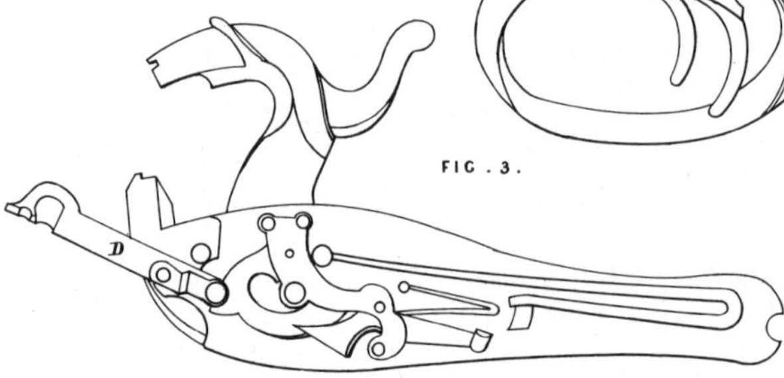

FIG. 4.

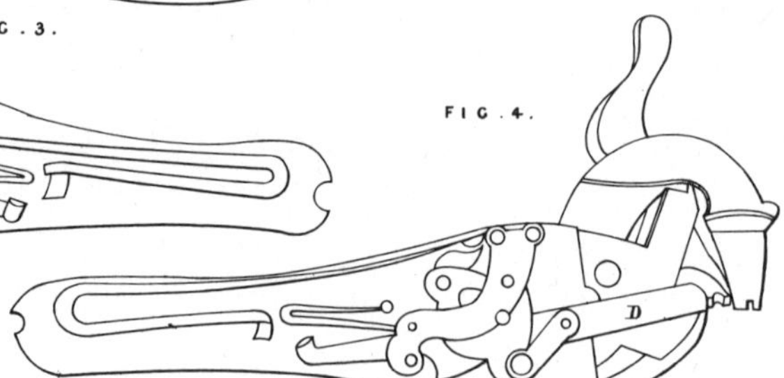

FIG. 5.

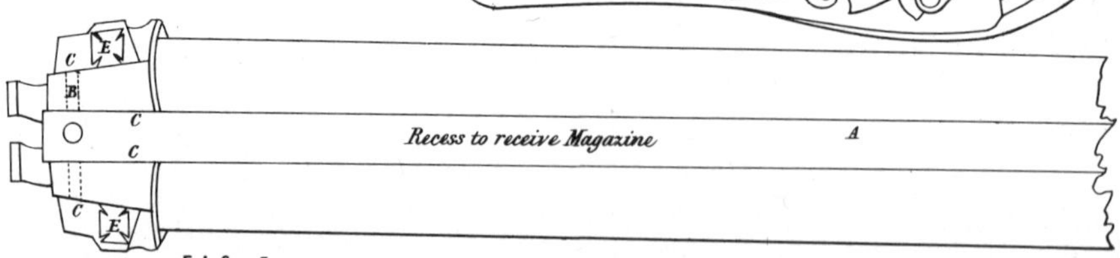

FIG. 7.

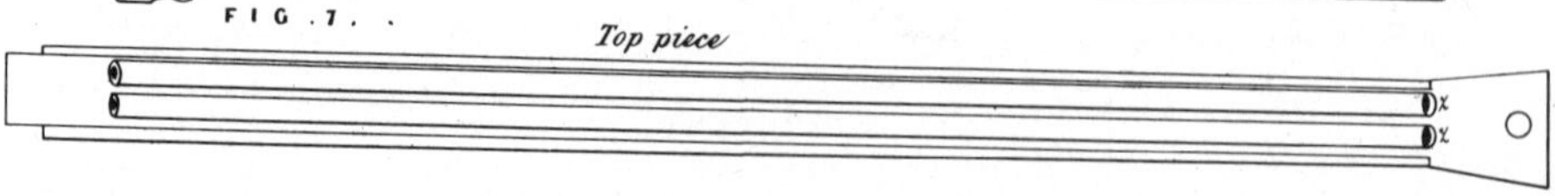

The enrolled drawing is colored.

Malby & Sons, lith.

London. Printed by George Edward Eyre and William Spottiswoode. Printers to the Queen's most Excellent Majesty. 1854.

A.D. 1835 . . . . N° 6909.

Revolving Fire-arms.

COLT'S SPECIFICATION.

TO ALL TO WHOM THESE PRESENTS SHALL COME, I, Samuel Colt, of Ludgate Hill, in the City of London, Gentleman, send greeting.

WHEREAS His present most Excellent Majesty King William the Fourth, by His Letters Patent under the Great Seal of Great Britain, bearing date at Westminster, the Twenty-second day of October, in the sixth year of His reign, did, for Himself, His heirs and successors, give and grant unto me, the said Samuel Colt, His especial license, full power, sole privilege and authority, that I, the said Samuel Colt, my ex̃ors, adm̃ors, and assigns, or such others as I, the said Samuel Colt, my ex̃ors, adm̃ors, or assigns, should at any time agree with, and no others, from time to time and at all times during the term of years therein expressed, should and lawfully might make, use, exercise, and vend, within England, Wales, and the Town of Berwick upon Tweed, my Invention of "Certain Improvements applicable to Fire-arms;" in which said Letters Patent is contained a proviso, that I, the said Samuel Colt, shall cause a particular description of the nature of my said Invention, and in what manner the same is to be performed, to be inrolled in His said Majesty's High Court of Chancery within six calendar months next and immediately after the date of the said in part recited Letters Patent, as in and by the same, reference being thereunto had, will more fully and at large appear.

NOW KNOW YE, that in compliance with the said proviso, I, the said Samuel Colt, do hereby declare that the nature of my said Invention of certain improvements applicable to fire-arms, and the manner in which the same is to be carried into effect, are fully set forth and explained in the following description thereof, as elucidated by the Drawings hereunto annexed, reference being had to the figures and letters marked on the Drawings in corrrespondence with the same figures and letters herein contained, that is to say:—

My Invention consists, First, in the adaptation and application of certain levers, lifters, slides, plates, connecting pieces, springs, a ratchet, a mandril or arbor, a shackle, and other apparatus, as herein-after particularly shewn and described, to that species of gun and pistol having a cylindrical breech piece revolving round an axis parallel to the axis of the barrel, in which breech piece six or more chambers are bored, each capable of containing a charge of powder and shot, and which charges by the revolution of the cylinder may be brought successively in a line with the bore of the barrel, through which they may be discharged one at a time; the object of my principles, adaptations, and applications being to cause the said cylinder to revolve the distance from centre to centre of two contiguous chambers by the action of cocking the gun or pistol, and by the same action to lock the cylinder firmly in its place when so brought, and thus that no care or attention on the part of the person using the fire arms is required in order to bring the charges into the proper place for being discharged through the barrel.

Secondly, in a peculiar mode of securing the revolving breech piece, that it may revolve freely and yet be discharged safely.

Thirdly, in protecting the lock and connecting works from wet and from the smoke of the explosions, by a peculiar shield which will allow the smoke to escape without annoyance to the person using the gun or pistol, which shield also receives and resists the recoil of the revolving cylinder.

Fourthly, in placing the axis of the aperture of the percussion tube in a right line with the axis of the chamber.

Fifthly, in a ready method of securing the barrel to the cylinder, and of detaching it at pleasure, so that the loading of the chambers may be performed with facility and dispatch.

Sixthly, in effectually separating the recesses in which the percussion tubes are placed from each other, so as to prevent fire communicating from the exploding cap to the adjoining ones.

And seventhly, (in the case of guns discharged from the shoulder) in the use of a ram or sliding piece to receive the blow of the hammer and communicate the same to the percussion cap, whereby more space is obtained between the revolving cylinder and the hammer for the apparatus. And the manner in which I carry into effect these my improvements upon fire arms is shown and explained as follows:—

Section 1, represents the parts necessary to the application of my improvements to fire arms discharged from the shoulder; section 2, represents the parts necessary to the application of my improvements to pistols. In order to a clear explanation, I will first describe the parts and their connections, shewn in both sections; then those peculiar to shoulder-arms, as shewn only in section 1; and after which, such parts as belong to pistols, as contained in section 2. The parts shewn in both sections are Figure 1, the main spring; Figure 2, the stirrup, which connects the main spring to the hammer; the stirrup is omitted in section 2, for the sake of shewing a different form of action of the spring on the hammer; Figure 3, the hammer; its fulcrum is at *a*; at *b*, is attached the stirrup (Figure 2); at *c*, are the notches for the trigger; *d*, is the place where the lifter (Figure 5) is attached; *e*, is a stud projecting from the side of the hammer that operates on the key. Figure 4, is the key that locks the cylinder (Figure 13); its fulcrum is at *a*; *b*, is the bolt at the end of the long arm which enters the mortice *b*, *b*, &c. in the cylinder (Figure 13.) The catch C, is attached to the arm *d*, of the key by the pin *e*, passing through its lower end to form a hinge, which permits the upper end of the catch to move laterally; *g*, is the plate acted upon by the stud *e*, on the hammer (Figure 3) to draw out the bolt *b*, of the key from the mortice in the cylinder (Figure 13); J, is a spring rivetted to the catch *c*, at *f*; the lower end of which spring acts below the hinge to bring the upper part back to its place, when the stud *e*, on the hammer (Figure 3) has passed down to the place *g*, of the catch C; L, is a spring (attached to the lock plate) that forces the arm *b*, into the mortices *b*, in the cylinder (Figure 13). Figure 5, is the lifter which turns the cylinder (Figure 13); the arm *a*, acts on the teeth of the ratchet (Figure 10); the round part of the arm C, passes through the hammer (Figure 3) at *d*, in which it is permitted to turn sufficiently to keep the arm *a*, in a proper situation to act under the teeth of the ratchet (Figure 10). The lifter is joined at *b*, with the forked end of the arm C, by the pin *e*, which forms a hinge to allow the arm *a*, some lateral motion in respect of the hammer, whereby the lifter may accommodate itself to the curvature of the ratchet and depth of its tooth. Figure 6, is the nutt which holds the arbor (Figure 8) in the shield (B, Figure 29). Figure 7, is an end view of the nutt (Figure 6); *a*, is the place where it screws on to the arbor. Figure 8, is the arbor on which the cylinder (Figure 13) and shackle (Figure 11) turn, and which arbor connects the barrel (Figure 18) with the cylinder (Figure 13); *a*, *a*, are the bearings of the cylinder; B, is a mortice, through which the wedge (Figure 19) enters to connect the barrel (Figure 18) with the cylinder (Figure 13); C, is the part on which the shackle (Figure 11) turns; *d*, is the screw on which the nutt (Figures 6 and 7) is screwed to fasten the arbor firmly to the shield (B, Figure 29); *e*, is a feather or ward which enters a groove in the shield (Figure 29) to prevent the arbor from turning; *f*, is a feather or ward which enters a groove in the plate (Figure 17) to prevent the plate from turning. The end G, passes through the plate (Figures 16 and 17), and through a projection under the barrel (Figure 18), to connect them to the cylinder (Figure 13). Figure 9, is the ratchet wheel which is placed on the shackle (Figure 11). Figure 10, is an end view of the ratchet; *a*, *a*, &c. are the teeth which are acted on by the arm *a*, of the lifter (Figure 5), to cause it to turn the shackle (Figure 11), which enters at *b*; and *c*, is a groove that receives the ward *a*, of the shackle. Figure 11, is the shackle which connects the ratchet (Figure 9) to the cylinder (Figures 13 and 14); *a*, is a ward which enters the groove *c*, in the ratchet (Figure 10), and causes the shackle to move with the ratchet; *b*, is a ward which enters a corresponding grove in the cylinder (Figure 14), to prevent it from turning except with the shackle and ratchet. Figure 12, is an end view of the shackle; the arbor (Figure 8) on which the shackle turns passes through its central aperture *a*. Figure 13, is the cylinder in which are chambers for the charges; *c, c,* &c. are tubes screwed into the bottom of the chambers on which the percussion caps are placed; *b, b,* &c. are mortices to receive the bolt *b*, of the key (Figure 4) which locks the cylinder when a chamber is directly opposite the barrel (Figure 18). The shield B (Figures 28 and 29) is fixed over the end A, of the cylinder, to protect the percussion cap from moisture. Figure 14, represents the back end of the cylinder (Figure 13). The shackle (Figure 11) enters the central apertures *a*, and the ward *b* of the shackle enters the mortice *b*, of the cylinder, which causes them to turn together; *c, c,* &c. are ends of the percussion tubes on which are placed the percussion caps; *d, d,* &c, are partitions between the tubes which prevent fire communicating from the exploding caps to the adjoining ones. Figure 15, represents the forward end of the cylinder (Figure 13); *a*, is the middle aperture through which the arbor (Figure 8) passes; *b, b,* &c. are the chambers for the charges. Figure 16, is the plate which covers the forward end of the cylinder (Figures 13 and 15) to protect the charges from moisture, and to which plate the barrel (Figure 18) is firmly secured. Figure 17, represents the forward view of the plate (Figure 16); *a*, is an aperture through which the end *g*, of the arbor (Figure 8) passes; *e*, is a groove to receive the ward *f*, of the arbor (Figure 8); at *b*, the barrel (Figure 18) is attached; *c*, is a hole through which the charges may be passed into the chambers *b, b,* &c. in the cylinder (Figure 15), when it is desirable to omit taking off the barrel and plate; and *d, d,* &c. are mortices to receive the ends *s, s,* of the guards or lock plates (Figures 28 and 29). Figure 18, is the barrel which gives direction to the charge; *a*, is the end that screws into the plate at *b* (Figure 17); *b, b,* is a projection under the barrel which receives the end G, of the arbor (Figure 8); C, is a mortice for admitting the wedge (Figure 19). Figure 19, is the wedge which passes through the mortice C, of the barrel (Figure 18), and the mortice B, in the arbor (Figure 8), to hold the plate (Figure 16), and the barrel (Figure 18), in contact with the cylinder (Figure 13), and the cylinder in contact with the shield (Figure 29); the small end of the wedge enters the mortices, and when it is pressed into them the edge *a*, acts on the end *d*, of the mortice C, in the barrel (Figure 18), the other edge *b*, acts on the end *h*, of the mortice B, in the arbor (Figure 8), which causes the plate (Figure 16), and the barrel (Figure 18), to be drawn to the fore end of the cylinder (Figures 13 and 15) with any requisite degree of tightness.

Parts shewn only in section 1, as not adapted to pistols:—Figure 20, is the lever used for drawing back the hammer; its fulcrum is at *a*; *b*, is the part that acts at *f*, on the hammer (Figure 3), and raises it when the end *c*, of the lever is drawn down; *f*, is a spring fastened in the lock plate (A, Figure 29), which spring acts at *e*, on the lever, and draws the end *c*, to its place against the lock plate, after the hammer has been drawn back to its proper position for being let off for exploding the percussion caps. Figure 21, is the trigger; its fulcrum is at *a*; *b*, is a spring fastened to the lock plate that forces the catch *c*, of the trigger into the notches *c*, of the hammer (Figure 3). Figure 22, is the spiral spring that draws the ram (Figure 23) from the tubes *c*, *c*, &c. in the cylinder (Figure 13), and permits the cylinder to turn when the end *g*, of the hammer (Figure 3) is drawn back. Figure 23, is the ram for conveying the force of the hammer to the percussion cap. The spiral spring (Figure 22) is placed on it at *a*; *b*, is a pin against which the spiral spring acts to force back the ram, the other end of which spring abuts against the back part of the shield; at the end *c*, of the ram there is a shoulder that comes against the shield (B, Figure 29) to prevent the coiled spring from drawing the ram too far back from the percussion cap. The end *g*, of the hammer (Figure 3) strikes the end *d*, of the ram, the other end *f*, of which discharges the percussion cap.

Parts peculiar to pistols shewn only in section 2:—Figure 24, is the connecting rod or sliding piece between the hammer (Figure 3), and the trigger (Figure 26); *a*, is a slot through which a pin passes to hold the connecting rod in its place, and to allow it a longitudinal motion The face of the end *b*, is oblique, so that when the part *d*, of the hammer (Figure 3) is raised, it commences to act at the bottom part *c*, of the oblique face of the connecting rod, and causes it to slide forward; the other end *f*, of which connecting rod operates at *b*, on the secret trigger (Figure 26). E, is a forked spring fastened to the lock plate (Figure 28), the upper end of which forked spring hooks into the end *b*, of the connecting rod to force its catch *d*, into the notches *c*, on the hammer (Figure 3), and hold it back when the notches are brought opposite the catch. Figure 25, is a view of the upper side of the connecting rod; *a*, *a*, the shoulders against which the prongs of the spring (E, Figure 24) press; *c*, *c*, the shoulders behind which the fork of the trigger acts. Figure 26, is the secret trigger to let off the lock; its fulcrum is at *a*; at *b*, the end *f*, of the connecting rod (Figure 24) acts when it is moved forward by the hammer (Figure 3), which causes the end *c*, of the trigger to be thrown into a suitable position for the finger. Figure 27, is a view of the upper side of the trigger (Figure 26); it turns on the pin *a*; *b*, is a spring (fastened to the lock plate) to hold the secret trigger in its place, by acting on the projection *c*; *d*, is the fork that acts behind the shoulders *c*, *c*, of the connecting rod (Figure 25), so that by pressing the finger on the end *c*, of the trigger, the catch *d*, of the connecting rod (Figure 24) is drawn from the notch *c*, of the hammer (Figure 3), and the main spring (Figure 1) forces the end *g*, of the hammer (Figure 3) against the percussion cap, which causes the load to be discharged, as is seen in the transparency (Figure 31). Figure 28, represents the handle, lock plate, and shield; the part colored yellow is the shield; the part colored red is the handle and lock plate; the main spring (Figure 1) is fastened at *a*; the hammer (Figure 3) turns at *d*; the key (Figure 4) at *c*; the connecting rod (Figure 24) slides on the pin at *b*; and the trigger (Figure 26) turns at *e*. The forked end of the lock plate *s*, enters the mortices *d*, *d*, &c. in the plate (Figure 17). At *g*, is a hinge to permit the back part of the handle to open when the screw *h*, is drawn. The screw *i*, connects the handle and lock plate with the shield; the forward part B, of the shield covers the hind part A, of the cylinder (Figure 13), and protects the percussion caps from moisture. The back part J, forms a complete cover for the lock, and protects it from the smoke of the percussion caps and from moisture. The form of the inside of the shield and lock plates must be such as to allow the parts their respective motions as seen in the transparent view (Figure 31). Figure 29, (section 1,) is the lock plate, shields, and guards, suited to guns discharged from the shoulder. A, is the lock plate to which the main spring (Figure 1) is attached, at *d*; the hammer (Figure 3) at *e*; the trigger (Figure 21) at *f*; the key (Figure 4), and the lever (Figure 26), at *g*. At *h*, is the guard for the trigger. *i*, is a bolt that passes through the lock plate and stock, and screws to the upper guard at *j*, to connect them together; the forked ends of the lock plate S, S, enter the mortices *d*, *d*, &c. in the plate (Figure 17). *l*, *l*, are screws which connect the lock plate and guards to the shield B. The forward part of the shield covers the end A, of the cylinder (Figure 13), and protects it from moisture. Between the ratchet (Figure 9) and cylinder (Figure 13), is a partition on the shield to prevent the smoke of the percussion caps getting to the lock. *r*, is a sliding shutter over an opening to the tubes for the purpose of putting on percussion caps. *t*, is a hole in the top of the shield for the smoke of the percussion caps to escape. The interior of the shield and lock plate must be suited to the lock, as is seen in the transparent view (Figure 31). Figure 30, sections 1, and 2, represent (by an end view) the relative position of the key (Figure 4); the lifter (Fig. 5); the ratchet (Fig. 10); and hammer (Fig. 3). The black line A, represents the hammer. B, in the centre is the position of the ratchet. C, is the key which is operated on by the stud *e*, and the hammer A. The end of the stud *e*, and the end *i*, of the back part of the key are both cut off obliquely, that they may, when brought in contact, cause the upper end of the catch of the key to move laterally and permit the stud *e*, to pass down to its place behind it again, when, by means of the spring *j*, which is fastened to the upper end *i*, of the catch, and acts below the joint at *k*, it is brought back to its former position before the stud *e*, on the hammer A. D, is the lifter. The arm *c*, passes through the hammer (A), where it is secured by a pin acting in a groove in the arm to prevent it from coming out. *b*, is the hinge that allows the upper end a lateral motion. *a*, is the arm which acts in the teeth *a*, *a*, &c. of the ratchet B. At *d*, a spring is attached that keeps the arm *a*, into the teeth of the ratchet by the action of its other end on the lower part of the hinge *b*. Figure 31, sections 1, and 2, shew, by a transparency, the relative position of all the parts combined; the colour and form of the outlines of each correspond with the colour and form of their separate parts. To work the lock of the gun draw on the lever at *c*, (Figure 31,) and as the arm *d*, of the hammer (Figure 3) is raised, the stud *e*, presses against the end *g*, of the key (Figure 4), and consequently its bolt *b*, is withdrawn from the mortice *b* of the cylinder (Figure 13), at which time the lifter (Figure 5) (being connected with the hammer (Figure 3) at *a*) is raised, and its arm *a*, catches in a tooth of the ratchet (Figures 9 and 10), which, being connected with the cylinder (Figure 13), by means of the shackle (Figure 11) causes the cylinder (Figure 13) to turn, and before the succeeding chamber is brought opposite the barrel (Figure 18), the stud *e*, on the hammer (Figure 3) passes the top of the catch of the key (Figure 4), and when the hammer (Figure 3) is drawn back to its proper position for discharging, then the succeeding chamber in the cylinder (Figure 13) has come directly opposite the caliber of the barrel (Figure 18). In this position the bolt *b*, of the key (Figure 4) is forced by the spring L, Fig. 4, into the succeeding mortice in the cylinder (Figure 13), and holds it until the pulling back of the hammer is repeated. By pulling the trigger the mainspring (Figure 1) forces the hammer (Figure 3) forward to its former position, and the upper end *g*, discharges the percussion cap. During which operation the lifter (by means of its lateral motion to the left) passes below the succeeding tooth on the ratchet, and by means of the lateral motion to the right of the catch C, of the key (Figure 4), the pin *e*, on the hammer (Figure 3) is permitted to pass below it again. A repetition of the same motions produce the same effects. To work the pistol lock the cock is drawn back, which, by its lower part, acts against the oblique back-end of the connecting rod, the front end of which presses against the shoulder of the trigger and throws its tail down into the position for discharging. The operation of the key, lifter, and other parts for unlocking, turning, and relocking the cylinder are similar to that for the gun, and the lock is let off by pulling the trigger, as in the gun, when the various parts will be restored to their places for a repetition of the working. Figure 32, represents a pistol when fitted for use. To load the gun or pistol it is only necessary to draw out the wedge *a*, (Figure 19,) which will liberate the barrel (Figure 18); then, by withdrawing the bolt *b* (Figure 4) from the mortice of the cylinder (Figure 13), which is effected by drawing back the hammer (Figure 3), then the cylinder (Figure 13) may be taken off from the arbor (Figure 8.) And, lastly, I hereby declare that I lay no claim to the various common parts of fire arms herein-before described, such parts having been introduced into the Specification merely for the purpose of more clear description; but I distinguish the parts and combinations constituting my Invention, and claim my right to the same as applicable to fire arms, by the following enumeration thereof:—First, the application of a key for locking and unlocking the revolving cylinder on the principle described at Figure 4. Secondly, the application of a lifter to act in turning the ratchet on the principle described at Figure 5. Thirdly, the application of a sliding connecting rod (Figures 24 and 25), for the triple purpose of throwing out the secret trigger (Figures 26 and 27), of holding the hammer at half and full cock, and of communicating with the trigger for letting off the hammer on the principle herein-before described. Fourthly, the combination of the key and lifter, or the key, lifter, and connecting rod, with the hammer, so that the backward motion of the hammer shall successively produce the unlocking, the rotary motion, and the relocking of the cylinderon the principle herein described. Fifthly, the application of the ratchet (Figures 9 and 10). Sixthly, the application of the shackle (Figures 11 and 12.) Seventhly, the application of the percussion tubes *c*, *c*, &c. (Figures 13 and 14,) in a right line with the axis of the chamber in the cylinder. Eighthly, the application of the partitions *d*, *d*, &c. (Figure 14) between the percussion caps. Ninthly, the application of the ram (Figure 23). Tenthly, the application of the shield (B, Figures 28 and 29.) Eleventhly, the manner of uniting the barrel with the cylinder. And, Twelfthly, the entire arrangement of the whole apparatus, as herein-before shewn and explained.

In witness whereof, I, the said Samuel Colt, have hereunto set my hand and seal, this Thirtieth day of October, One thousand eight hundred and thirty-five.

SAMUEL (L.S.) COLT.

A.D. 1835. October 22. N.o 6909.

COLT'S SPECIFICATION.

(1 SHEET.)

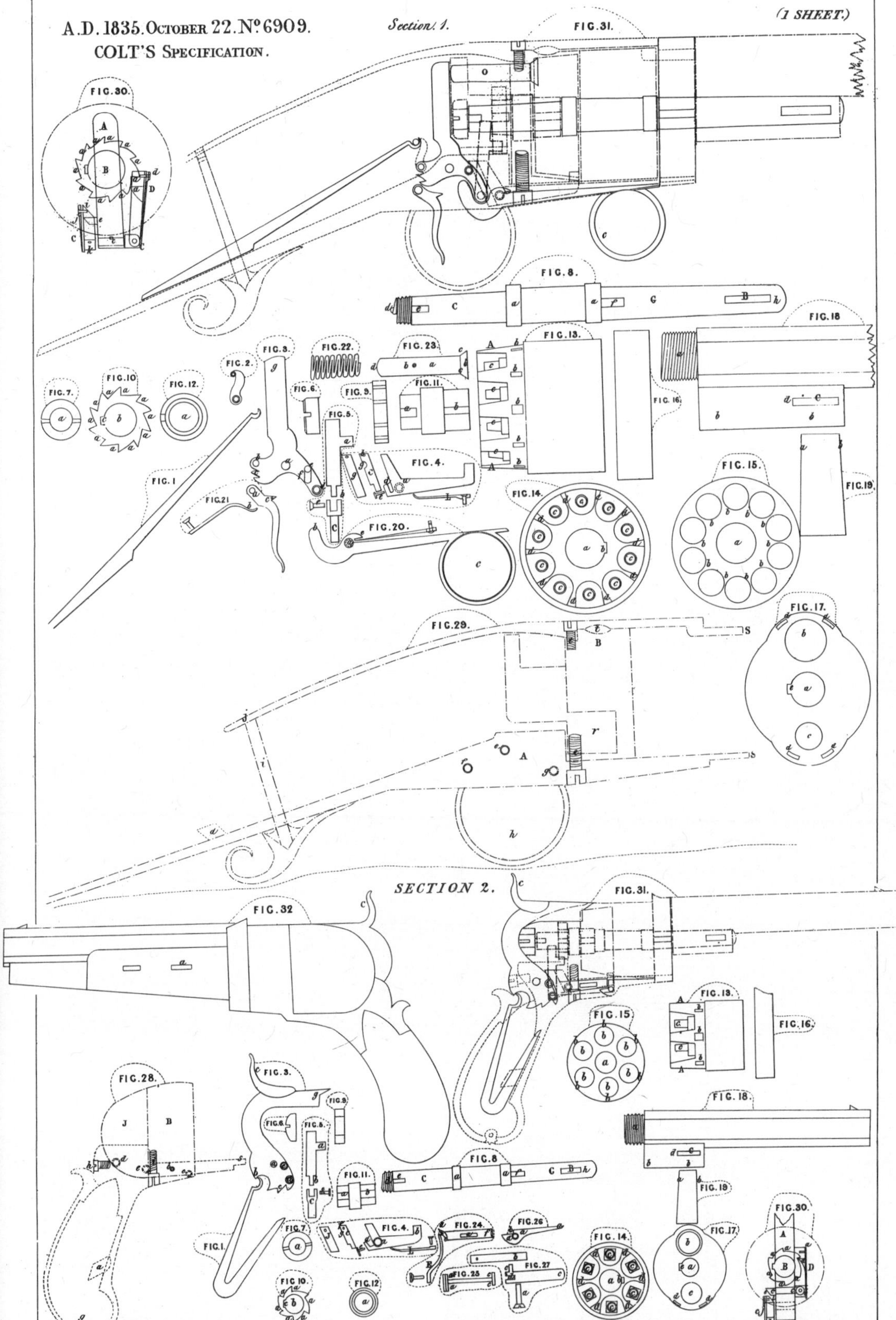

The Enrolled Drawing is partly Colored

Malby & Sons. Lith.

London: Printed by George Edward Eyre and William Spottiswoode. Printers to the Queen's most Excellent Majesty. 1854.

A.D. 1839 . . . . Nº 7965.

Locks of Fowling Pieces and Fire-arms.

MANTON'S SPECIFICATION.

TO ALL TO WHOM THESE PRESENTS SHALL COME, I, GEORGE HENRY MANTON, of Dover Street, Piccadilly, in the County of Middlesex, Gun Maker, send greeting.

WHEREAS Her present most Excellent Majesty Queen Victoria, by Her Letters Patent under the Great Seal of Great Britain, bearing date at Westminster, the Eleventh day of February, in the second year of Her reign, did, for Herself, Her heirs and successors, give and grant unto me, the said George Henry Manton, Her especial licence, full power, sole privilege and authority, that I, the said George Henry Manton, my ex̃ors, adm̃ors, and assigns, or such others as I, the said George Henry Manton, my ex̃ors, adm̃ors, or assigns, should at any time agree with, and no others, from time to time and at all times thereafter during the term of years therein expressed, should and lawfully might make, use, exercise, and vend, within England, Wales, and the Town of Berwick-upon-Tweed, my Invention of "CERTAIN IMPROVEMENTS IN FOWLING PIECES AND OTHER FIRE-ARMS;" in which said Letters Patent is contained a proviso that I, the said George Henry Manton, shall cause a particular description of the nature of my said Invention, and in what manner the same is to be performed, to be inrolled in Her said Majesty's High Court of Chancery within six calendar months next and immediately after the date of the said in part recited Letters Patent, as in and by the same, reference being thereunto had, will more fully and at large appear.

NOW KNOW YE, that in compliance with the said proviso, I, the said George Henry Manton, do hereby declare that the nature of my said Invention, and the manner in which the same is to be performed, are particularly described and ascertained in and by the following statement thereof, reference being had to the Drawings hereunto annexed, and to the figures and letters marked thereon (that is to say):

My Invention of Certain Improvements in Fowling Pieces and Other Fire-arms consists in the mode of constructing and applying a circular magazine, containing a series of copper caps, one of which is constantly kept at the point of the neck by a spring, as is well understood, to the lock of a fowling piece or pistol, (single or double,) which lock is so constructed that in bringing it to the full cock the cap discharged is removed and its place supplied by another; which movements are effected by a spring (1) at the back of that part of the cock which inflames the cap having a projection, which removes the cap above named, as shewn by the Drawing; and the other cap is placed on the nipple by a projection (2) on the lower part of the cock acting in a hollow (3) cut in the top or cover of the magazine, as also shewn in the Drawing, so that the party using the same have no trouble in effecting an exchange of the cap thus discharged until those contained in the magazine are all consumed. The cap to be next supplied to the nipple is kept in its place, and protected from wet in the short neck of the magazine by a slide (4) pressed outward by a spring (5), and the slide (4) is removed out of the way by the spring (5) coming against a projection (6) at the back of the breech.

In witness whereof, I, the said George Henry Manton, have hereunto set my hand and seal, this Second day of August, in the year of our Lord One thousand eight hundred and thirty-nine.

GEORGE HENRY (L.S.) MANTON.

A.D. 1839, Feb. 11. N°. 7965.
MANTON'S Specification.

(1 SHEET)

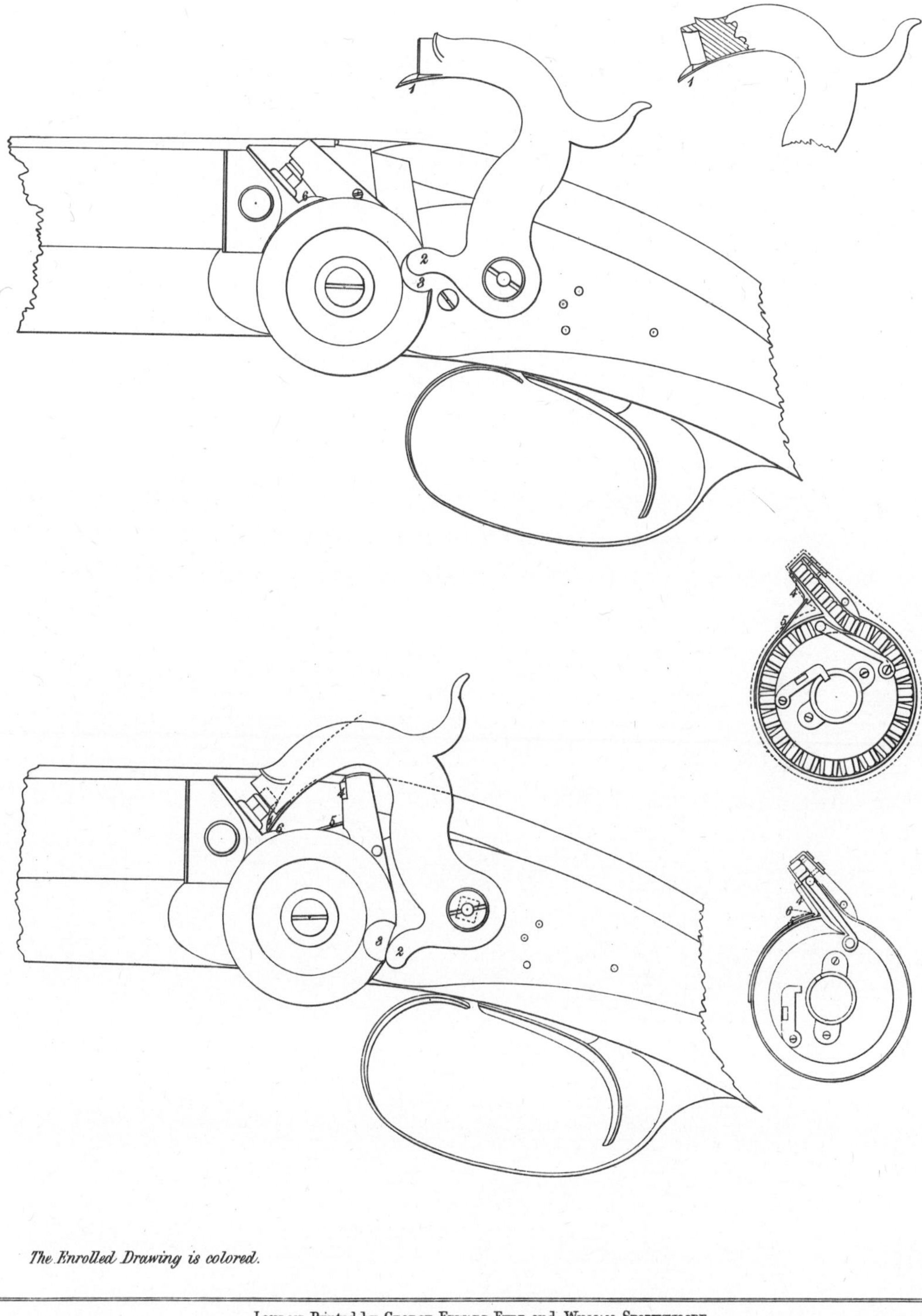

The Enrolled Drawing is colored.

London. Printed by George Edward Eyre and William Spottiswoode.
Printers to the Queen's most Excellent Majesty. 1854.

J.R. Jobbins.

A.D. 1839 . . . . N° 7980.

Fire-arms, and Balls to be used therewith.

HEURTELOUP'S SPECIFICATION.

TO ALL TO WHOM THESE PRESENTS SHALL COME, I, Charles Louis Stanislas Baron Heurteloup, of Queen Ann Street, in the County of Middlesex, send greeting.

WHEREAS Her present most Excellent Majesty Queen Victoria, by Her Letters Patent under the Great Seal of Great Britain, bearing date at Westminster, the Twenty-third day of February, in the second year of Her reign, did, for Herself, Her heirs and successors, give and grant unto me, the said Charles Louis Stanislas Baron Heurteloup, Her especial licence, full power, sole privilege and authority, that I, the said Charles Louis Stanislas Baron Heurteloup, my ex̃ors, adm̃ors, or assigns, or such others as I, the said Charles Louis Stanislas Baron Heurteloup, my ex̃ors, adm̃ors, or assigns, should at any time agree with, and no others, from time to time and at all times during the term of years therein expressed, should and lawfully might make, use, exercise, and vend, within England, Wales, and the Town of Berwick-upon-Tweed, my Invention "**Certain Improvements in Fire-arms, and in the Balls to be used therewith**;" in which said Letters Patent is contained a proviso, that I, the said Charles Louis Stanislas Baron Heurteloup, shall cause a particular description of the nature of my said Invention, and in what manner the same is to be performed, to be inrolled in Her said Majesty's High Court of Chancery within six calendar months next and immediately after the date of the said in part recited Letters Patent, as in and by the same reference being thereunto had, will more fully and at large appear.

NOW KNOW YE, that in compliance with the said proviso, I, the said Charles Louis Stanislas Baron Heurteloup, do hereby declare the nature of my said Invention, and the manner in which the same is to be performed, are fully described and ascertained in and by the following statement thereof, reference being had to the Drawing hereunto annexed, and to the figures and letters marked thereon (that is to say):

The improvements which constitute the object of the present Patent are applicable to portable fire-arms, single or double barrelled guns, rifled or not, and may be comprised under four heads. First, comprises improvements made in the mechanism of the lock, adapted to the use of the particular priming called continual prime (amorce continue). Second, comprises improvements in the construction of fire-arms with reference to the interior of the barrel, and also in the ramrod, in order to facilitate the loading of rifled fire-arms with ball. Third, comprises improvements in the construction of fire-arms in order to enable the ramrod to be also used as a bayonet. Fourth, comprises improvements in the balls used to load rifled or not rifled fire-arms.

Description of the Drawing.

Figure 1, a side elevation of a single rifled fire-arm; Figure 2, a longitudinal section of the same, shewing the interior of the mechanism of the lock and the grooves of the single rifled fire-arm; Figure 3, a view of the under part of the same; Figure 4, a section of a portion of a doubled barrelled gun with the application of the system of the continual prime; Figure 5, shews the under part of the same; Figure 6, shews the ramrod bayonet belonging to the single rifled fire-arm; Figure 7, an outside view of the thin piece, whereby the box containing the whole mechanism of the lock is closed; Figure 8, an inside view of the same piece; Figure 9, an inside view of the thicker and principal piece forming the box, which contains the whole mechanism of the lock; Figure 10, an outside view of the last-mentioned piece;

Figure 11, shows the inner side of the compressor; Figure 12, the under side of the box containing the mechanism in which the continual prime is moved; Figure 13, a detached view of the spring with lips; Figure 14, a transverse section of the barrel of a rifled gun, showing the form of the border of the patent breech; Figure 15, shows the extremity of the ramrod bayonet for the rifled gun; Figure 16, different views of the shape of the ball before and after being pressed between the ramrod and the border of the patent breech; Figure 18, detached views of the moveable box designed to receive the smoke and flame from the priming; N°s 1, 2, and 3, shew the plate carrying the springs with lips for the double barrelled gun; Figure 19. a longitudinal section of a portion of a double barrelled gun with a particular disposition of the compressor, and the box which receives the smoke and flame of the ignited prime; Figure 20, the under view of the same double barrelled gun; Figure 21, the particular manner to prepare the balls for their being attached to the cartridges without much increasing the volume of these balls.

As to the first head, improvements made in the mechanism of the lock adapted to the use of the particular prime called continual prime (amorce continue). The lock which I discover to be best adapted to the system of the continual prime consists of a box composed of two pieces (see Drawing). Figures 9, and 10, shew different views of the first piece; Figures 7, and 8, show different views of the second piece. These two pieces, joined together by means of two screws passing through the holes *a*, and *b*, Figure 2, leave between them two principal cavities; the one *c*, is towards the extremity of the stock; the other *d*, towards the extremity of the barrel. The first cavity *c*, contains the main spring *e*, the sear *g*, the trigger *h*, and the cock *i*. The second cavity *d*, Figure 2, is intended to receive the gas and flame produced on firing, in order, first, to prevent the dispersion of the gas in the interior of the lock, and, second, that the fire from the exploded piece of priming which has just been cut off at the point z^1, may be kept separate from the rest of the priming tube. The piece, Figure 9, in which more particularly the before-mentioned cavities are formed, has a prolongation *j* intended to carry a particular spring, to be described hereafter. This spring is important to the proper motion of the prime. To produce the motion of the prime, to protect the prime, and to fix it, there is another piece, represented separately in Figure 11; the same is shewn by *k*, Figures 2 and 3. This piece is jointed to another long one *l*, Figures 2 and 3. This piece *k*, which I have called a compressor (compresseur), because it presses upon the priming tube, has an aperture *o*, Figure 2, in which a small wheel is inserted, the use of which is shewn hereafter. This piece has also two oblong narrow cavities or grooves *a*, Figure 11, intended to receive, first, two oblong prominences made to fit into such cavities or grooves, shown by *x*, Figure 9, and by *x*, Figure 12; and, second, two prominences *y*, Figures 12, and 13. This compressor has two shields, one for preserving from rain the extremity of the prime, the other to protect the aperture containing the progression wheel. The shield guarding the prime is marked *f*, Figures 1, 2, 3; that which guards the aperture for the wheel is marked *m*, Figures 1, 2, 3.

Now, as an important part of the improvements in the lock adapted to the use of the continual prime consists in the new means of bringing forward successive portions of prime, one after the other, in regular quantities or lengths, I will give a particular description of this part of the lock. In the aperture *o*, in Figures 2 and 11, of the compressor, is a small wheel *v*, Figure 2 (colored yellow in the Drawing); it turns upon a pivot. This wheel has six divisions. In a groove shown in the prolongation *j*, Figure 2, marked by *r*, on Figures 10 and 12, is placed a spring, fixed by a screw *n*, Figure 2. This spring is drawn separately in Figure 13. It has a gutter *c*, Figure 13, bordered with two projecting lips *y*, Figures 12 and 13. The lips are received when the compressor is closed in the hinder part of the grooves *a*, Figure 11, by means of the above mentioned small wheel placed in the compressor and by the last described spring. I move the prime forward or backward in certain lengths or quantities, and the spring pressing upon the prime keeps it fixed and steady. Thus by turning the wheel with the finger either backwards or forwards, I push forward or draw back the priming tube. This being done, it is evident that two of the projections of the wheel resting upon the priming tube in the groove fix and keep it steady in its place. The prime priming tube or continual prime (amorce continue), for it may be called by either of these names, is shewn by a red line on the Drawing, Figure 2; and being placed in a channel or conduit, formed by the stock by the piece *l*, Figure 2, of the compressor, when closed by the compressor *k*, Figure 2, by the groove *r*, Figures 10 and 12, and by the spring *c*, Figures 12 and 13, must obey the impulse given it by the wheel *v*, Figure 2; and thus the priming is freely drawn back or pushed forward to be cut and struck upon the touch-hole, as represented in Figure 2. Thus the compressor having grooves *a*, Figure 11, to receive the projections *x*, Figure 12, which are lateral to the groove *r*, Figure 12, in which groove the prime is fixed, forms a complete channel from which the prime cannot deviate. This channel is the more complete as the prominence *q*, Figure 11, left between the two grooves of the compressor, keeps the priming tube correctly at the bottom of the groove.

The application of this mechanism to a double barrelled gun:—I apply to a double barrelled gun the same mechanism before described for moving the prime forward or backward in a certain length or quantity, with the exception, however, that the necessity of supplying with primes two barrels, obliges me to make certain alterations of sufficient importance to render a detailed account necessary that they may be secured by the present Patent. I have not united, as in the single gun, the mechanism of the lock, the mechanism moving the prime and the chamber receiving the smoke and flame of the prime, but in the double barrelled gun I have divided these parts into several divisions. Thus *a*, Figures 4 and 5, shew the frame bearing the double lock *c*, the cocks each provided with a sharp blade to cut off the prime; *b*, Figure 4, the sears; and *e*, Figure 4, the triggers. Figure 4, *d* is a box receiving the smoke and fire from the prime; it can be taken out at will. It is kept in its place by the compressor *x*, Figure 4, and has two holes to allow the nipples to pass through, as shown by Figure 18, in *a*. It shews also the projections *b*, Figure 18, which fall into the grooves made on the compressor in *p*, Figure 4, in order to form a complete channel for the prime. Figure 4, *g* represents a piece fixed on the stock supporting the springs in which the prime is to slide; this piece is separately drawn in Figure 18, N°s 1, 2, and 3. Finally, the compressor *x*, falls down, and is caught by the spring *i*, Figure 4, which fastens the piece *d*, Figure 18, by means of the hook *w*, Figure 4. The continual prime, which is marked by a red line, Figure 4, slides in a channel formed in front by a groove shewn on the box *e*, Figure 18, by the compressor *x*, Figures 4 and 5, by the springs *g*, Figure 18, N° 3, by the pieces Figure 18, N° 1, in *o*, by the stock (videlicet) by means of a channel cut in it, marked *q*, Figure 4, and, finally, by the stock and the barrel by a groove made in the stock in *m*. Such is the arrangement best adapted to the application of the continual prime to the double barrelled gun; the elements are the same as those pointed out in the description of the single gun. Thus my double barrelled gun represents the mechanism moving the cock, the chambers receiving the smoke and flame that propelling the prime, and, lastly, the compressor, which not only acts with the other parts, but protects the prime, with the difference that, instead of their being united as in the single gun, they are separated in the double barrelled one. To give more compactness to the system of applying the continual prime to the double barrelled gun, I have joined the compressor to the box, which is intended to receive the products of the powder, as represented by Figures 19, and 20, Thus *a*, shows the box designed to receive the products of the ignition of the powder; *b*, shows the compressor joined with the box by means of a hinge *c*. The compressor is shut by a spring *d*, fixed to the box by the screw *e*. *f*, is a long pin adapted to the spring, and which passes through the partition which separates the two chambers of the box; the extremity of that pin keeps the compressor closed. The traverse *d*, enables me to draw the spring backwards by the points *o*, Fig. 20. The box is kept in its place in the gun by a small spring *g*, fixed on the box towards the extremity of the barrel by a screw *v*. This small spring has a hole, in which a pin *h*, enters, placed on the fixed piece *x*.

As to the second head, improvements in the construction of fire-arms with reference to the interior of the barrel, and also in the ramrod, in order to facilitate the loading of the rifled fire-arms. A new system of loading rifled guns has lately attracted public attention, particularly in France. This system consists in putting in the barrel a ball of a size sufficiently small to allow it easily to enter into the barrel and fall to the bottom without having recourse to a hammer or any other means to force it down; thus the ball having reached the bottom is struck with the ramrod upon the edge of the breech, the circumference of which is perfectly even, and the ball being thus flattened it is increased in its transversal diameter, and the circumference of that diameter takes the shape of the grooves of the rifled barrel. The ball thus expanded in the grooves is fit to follow them when discharged, by which means a more rapid loading of the rifle gun has been attained. This ingenious system in its principle appears to me to have been carried out in an imperfect manner, and I have therefore invented means to remedy the defects which are connected therewith, and which are as follows:— First, it does not flatten the ball enough to make its projections sufficiently enter the grooves, and therefore force it to follow them correctly, and consequently to receive the necessary rotatory motion. Second, it does not sufficiently press the ball upon the sides of the barrel so that the contact of the ball with the sides of the barrel be sufficiently close to determine a pressure upon the barrel its whole breadth, to prevent the ball from moving and from keeping its axis (when issuing) parallel to the axis of the barrel. In order to obtain the requisite figure of the ball to the grooves of a rifled gun I have shaped the bottom or breech of the barrel and the ramrod in a manner that the parts of the ball which correspond with the grooves are pinched between the projections placed at the bottom of the barrel, corresponding with the grooves and the projections of the thick extremity of the ramrod, whereby I force these pinched parts of the ball into the interior of the grooves. I effect by this means not only of forcing the ball to follow the grooves, whatever be the strength of the charge, but also of placing the protruding parts of the ball in contact with the barrel sufficiently close to insure at all times its retaining the same horizontal axis. Thus on the borders of the breech B, Figure 2, of the patent breech A, Figure 2, I have left projections C, Figures 2 and 14. These projections are separated by hollow intervals D, Figures 2 and 14. These projections C, Figures 2 and 14, correspond with the grooves E, Figure 2, made in the barrel, so that there are as many projections as grooves. This being done it will be easily conceived that the ball introduced into the barrel and resting on the borders of the breech, will, when the blow of the ramrod has been given, assume the figure of the outer end of the breech, and thus produce projections which enter the grooves in the barrel. If the extremity of the ramrod presents projections which are calculated to act conjointly with the projections of the borders of the breech, in order to effect the pinching of the ball, I shall attain the desired result in a manner still more complete. Figure 15, shows how the extremity of the ramrod is shaped. Figure 16, N° 1, shows the round ball as it is introduced into the barrel; and 2, 3, and 4, show various views of the ball after having been pressed between the projections of the breech and those of the ramrod. In *a*, Figure 16, the projections of the ball thus shaped are represented, and it will be easily understood that these projections, once fitted into the grooves of a rifled gun, cannot swerve from the proper course. 3, Figure 16, shows that this effect produces on the ball a large belt which prevents it from swerving from its proper course. Note, I prefer the projections to be formed on the borders of a patent breech, but such projections may be applied otherwise than on the borders of a patent breech, the object being suitably to apply projections corresponding with the grooves in order to obtain the result above described.

Improvements in the construction of fire-arms, in order to enable the ramrod to be also used as a bayonet. I have made the ramrod, which I intend to use also as a bayonet, from an iron tube, shown by 2, in Figure 2; the interior of it is marked with 3. At one of the ends of the tube there is a projecting swell 4, Figures 2 and 6, the interior part of which is shown by Figure 15. At the other end of the tube there is a sharp point 5, Figure 6, which serves as a bayonet. 7[1], 7, are projecting collars slopingly shaped in the direction of the stock, and abrupt in the direction of the outer extremity of the barrel. An iron tube 8, Figures 2 and 3, is soldered to the barrel; it shows in 9, a flat piece pierced by a hole 10. 12, is a piece or spring latch, of which the extremity passes through the hole 10, and rests against the abrupt part of the projecting swell 7^1. The ramrod bayonet thus fastened cannot issue during the recoil of the fire-arm. By pressing upon the outer extremity of the piece or spring latch 12, I bend the spring 14, and thus withdraw the extremity of that spring latch 12, which, leaving its hold of the projection, allows the bayonet to be taken out and used as a ramrod.

Improvements in the balls used to load rifled or not rifled fire-arms. My improvements applicable to balls for fire-arms consist in certain additions thereto, whereby the case for the powder is more easily and advantageously applied in making ball cartridge; and my improvements consist in enveloping the ball in muslin or other suitable material, in such manner as to leave a tuft to which the case for the powder is affixed, by tying in the manner I will now explain:—I envelope the ball in a piece of muslin, thin and strong, or any other suitable material which stretches easily upon the ball without making any folds or creases on the transversal diameter of that ball. I bind this wrapping with a tie *a*, Figure 21, taking care to leave beyond the tie a tuft or length *b*, Figure 21, of the wrapping, sufficient to bind or attach the paper thereto which is to contain the powder of the cartridge *c*, Figure 21. The ball thus prepared can be greased without injuring the powder of the cartridge, and can be introduced into the barrel at the same time as the powder, with a diameter nearly equal to the interior of the barrel, for the thin and well stretched material upon the ball does not increase the latter but very slightly.

Having thus described the nature of my said Invention, and the manner of carrying the same into effect, I would wish it to be understood that what I claim as my Invention is,—

First, the mode of making the box, Figures 7 and 9, which contains the whole mechanism of the lock, of two pieces only. One of the side parts, Figure 9, of this box contains, first, the cavity in which the parts of the lock play; secondly, the space receiving the smoke and flame produced by the detonation; and, thirdly, the particular prolongation which supports a part of the mechanism producing the progression of the continual prime. The hole for the pivot of the cock, and the projection upon which the continual prime is cut, are included in this piece, and these two points being thus retained always at the same distance the one from the other, ensures the continual prime being cut in the most accurate manner.

Secondly, the mode of making a complete channel, to be opened at will, to contain the continual prime, so as to prevent its deviating when moved forward or drawn back. Also a spring with a gutter, which forms a part of and is placed in the channel, and presses towards the wheel, which moves forward or draws back the continual prime.

Thirdly, the mode of applying a wheel so arranged as to enable me to move forward or to draw back the continual prime in regulated and limited lengths by means of the finger. Also the use of a shield to protect the mechanism intended to move the continual prime.

Fourthly, making in the double barrelled gun a small box, Figure 18, separated from the mechanism which effects the action of the cock, and which box, being moveable, allows the cleaning of the nipples. This box is intended to receive the products of the ignition and to guide the flame, in order to prevent its communication with the remaining prime.

Fifthly, the mode of constructing a channel between the barrel and the stock to contain the continual prime, as described.

Sixthly, the mode of constructing the breech of a rifled gun with projections at regular intervals, and also the extremity of the ramrod, in order that the ball resting on such projections and being struck by the ramrod may be spread out with more completeness to fill up and be governed by the grooves of the barrel.

Seventhly, the mode of preparing balls in the making of ball cartridges, by wrapping them in muslin or other suitable material, and leaving tufts to which the paper cases for the powder are to be attached thereto, and whereby the balls of ball cartridges may be greased without predudice to the powder.

And, eighthly, the mode by which a bayonet used as a ramrod may be firmly fixed when used as a bayonet, and at the same time easily released when used as a ramrod.

In witness whereof, I, the said Charles Louis Stanislas Baron Heurteloup, have hereunto set my hand and seal, this Twenty-third day of August, One thousand eight hundred and thirty-nine.

CHARLES LOUIS STANISLAS (L.S.) B^N HEURTELOUP.

FIG. 4.

FIG. 18.

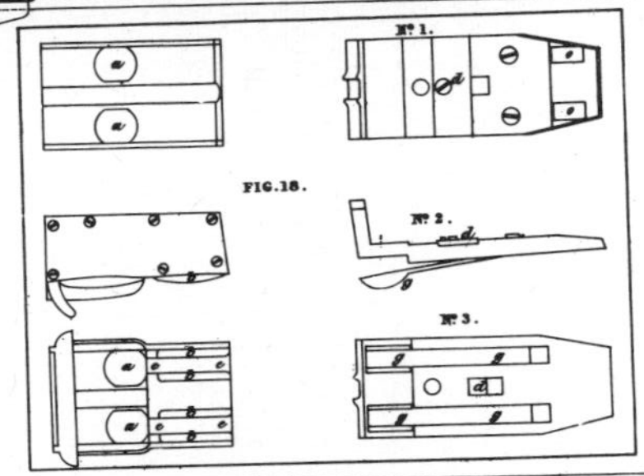

FIG. 5.

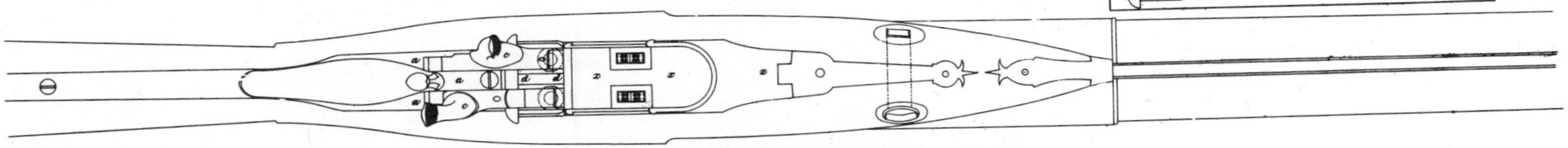

FIG. 19.

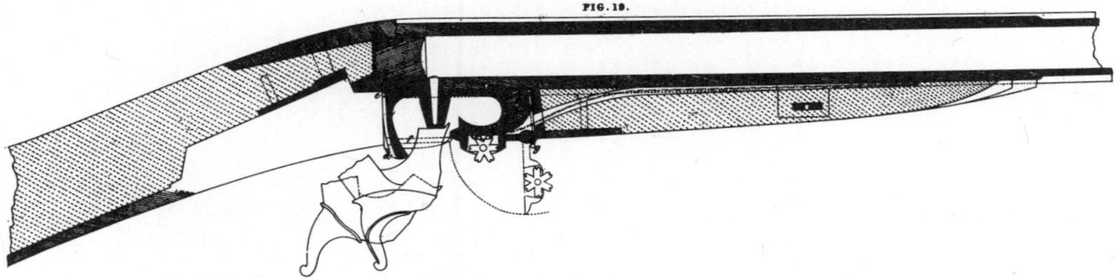

FIG. 20.

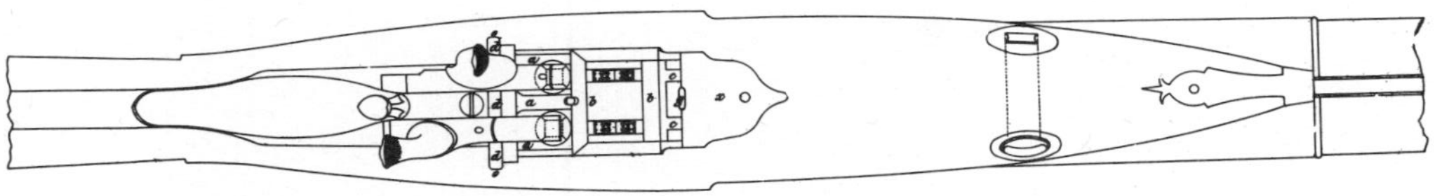

The Bordered Drawing is colored.

A.D. 1839 . . . . N° 8119.

Locks of Fire Arms.

WILKINSON'S SPECIFICATION.

TO ALL TO WHOM THESE PRESENTS SHALL COME, I, Henry Wilkinson, of Pall Mall, in the County of Middlesex, Gun Maker, send greeting.

WHEREAS Her present most Excellent Majesty Queen Victoria, by Her Letters Patent under the Great Seal of Great Britain, bearing date at Westminster, the Twenty-second day of June last, did, for Herself, Her heirs and successors, give and grant unto me, the said Henry Wilkinson, Her especial licence, full power, sole privilege and authority, that I, the said Henry Wilkinson, my ex̃ors, adm̃ors, and assigns, or such others as I, the said Henry Wilkinson, my ex̃ors, adm̃ors, or assigns, should at any time agree with, and no others, from time to time and at all times during the term of years therein expressed, should and lawfully might make use exercise, and vend, within England, Wales, and the Town of Berwick-upon-Tweed, my Invention of "**An Improvement in Fire-arms**;" in which said Letters Patent is contained a proviso, that I, the said Henry Wilkinson, shall cause a particular description of the nature of my said Invention, and in what manner the same is to be performed, to be enrolled in Her said Majesty's High Court of Chancery within six calendar months next and immediately after the date of the said in part recited Letters Patent, as in and by the same, reference being thereunto had, will more fully and at large appear:

NOW KNOW YE, that in compliance with the said proviso, I, the said Henry Wilkinson, do hereby declare the nature of my said Invention, and the manner in which the same is to be performed, to be as follows (that is to say):

The improvement in fire-arms consists in a new arrangement of the mechanism by which fire-arms may be discharged, herein-after described, either in combination with or without a new method of priming, also herein-after described. The advantages of this improvement, in my opinion, consist,—

1st. In the simplification of the whole arrangement of the lock, which according to my plan consists of three or four separate principal parts instead of twenty-five or thirty pieces as usual in a common flint musquet lock; and these three or four parts (screws not included) perform all the offices or functions of an ordinary lock, including the trigger and guard also. 2nd. The stock is much stronger, in consequence of there not being either lock, mechanism, or side piece to let in on the sides, and no screws through the sides as usual. 3rd. My improved fire-arms being much simpler in all its parts and easier to manufacture, must be, of course, cheaper. 4th. This plan admits of adopting the common copper caps, and greatly facilitates the application of my new method of priming, which offers great advantages to the soldier and to all persons employing it. The Drawings will explain all the particulars of my Invention, and the same letters are employed to distinguish the same or corresponding parts in all the Figures.

Explanation of the Drawings.

Fig. 1; A, the stock; B, the barrel; C, the plug or breeching, represented by dotted lines, which also shew the section of the whole. *a*, the mainspring and cock combined in one piece and represented at half cock; *b*, the trigger, sear, and tumbler, also in one piece; *c*, the guard cast in brass; *d*, the trigger spring held by the screw *m*; *e*, the nipple screwed into the under side of barrel; *f*, *f*, a long strap of iron forming the trigger and lock plate, concave at that part which surrounds the nipple *e*, to give more room, and on which the whole of the action is fixed, so as to work or act quite independantly of the wood, the said iron strap being bent to suit the curve of the stock and let flush into the wood; *i*, *i*, two screws to hold on the front part of guard; *g*, *g*, are screws to hold the plate *f*, *f*, in the wood, one of which may hold the mainspring and the other the guard also. When this strap is required to be used with the new method of priming, to be hereafter described, there must be a groove down the middle of the under side capable of receiving the priming magazine, in which case one of the screws *g*, in the fore part of the stock cannot be used, but the strap *f*, *f*, must be secured by screws on the side so as to leave the end of priming tube open. Fig. 2, and Fig. 4, side view of the mainspring and cock *a*, taken off the plate, and Fig. 4, the under side of the same, and lettered in the same way as in Fig. 1. *h*, is a conical raised piece of steel, either forged solid with the spring, or better if made to screw in so as to be renewed when worn by striking on the nipple. Fig. 3, the strap or trigger plate *f*, *f*, the holes being all lettered to correspond with the screws on parts which enter them, as described in Fig. 1; *e*, is the nipple in the centre of the concave part. Fig. 5, the trigger, &c. *b*, shewn separately, with its spring *d*, and pin *m*. Fig. 6, front view of the guard within the sides of which the triggers, &c. works, and is fixed to the plate *f*, *f*, by two screws *i*, *i*; Fig. 7, another arrangement complete, as ready to let into the stock, in which the action is reversed, the parts corresponding with the action or mechanism, Fig. 1, being marked the same; the mainspring *n*, is here separate from the cock, and acts on the tail of it; *h*, the part which strikes the priming on the nipple; *o*, is a thin piece of steel projecting beyond *h*, and intended to separate the primers in the tube or magazine, herein-after described, from the fire of that which is discharged, and which piece or projection *o*, passes between the said priming tube and the parts Fig. 10, which supplies the place of the nipple *e*, Fig. 1, when the priming magazine is employed, but is not required for copper caps; this piece *o*, may be applied to either of the arrangements Fig. 1, or Fig. 7. Fig. 8, a rectangular brass tube having a narrow slit *p*, *p*, throughout its whole length, and made to fit the primers Fig. 9, so as to allow them to slide easily along. A small spring *x*, enters the tube a little at *z*, to prevent them from falling out and to deliver one at a time as pressed forward by the button *r*, as required, which button fits the tube and slides through it, being retained in its place either by its own fitting or by a light spring. Fig. 9, two views of the primer intended, which may be made of card punched out with a small hole in the centre, into which a pellet of fulminating powder is placed, and both sides covered with a disc or circle of thin paper, and then varnished to render the whole waterproof, or they may be made of copper or other suitable metal. These primers are introduced into the tube or priming magazine by merely pulling out the slider *r*, and chopping them in when required. Fig. 10, the tube or nipple suitable for these primers out of the barrel, the edges at top being slightly undercut or dovetailed, to prevent the priming from falling out until it has been exploded by the action of the cock, when it is either blown to pieces, or sufficiently altered in form to fall out when the cock is raised. Fig. 11, part of the under side of the shaft of stock A, into which the priming magazine *p*, *p*, is inserted by making a groove in the wood; one end is entered under a small plate *s*, and the other being pressed down is retained in its place by a spring *t*, screwed into the wood under it, so that the priming magazine may be easily removed and refilled or replaced by another full one. If, however, the arrangement Fig. 1, be employed, which is preferable to Fig. 7, the priming magazine has no slit. The spring must be placed on one side to leave the end open, and the primers must be pushed forward by means of a narrow long strip of metal fitting the tube and sliding in a groove in the stock in a line with the magazine, so as to enter it when pressed into its place. Fig. 12, a side view of the priming magazine out of the wood, with the spring and pin *t*. N.B.—The ramrod must be placed a little on the right side of the shaft of stock looking towards it, to avoid the screws or magazine in the centre, which is rather an advantage than otherwise.

Now I do not claim as new the placing of a lock on the under side of the barrel, which has been done before; nor any of the springs, screws, straps, tubes, or pieces aforesaid, separately; but I claim as my Invention the general combination and arrangement substantially herein-before described in the mechanism for discharging fire-arms; and whether the same is used in whole or in part, it being obvious that my new method of priming herein-before described may be used either as a part of my said general improvement, or separately applied to fire-arms of the ordinary construction. And such my Invention being, to the best of my knowledge and belief, new, I declare this to be a true Specification thereof.

In witness whereof, I, the said Henry Wilkinson, have hereunto set my hand and seal, this Twenty-first day of December, in the year of our Lord One thousand eight hundred and thirty-nine.

HENRY (L.S.) WILKINSON.

AND BE IT REMEMBERED, that on the Twenty-first day of December, in the year of our Lord 1839, the aforesaid Henry Wilkinson came before our said Lady the Queen, in Her Chancery, and acknowledged the Specification aforesaid, and all and every thing therein contained and specified, in form above written. And also the Specification aforesaid was stamped according to the tenor of the Statute made for that purpose.

WINGFIELD.

Inrolled the Twenty-first day of December, in the year of our Lord One thousand eight hundred and thirty-nine.

LONDON:
Printed by GEORGE EDWARD EYRE and WILLIAM SPOTTISWOODE,
Printers to the Queen's most Excellent Majesty. 1854.

A.D. 1839, June 22. Nº 8119.
WILKINSON'S Specification.

(1 SHEET)

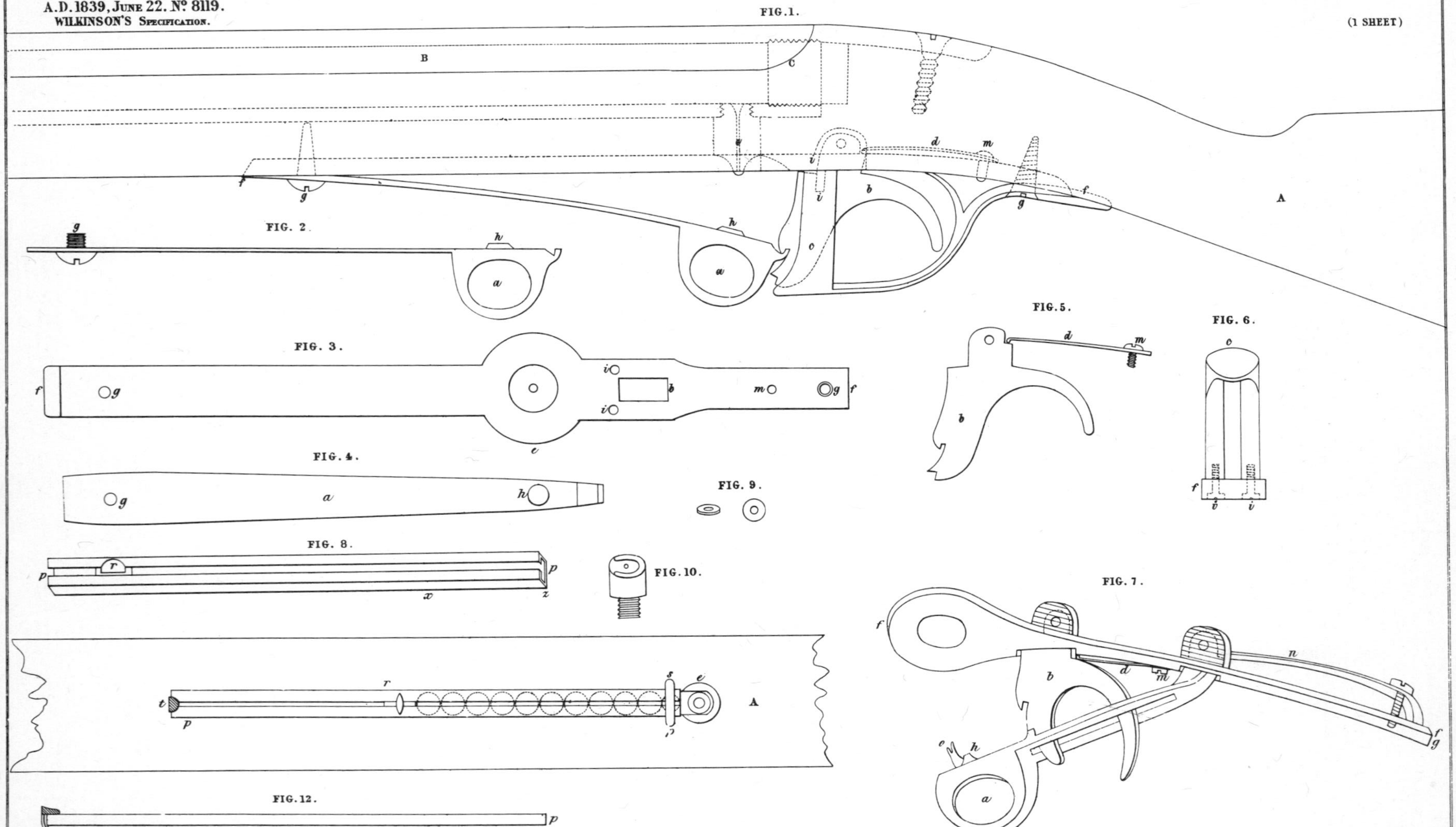

The Enrolled Drawing is colored.

J. R. Jobbins

London. Printed by George Edward Eyre and William Spottiswoode.

A.D. 1840 . . . . N° 8704.

Primer and Percussion Caps for Fire-arms.

SHAW'S SPECIFICATION.

TO ALL TO WHOM THESE PRESENTS SHALL COME, I, Joshua Shaw, of Goswell Street Road, in the County of Middlesex, Artist, send greeting.

WHEREAS Her present most Excellent Majesty Queen Victoria, by Her Royal Letters Patent under the Great Seal of Great Britain, bearing date at Westminster, the Seventeenth day of November, in the fourth year of Her reign, did, for Herself, Her heirs and successors, give and grant unto me, the said Joshua Shaw, Her especial license, full power, sole privilege and authority, that I, the said Joshua Shaw, my executors, administrators, or assigns, and such others as I, the said Joshua Shaw, my executors, administrators, or assigns, should at any time agree with, and no others, from time to time and at all times during the term of years therein mentioned, should and lawfully might make, use, exercise, and vend, within England, Wales, and the Town of Berwick upon Tweed, and in all Her Majesty's Colonies and Plantations abroad, my Invention of "Certain Improvements in Discharging Ordnance, Muskets, Fowling Pieces, and other Fire-arms;" in which said Letters Patent is contained a proviso, obliging me, the said Joshua Shaw, by an instrument in writing under my hand and seal, particularly to describe and ascertain the nature of my said Invention, and in what manner the same is to be performed, and to cause the same to be inrolled in Her Majesty's High Court of Chancery within six calendar months next and immediately after the date of the said in part recited Letters Patent, as in and by the same, reference being thereunto had, will more fully and at large appear.

NOW KNOW YE, that in compliance with the said proviso, I, the said Joshua Shaw, do hereby declare that the nature of the said Invention, and the manner in which the same is to be performed, is particularly described and ascertained in and by the following description thereof, reference being had to the Drawings hereunto annexed, and to the letters and figures marked thereon (that is to say):

In making the discharge I ignite the powder constituting the charge by means of a percussion cap charged with fulminating powder in the ordinary way, but when used for the discharging of small arms, I employ percussion caps of a size considerably smaller than those in ordinary use, and for the discharge of cannon they need not be larger than those now employed with small arms. A distinguishing feature of my Invention is the manner of employing or using these caps, which, under all the modifications thereof, is by placing them upon one end of a cylindrical rod or wire of steel or other metal, which rod or wire, for distinction sake, I will call a piston, and inserting the said piston, with the cap on the end thereof, into a cylindrical opening, occupying the place of the ordinary nipple, or otherwise conveniently situated, the said opening being adapted in size to the piston which is to slide freely but closely within it. The cylindrical piston is to be a trifle larger in its diameter than the outside diameter of the cap which is to be placed upon it, its end being turned down or reduced in size, so as to pass into and to be embraced by the cap. The hole or opening into which the piston is to pass is to contain in some part of it a piece or pieces of leather, cork, or other elastic substance or material, which is to embrace the piston, and thus render the passage air and water tight, whilst at the same time the said material will serve by its elasticity to hold the piston in its place, or the piston may be retained by a spring. In small arms the hole or opening through which the piston passes may be of equal bore or diameter throughout, as the piston is, in some cases, to be of sufficient length to enter the chamber of the barrel, and to extend across it, so that the percussion cap which is placed upon its end may be brought into contact with the interior of the chamber on the side opposite to that at which it entered; and it is manifest that if the outer end of the piston be then struck by a hammer or mallet the percussion cap will explode, and the gunpowder will be ignited. In pieces of ordnance there are sundry objections to the allowing of the piston to extend across the chamber of the gun, and from the thickness of the breech in arms of this description a sufficient depth is obtained in the solid metal for the operation of the piston, without its entering the chamber of the gun In using my piston, with the percussion cap at its end, for the discharging of pieces of ordnance, I bore a hole in the manner and generally in the place of the ordinary touch hole, such hole being of the size of the piston intended to be used, say of three sixteenths of an inch in diameter, more or less, and extending down nearly to the chamber of the gun, say within a fourth or half of an inch thereof, and through this remaining part I drill a hole of much smaller size, say one sixteenth of an inch, more or less, in diameter until it enters the chamber. The piston must be of such length as to extend down to the bottom of the larger portion of the bore, and to rise to a sufficient height above the top of it for the action of the hammer by which it is to be struck. Its lower end, or that which enters the bore, is, as above stated, to be turned down, so as to form a nipple for the reception of the percussion cap. The percussion cap which I use under this modification of my Invention I have improved in such a manner as adapts it especially to the object in view This improvement consists in the making a small hole, say of one sixteenth of an inch in diameter, through the centre of what in the ordinary cap is its closed end, this being done before the cup is charged with percussion powder. When the piston is inserted in the bore, having a percussion cap of the kind just described on its lower end, if the piston be struck by a mallet or hammer on its upper end the cap will explode, and the ignited percussion powder will pass through the hole in the head of the cap and through the small part of the bore under it into the chamber of the cannon, where it will ignite the powder. I sometimes drill a small hole in the end of the piston in the direction of its axis, which hole may be from an eighth to a fourth of an inch in depth, and of the same size with that in the percussion cap and also with that leading into the chamber of the gun; this hole is to be filled with common gunpowder which will be ignited by the percussion powder. By using this device the percussion caps may be very lightly charged with percussion powder, the gunpowder greatly aiding in causing the discharge of the cannon. This is a is a point of considerable importance, as the destructive influence of the percussion powder upon the parts with which it is in direct contact is thereby to a considerable extent obviated. It may also be remarked that the percussion cap which remains upon the piston after the discharge, has the effect of protecting it and the adjacent parts from injury by the percussion powder. I will now exemplify, by reference to the accompanying Drawings, the manner in which my improvements may be carried into actual operation. Figures 1, and 2, represent modes in which it may be conveniently applied to small arms. In Fig. 1, *a*, is a perforated screw nut which screws into the barrel at the place usually occupied by the nipple, or in any other preferred situation; the end of this screw is to be flush with the inside of the chamber; *b*, is a screw on the upper part of this nut, upon which there is fitted a screw cap *c*; the cavity in this cap is somewhat deeper than the length of the screw *b*, in order to admit of the insertion of a piece of leather, or of some other elastic substance, through which the stem of the piston *d*, *e*, is to pass. This stem is to be made perfectly cylindrical, and the holes in the nut and cap through which it passes are to be made perfectly true and adapted nicely to it in size, so that such pistons may slide in and out easily, and yet have no shake. Fig. 6, is a section of Fig. 1, shewing the piston *d*, and *e*, and the piece of leather, cork, or other suitable material *g*, for the purpose of retaining the piston in its proper place. Fig. 2, is another method of applying my Invention, but more simple in its construction than Fig. 1. *a*, is a screw nut which screws into the barrel, but does not pass through into the chamber, there being a solid portion of the barrel below it, with the exception of an opening of such size only as to admit the piston to pass through it. The elastic material *g*, under this arrangement is placed beneath the screw *a*, and is pressed between it and the part of the barrel beneath it. The head *f*, of this screw nut, together with its screw part *a*, receives the piston *d*, *e*, as in Fig. 1, its whole construction and operation being substantially the same. The length of this piston is under this arrangement to be such as to extend across the chamber of the piece to be discharged. The small percussion cap is received on its inner end *e*, and this must be, as before remarked, somewhat smaller in diameter than the piston; these percussion caps are in other respects constructed in the ordinary way. Any required number of these pistons, armed with their caps, may be conveniently carried by arranging them in holes prepared for their reception in the stock or breeching part of the gun or pistol, or they may be disposed of in any way that will admit of their being readily applied when wanted. Fig. 3, is a section through the touch hole part of the breech or vent field of a piece of ordnance, showing the manner in which I drill or bore through this part for the insertion of the piston and the exploding of the cap; and Fig. 4, is the piston; *g*, *g*, is the part of the bore which is to receive the cylindrical piston, Fig. 4; and *h*, the lessened aperture leading from the lower end of *g*, *g*, into the chamber of the gun, thus leaving a shoulder at *i*, *i*, against which the percussion cap is to be forced when the discharge is to take place. I have said that the bore *g*, *g*, should be perfectly cylindrical, and this is true, excepting that there is a slight contraction at its lower end, in the part which embraces the percussion cap at the time of the explosion, this part being made slightly conical to the height of from an eighth to a fourth of an inch, which causes it to embrace the cap the more closely after the blow, and still allows it to be readily withdrawn, as the slightest motion relieves it from the conical termination. I have, in the application of my improvement to cannon, devised a very simple mode of using an elastic substance through which the piston is to pass, and by which the entrance through the opening *g*, *g*, is rendered air and water tight. I make a countersink *l*, *l*, at the upper end of the bore *g*, *g*, in which I insert a piece of cork having a hole through the middle of it, through which hole the piston will pass and by which it will be closely embraced. This cork may be firmly held in its place by small asperities on the sides of the countersink by widening the countersink a little towards its bottom, or in various other ways. If preferred, an elastic substance may be held down by means of a perforated screw nut inserted in the vent field, but I think that the cork, inserted as above described, is preferable, as I have found it to be perfectly efficient, and it may be readily and instantaneously renewed. I prefer in all cases to make the piston of tempered steel, but hard drawn brass wire, or iron, will do very well, and they are accurately turned and ground and fitted to the perforations prepared to receive them. In Fig. 4, *n*, is a section of the copper cap which is to be placed on the lower end of the piston *m*, *m*; this cap is perforated at *o*, prior to its being charged with percussion powder, the opening being equal in size, or nearly so, to the perforation at *h*; the lower end *p*, of the piston is turned down to receive the cap, and it is diminished in diameter at the part *q*, so as to leave a shoulder at *r*, above which the copper cap rises, and the upper edge of the cap is to be slightly bent inwards, so as to embrace this shoulder, and this may be easily effected by holding a piece of steel, say one of the pistons, or a tool kept for the purpose, upon the edge of the cap after it is in place, and rolling it over upon a table; by this means it is made to embrace the piston so firmly as to ensure its being withdrawn with it after the discharge, whilst without this precaution it would be liable to be left behind. The dotted lines at the lower end of the piston at *p*, shew the perforation sometimes made for containing gunpowder. The discharge is made by allowing a hammer to strike on the head *d*, of the piston, and this may be done by hand, but it will usually be effected by means of a lock, constructed in any convenient manner and adapted to the piece to which it is to be applied. Having described the manner in which I apply my piston and its percussion cap for the discharge of pieces of ordnance, I will now observe that I intend sometimes, and shall probably prefer, to adopt a similar arrangement in small arms (that is to say), to use a piston which shall not enter the chambers of such arms, but which shall be contained entirely within the tubular screw nuts represented in Figures 1, and 2, or within tubular pieces analogous thereto. In this case I shall use also the perforated percussion cap, like those used for the discharge of cannon, and allow the percussion powder to pass therefrom into the chamber of the pistol or gun, through a perforation smaller than that which receives the piston, the whole arrangement of the apparatus being in this case substantially the same with that described in my application of it to cannon, the main difference being in the size only of the respective parts. From the smallness of the openings through which the discharge is made from the percussion cap into the powder forming the charge, or from the smallness of the section of the piston used in small arms, when the said piston is allowed to pass through the chamber, the reaction from the discharge tending to throw the piston out will be but small, and it will be held down by the force of the spring acting upon the hammer of the lock when a lock is used. When the piston is to be struck by hand, as is the case in ordnance, a button attached to the vent field of the gun, and bearing on the piston near its upper end, may be used for checking the rise of the piston from the force of the discharge. Fig. 5, shews the manner in which this button may be affixed; *d*, is the head of the piston, and immediately under this head there is a shoulder r^1, made by turning the shank of the piston so as to form a neck smaller than the shank; *s*, is a steel button which turns on a screw at *t*, and passes under a catch *u*, at its opposite end. The neck of the piston is embraced by the notch in the button *s*, and is thus held from being blown out. The neck is of sufficient length to give the requisite play to the piston, which can be drawn out when the button is turned back. I now proceed to describe a third modification of my Invention, in which I dispense with the cap as constructed and applied in the manner hereinbefore described, in place of which I make a cavity in the lower end of the piston, about the twentieth part of an inch in depth, and nearly as large as the diameter of the piston will admit of, leaving but a slender rim to receive and retain the fulminating powder with which it is filled, and which is further secured in its place by wax, varnish, or other means convenient for the purpose; this piston when applied to small arms is about one eighth of an inch in diameter, and about five eighths, more or less, in length, and for distinction sake I now denominate it the priming cap. When this priming cap is placed in the priming tube or nut pre-paratory to discharging the priming at its lower end resting immediately over the vent, the instant it receives the blow of the cock on its upper extremity the slender edge or rim retaining the fulminating powder gives way, bulging upwards, and as the matter ignites and the intensity of its fire increases, the bulging of the metal increases, so as to be capable of resisting any escape of fire otherwise than by passing through the vent in the bottom of the nut or priming tube into the chamber of the gun, whereas in the ordinary mode in which the priming caps are discharged, it is effected by projecting a nipple or a slightly projecting fillet of metal into their cavity, which allows in most cases the escape of one half of all the fire generated, driving off portions of the cap, to the imminent danger of the eyes, or otherwise creating dirt and rust about the lock and breech of the gun. The nut or priming tube employed to receive and to operate in combination with this priming cap I construct in the same way as the one previously described, Fig. 1, with this difference, that the cylindrical bore is not continued entirely through it, but only to about four fifths of its entire length, or to within an eighth or the sixteenth of an inch of its lower end, and leaving a flat bottom for the priming cap to rest upon, below which a smaller perforation is continued, and which constitutes the vent leading into the chamber of the gun, which vent is made conical, terminating in the smallest possible point where it enters the chamber of the gun, by which means the return of fire of the charge is cut off, as shewn in Fig. 7, which is a section taken through the vent hole. On applying these improvements to practical use the nut or priming tube is screwed into the breech of the gun, the priming cap or piston is then introduced, and on being struck on its upper extremity by the cock of the lock the cap is exploded and the gun discharged. Fig. 8, is a sectional view of the several parts of my third modification, constructed in the manner just described, the parts being shown detached to avoid confusion. This construction of primer somewhat resembles that shewn in Fig. 1, except that in Fig. 1, that end of the piston which contains the detonating powder is passed through the nipple into the magazine, whereas in the present instance such is not the case, the detonating powder being exploded at the lower end of the tube and the fire generated thereby blown directly into the magazine through the small aperture formed in the lower end of the tube for that purpose, as seen in Fig. 7. Fig. 9, represents a section of the piston, showing the recess formed at the lower end thereof, containing the detonating powder. Sometimes, instead of the disc of leather or cork *g*, used for the purpose of keeping the piston in its proper place and preventing the escape of fire or smoke from the nipple, I employ a small spring formed on or attached to the nipple, and which as it presses against the piston prevents it from falling out, or I raise a burr upon the side of the primer, which effects the same object. This, if preferred, may of course be applied to either of the modifications above described. In cannon and other fire-arms, constructed as herein described, the vent is not only rendered air and water tight, but the escape of fire and smoke through it is effectually prevented, as has been proved by numerous trials.

Having thus fully described the nature of my improvements, and shewn the manner in which I carry the same into operation, what I claim as constituting my Invention, and desire to secure by Letters Patent, is, the effecting of the discharging of fire-arms,—Firstly, by placing a percussion cap on the end of a cylindrical piston or rod, formed for the purpose in the manner herein set forth, such piston with its cap being made to pass through an opening adapted thereto, either in the body of the breech of the piece to be discharged or through tubular pieces screwed into or otherwise attached thereto, the said piston being of such length as shall adapt it to the respective modifications of my apparatus, as above described, (that is to say,) I claim the passing of such a piston through a tubular opening, allowing it to reach across the chamber, as in my first described modification, or I pass it into a cylindrical opening extending nearly to the chamber, where a shoulder is formed by diminishing the size of the aperture in the manner and for the purpose set forth, the respective parts concerned in the discharges being constructed and operating substantially as herein fully made known. I claim likewise the improvement in the percussion cap, consisting of the perforation through the middle of what is usually its closed end in the manner and for the purpose described. And, secondly, I claim the method shown in Figs 8, and 9, of applying, adapting, or containing the detonating powder in a recess formed at the lower end of the piston, on which the slender edge or rim inclosing the fulminating powder is made so thin that on receiving the blow of the cock it swedges upwards, and progressively giving way as the matter becomes ignited, drives all the fire before it into the chamber of the gun, allowing none of it to escape, by which means any accumulation of dirt or rust about the lock or breech is entirely prevented, insuring quickness and safety, at the same time the fire being generated within the sixteenth of an inch of the magazine and upon a level with it. And I also claim as new and useful the nut or priming tube, Fig. 8, made to receive and secure the primer containing the detonating powder in its proper place, and from the effects of rain or moisture, and for the purposes herein-before set forth and described; but more especially do I claim as my Invention the priming cap, constructed to operate in the way above stated, and acting in combination with the priming nut or tube.

In witness whereof, I, the said Joshua Shaw, have hereunto set my hand and seal, this Twenty-ninth day of April, in the year of our Lord One thousand eight hundred and forty-one.

JOSHUA (L.S.) SHAW.

RUSSELL.

AND BE IT REMEMBERED, that on the Twenty-ninth day of April, in the fourth year of the reign of Her Majesty Queen Victoria, the said Joshua Shaw came before our said Lady the Queen, in Her Chancery, and acknowledged the Instrument aforesaid, and all and every thing therein contained and specified, in form above written. And also the Instrument aforesaid was stamped according to the tenor of the Statute made in the fifty-fifth year of the reign of His late Majesty King George the Third.

Inrolled the Seventeenth day of May, One thousand eight hundred and forty-one.

LONDON:
Printed by GEORGE EDWARD EYRE and WILLIAM SPOTTISWOODE,
Printers to the Queen's most Excellent Majesty. 1854.

A.D. 1840. November 17. Nº 8704.

(1 SHEET)

SHAW'S Specification.

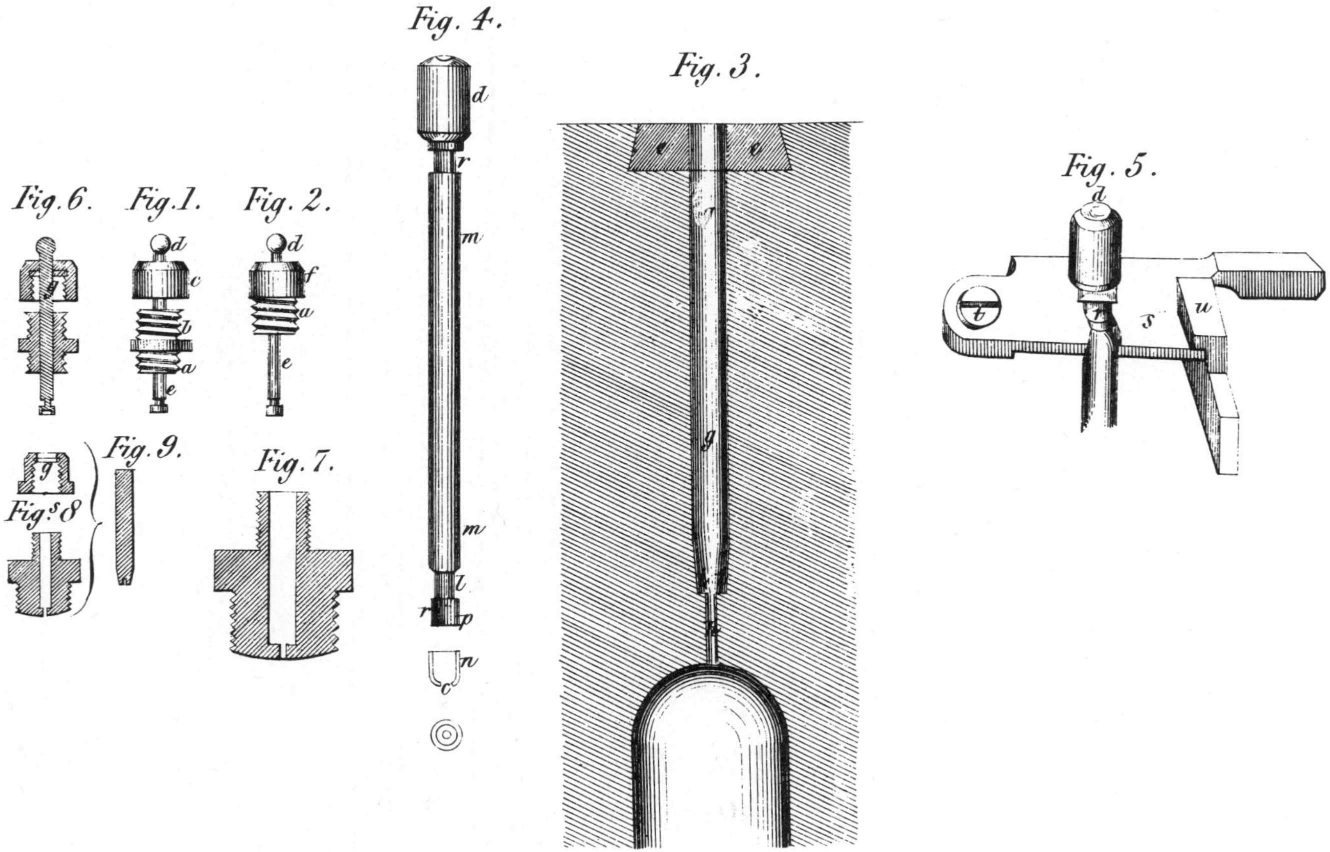

Fig. 1. Fig. 2. Fig. 3. Fig. 4. Fig. 5. Fig. 6. Fig. 7. Figs. 8. Fig. 9.

The Enrolled Drawing is not Colored.

Malby & Sons. Lith.

London: Printed by George Edward Eyre and William Spottiswoode. Printers to the Queen's most Excellent Majesty. 1854.

A.D. 1841 . . . . N° 9084.

Continuous Priming for Fire-arms, and Mechanism for applying the same.

HEURTELOUP'S SPECIFICATION.

TO ALL TO WHOM THESE PRESENTS SHALL COME, I, Charles Louis Stanislas Baron Heurteloup, of Albany Street, Regent's Park, in the County of Middlesex, send greeting.

WHEREAS Her present most Excellent Majesty Queen Victoria, by Her Letters Patent under the Great Seal of Great Britain, bearing date at Westminster, the Ninth day of September, in the fifth year of Her reign, did, for Herself, Her heirs and successors, give and grant unto me, the said Charles Louis Stanislas Baron Heurteloup, Her especial licence that I, the said Charles Louis Stanislas Baron Heurteloup, my exors, admors, and assigns, or such others as I, the said Charles Louis Stanislas Baron Heurteloup, my exors, admors, or assigns, should at any time agree with, and no others, from time to time and at all times during the term of years therein expressed, should and lawfully might make, use, exercise, and vend, within England, Wales, and the Town of Berwick-upon-Tweed, the Islands of Jersey, Guernsey, Alderney, Sark, and Man, and also in all Her Majesty's Colonies and Plantations abroad, my Invention of "**An Improved Manufacture of Continuous Priming for and Improved Mechanism for**

THE APPLICATION OF THE SAME TO CERTAIN DESCRIPTIONS OF FIRE-ARMS;" in which said Letters Patent is contained a proviso, obliging me, the said Charles Louis Stanislas Baron Heurteloup, by an instrument in writing under my hand and seal, particularly to describe and ascertain the nature of my said Invention, and in what manner the same is to be performed, and to cause the same to be enrolled in Her said Majesty's High Court of Chancery within six calendar months next and immediately after the date of the said herein-before in part recited Letters Patent, as in and by the same, reference being thereunto had, will more fully and at large appear.

NOW KNOW YE, that in compliance with the said proviso, I, the said Charles Louis Stanislas Baron Heurteloup, do hereby declare the nature of my said Invention to consist,—

First, in an improved material or composition of the detonating materials of which the continuous priming is manufactured. The material being such, or the relative proportions of the said materials when more than one material is used, being so regulated, that though the explosion caused by the percussion on the piece of priming, which is immediately over the nipple of the fire-arm, is sufficient to ignite the charge in the barrel, it will not cause the remainder of the priming to ignite, or if it should the priming will only burn along a very short length, and will extinguish itself immediately, without the possibility of damage accruing from such last-mentioned ignition of so small a portion.

And, secondly, my Invention consists in improved mechanism for the application of the said priming, in which the priming is coiled round, within a hollow case in a very small space, without interfering in any way with the stock of the fire-arm, and carefully protected from rain, and such moisture as fire-arms are subjected to in ordinary use.

And in further compliance with the said proviso, I, the said Charles Louis Stanislas Baron Heurteloup, do hereby describe the manner in which my said Invention is to be performed, by the following statement thereof, reference being had to the Drawings annexed, and to the figures and letters marked thereon, similar letters being used to denote similar parts in all the Figures and Drawings. And first, of the improved manufacture of continuous priming, (which I have before said depends upon the material or the relative proportions of materials used,) I have found that continuous priming in straight narrow tubes, heretofore used, and which priming is composed of chlorate of potash, sulphur, and charcoal, in proportions of about five parts chlorate of potash, one sulphur, and one and a half charcoal, is liable to ignite beyond the part struck by the hammer, and necessary for exploding the charge in the barrel of the fire-arm, and then often continues to burn like a slow match, until the whole tube of priming is consumed; and thus, in the event of any portion of a stored stock of such priming being by accident set fire to, the whole would probably be consumed, though the magazine in which it was stored might escape injury. This liability to loss may be entirely prevented, as I have discovered, by either using chlorate of potash by itself, or manufacturing the materials usually compounded for the purpose in the following proportions, that is to say, to forty-eight parts of chlorate of potash I add only one part of sulphur, and one part of charcoal; these, being finely powdered, I shake into small soft metal drawn tubes of pewter or other soft metal, of about one tenth of an inch in diameter, by means of a hopping or jolting apparatus, as shown at Figure 1, on the Drawing marked No. 1; the parts marked *a*, being a framing mounted upon a platform *b*, and provided with guide pieces *c*, *c*. *d*, is a spindle having two rollers *e*, *e*, mounted upon it, with notches or teeth on their peripheries, and which rollers are turned by a handle *f*, at the end of the spindle. *g*, *h*, are two boards, the former being smooth on both sides, and the other having one side cut into grooves *i*, *i*, the grooves being sufficiently deep to allow the tubes to be placed in them without projecting beyond the grooves, so that the board *g*, when laid against the grooved surface, shall only touch the spaces between the grooves, and also be clear of the tubes placed therein. Upon the top of the grooved board is fixed a row of hoppers *k*, corresponding in number with the grooves in the board; the hoppers are larger in diameter than the grooves, but are tapered at bottom to the same size as the grooves. *l*, *l*, *l*, are screw bolts for fastening the boards *g*, and *h*, together when the tubes are placed between them; at the bottom of the board *g*, is screwed a strip of metal *m*, (see more clearly in the enlarged Figure

31,) for the purpose of pressing against the end of the tube, and keeping it tight whilst being jolted upon the rollers. The tubes, previously to being placed in the grooves, are closed at their lower ends by being pinched, which keeps the powder from running out, and the upper end is widened out as in Figure 31, where it rests in the conical part of the hoppers. As soon as the tubes are placed in the grooves the flat board *g* must be placed against them, and tightened up by the thumb screws *l*, *l*; the hoppers may then be supplied with the priming powder, and placed between the guides of the frames, resting at bottom upon the rollers *e*, *e*; the handle being turned will cause the boards to be jolted up and down, and shake the powder into the tubes; after which the boards may be unscrewed, and the tubes taken out by hand; and by drawing them through the conical part of the hoppers they will become compressed, and thus enclose the powder within them; the tubes must then be passed through a pair of rollers, which will flatten them, and cause them to assume the form of tapes shewn in Figures 32, 33, and 34. Figure 28, is a plan of Figure 1; Figure 29, is a front view of the groove board *h*; and Figure 30, is a side view of one of the rollers *e*. In the aforesaid shape of long narrow tapes they are introduced into my said improved mechanism, from which they are to be used. And I wish it to be distinctly understood here, that I lay no claim, as regards the said Letters Patent, to the mode of filling the tubes, though of my own Invention, the same having been used and patented by me in France previously to the date of the said herein-before in part recited Letters Patent, and therefore probably previously known in England, through the medium of the foreign periodical publications; but as I do herein-after claim chlorate of potash by itself, or the composition in the proportions aforesaid, which is shaken into the said tubes, I have thought it right, in order to leave nothing unexplained, to enable competent workmen to carry out my Invention from the instructions contained in the Specification alone, to insert the foregoing description of the manner of filling the tubes with priming powder. Thus, it will be seen that chlorate of potash by itself, inclosed as aforesaid, or the new relative proportions of the materials of which the powder is composed, (as explained in the claim hereafter,) enclosed as aforesaid, constitutes my said improved manufacture of continuous priming. And I will now proceed to describe the improved mechanism in which it is to be placed for use. Figure 2, on Drawing No. 2, represents the appearance of the said mechanism complete, as applied to an ordinary single-barrelled gun with a common breech. To the parts here visible I have given the following names, a statement of which will facilitate the comprehending of the other Figures. The names are as follows:—A, the hood; B, the side shield; C, the foundation plate; D, the securing screw, that holds the three last-mentioned pieces, and also a fourth piece (not here seen), which they enclose firmly together; E, a thumb lever or priming guage for determining the length of priming tape to be brought over the nipple; F, the hammer for cutting off and detonating the piece of priming tape required for exploding the charge in the barrel of the fire-arm, being furnished with a nib or blade, which, passing close to and under the end of the beak, cuts off the piece of tape required. Figure 3, is a side elevation of the foundation plate C, and Figure 4, is a similar elevation, showing its position on the barrel of the gun to which it is to be brazed. Figure 5, is a section of the foundation plate and barrel brazed together; and Figure 6, is a perspective view of the foundation plate, in order to give a more perfect idea of its formation. The Figures 4, 5, and 6, also shew the direction of the touch-hole. Figure 7, is a side elevation of the hood A; Figure 8, a plan of its under side; and Figure 9, a front elevation of the same. Figure 10, is a side elevation of the side shield B; and Figure 11, an edge view of the same. Figure 12, is the securing screw D, shown separately. Figure 13, is a side elevation of the fourth piece before alluded to when describing Figure 2, and which in that Figure is concealed under the hood A. This fourth piece G, Figure 13, I call the magazine; and Figure 14, shows it placed in its proper position in the foundation plate C. This magazine contains the priming tape coiled round within it, a portion of which is seen protruding at K, from the mouth, or (as I call it) the beak of the magazine. The thumb lever or priming guage E, it will be seen, is attached to the magazine, and its action in bringing the priming tape over the nipple will be explained hereafter. Figure 15, is an end view of the magazine; and Figure 16, is a separate view of the thumb lever or priming guage, which it will be perceived is furnished at its inner end with a triangular boss, the apex of each angle of the boss being roughed or toothed for a purpose herein-after explained. Figure 17, is a side elevation of the magazine G, with the priming tape, which is colored red, coiled round within it; the side here shown is the side which is placed next the foundation plate when in use, and thus when in its position on that plate the whole of the priming may be said to be inclosed within a complete case. But now more particularly to describe Figure 17. J, is a small spring on the top of the magazine, and screwed on to it at L; this spring is furnished with a boss at its end J, which presses upon the priming tape, and keeps it down upon the triangular boss H of the thumb lever below it; the tape being thus pressed between the two bosses is acted upon by the teeth on the apex of each angle of the boss H, and drawn forward or backward as this boss is turned one way or the other by the thumb lever E. If turned in the direction of the arrow marked in this Figure it will advance the tape out of the beak of the magazine; and this Figure shows the teeth of the boss H, biting on the tape, while Figure 18, shows the tape advanced far enough to be over the nipple, and the tape held upon one of the flat sides of the boss H. Figure 19, is a transverse section through the magazine, from *r* to *s*, and Figure 20, a similar section through the beak of the magazine, while Figure 21, is a transverse section through the beak, side shield, and hood, showing the space occupied by the foundation plate when the magazine is placed in its proper position on that plate. It is only further necessary, in explanation of this Sheet of Drawings No. 2, to observe that the nib shown at F, on the hammer in Figure 2, answers the purpose of the steel blade formerly used, giving a scissars cut to divide the tape, instead of a knife cut as formerly. In order to charge this magazine it is only necessary to introduce an end of the priming tape at the mouth or beak of the magazine until it reaches the bosses, when, if the thumb lever is turned backwards, it will draw the tape into the magazine, and the tape will coil itself round the central part of the circular portion of the magazine as shown in

A.D. 1841. SEPTEMBER 9. Nº 9084.

HEURTELOUP'S SPECIFICATION.

(3 SHEETS)

FIG. 22.

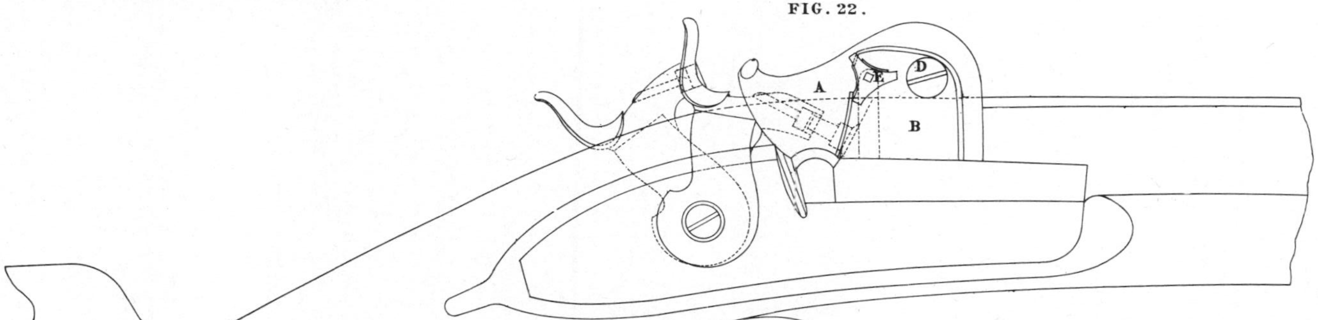
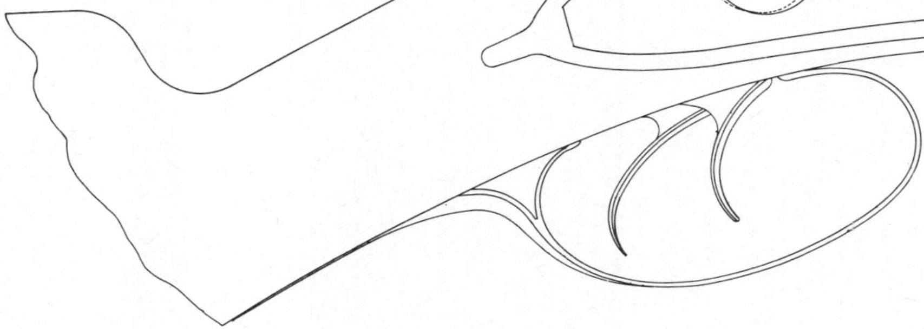

FIG. 24.

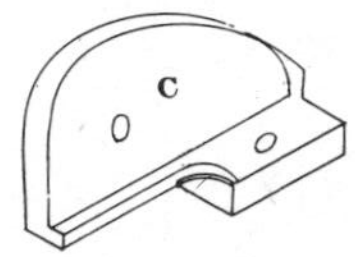

FIG. 26.

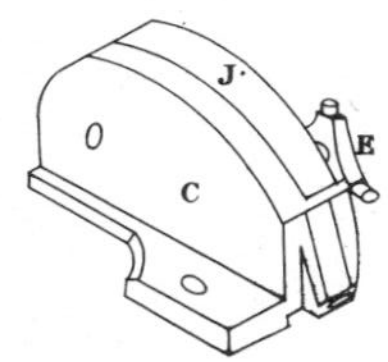

FIG. 25.

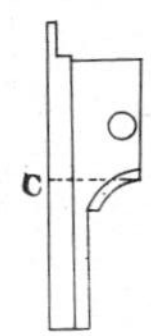

FIG. 23.

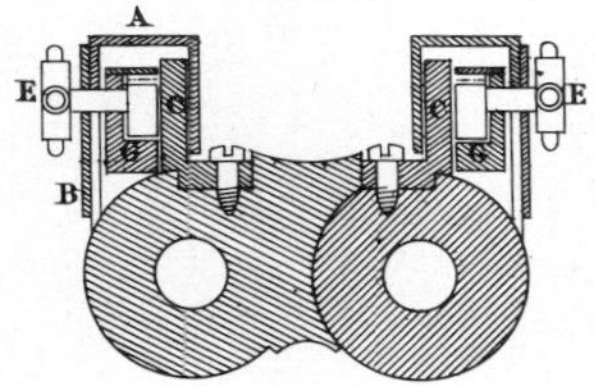

FIG. 27.

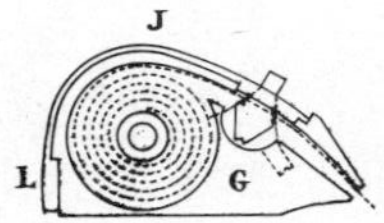

The Enrolled Drawing is partly colored.

LONDON. Printed by GEORGE EDWARD EYRE and WILLIAM SPOTTISWOODE.
Printers to the Queen's most Excellent Majesty. 1854.

J.R. Jobbins.

Figures 17 and 18, and when full the tape may be cut off even with the mouth of the magazine. Sheet No. 3, represents the mode of applying my said Invention to a double-barrelled fowling piece having patent breeches. A mere inspection of Figures 22, 23, 24, 25, 26, and 27, on this Sheet, will be sufficient to explain it, after what has been already described; similar letters being used to denote such parts as correspond with those similarly marked in Sheet No. 2.

Now whereas I have felt it necessary, in order to a clear comprehension of the details of my said Invention, to describe several things which I do not claim as any part of the Invention for which the said herein-before in part recited Letters Patent were granted: And whereas I therefore declare that I hereby only claim as my Invention,—first, the improved manufacture of continuous priming herein-before and herein-after described, that is to say, continuous priming made of chloride of potash only, or chloride of potash, sulphur, and charcoal, either or both, in such proportions, dependant on the particular strength and condition of the materials used, as shall be sufficient to explode such parts thereof as shall be exposed in tubes rolled into tapes to the action of a blow, for the purpose of continuous priming, without thereby at the same time exploding, burning, or otherwise destroying any other or material part of the same in immediate contact with the part so struck, such improved continuous priming having the inherent property of self-extinguishment, in consequence of chloride of potash only being used, or when the compound of materials is used in consequence of such attention to relative proportions as aforesaid. And, secondly, the improved mechanism by which such improved continuous priming is applied to muskets, fowling pieces, and such other fire-arms as are from their form and nature susceptible of such application; that is to say, the covering marked A and B, the foundation plate C, and the magazine G, and the mode of applying the same to the fire-arm with which it is to be used, either as shown in Sheet No. 2, or in Sheet No. 3.

And such my Invention being to the best of my knowledge and belief entirely new, and never before used within that part of Her said Majesty's United Kingdom of Great Britain and Ireland called England, Her said Dominion of Wales, or Town of Berwick-upon-Tweed, the

Islands of Jersey, Guernsey, Alderney, Sark, and Man, or in any of Her Majesty's Colonies and Plantations abroad, I do hereby declare this to be my Specification of the same, and that I do verily believe this my said Specification doth comply in all respects fully and without reserve or disguise with the proviso in the said herein-before in part recited Letters Patent contained; wherefore I hereby claim to maintain exclusive right and privilege to my said Invention.

In witness whereof, I, the said Charles Louis Stanislas Baron Heurteloup, hath hereunto set my hand and seal, this Seventh day of January, in the year of our Lord One thousand eight hundred and forty-two.

B^ON^ HEURTELOUP. (L.S.)

RICHARDS.

AND BE IT REMEMBERED, that on the Seventh day of January, in the year of our Lord 1842, the aforesaid Charles Louis Stanislas Baron Heurteloup came before our said Lady the Queen in Her Chancery, and acknowledged the Specification aforesaid, and all and every thing therein contained and specified, in form above written. And also the Specification aforesaid was stamped according to the tenor of the Statute made for that purpose.

Inrolled the Ninth day of March, in the year of our Lord One thousand eight hundred and forty-two.

LONDON:
Printed by GEORGE EDWARD EYRE and WILLIAM SPOTTISWOODE,
Printers to the Queen's most Excellent Majesty. 1854.

A.D. 1841 . . . . N° 9177.

Locks and Primers for Fire-arms.

RICHARDS' SPECIFICATION.

TO ALL TO WHOM THESE PRESENTS SHALL COME, I, William Westley Richards, of Birmingham, in the County of Warwick, Gun Maker, send greeting.

WHEREAS Her present most Excellent Majesty Queen Victoria, by Her Royal Letters Patent under the Great Seal of Great Britain, bearing date at Westminster, the Fourteenth day of December, in the fifth year of Her reign, did, for Herself, Her heirs and successors, give and grant unto me, the said William Westley Richards, Her especial licence, full power, sole privilege and authority, that I, the said William Westley Richards, my executors, administrators, and assigns, and such others as I, the said William Westley Richards, my executors, administrators, or assigns, should at any time agree with, and no others, from time and at all times during the term of years therein mentioned, should and lawfully might make, use, exercise, and vend, within England, Wales, and the Town of Berwick upon Tweed, my Invention of "Improvements in the Construction of Gun and Pistol Locks and Primers for the Discharge of Fire-arms," in which said Letters Patent is contained a proviso, obliging me, the said William Westley Richards, by an instrument in writing under my hand and seal, particularly to describe and ascertain the nature of my said Invention, and in what manner the same is to be performed, and to cause the same to be inrolled in Her Majesty's High Court of Chancery within six calendar months next and immediately after the date of the said in part recited Letters Patent, as in and by the same, reference being thereunto had, will more fully and at large appear:

NOW KNOW YE, that in compliance with the said proviso, I the said William Westley Richards, do hereby declare that the nature of my said Invention, and the manner in which the same is to be performed, are particularly described and ascertained in and by the following description thereof, reference being had to the Drawings hereunto annexed, and to the letters and figures marked thereon, that is to say:

My improvements in the construction of gun and pistol locks and primers for the discharge of fire-arms consist,—Firstly, in an improved form given to the main spring, whereby without the employment of a swivel for connecting the end of the main spring to the tumbler I am enabled to do away with the eye or hooked end of the spring, and thereby afford a greater length and elastic action than has heretofore been found practicable in the construction of the locks of muskets and other fire-arms; and secondly, in enclosing the priming or detonating composition by which the gunpowder is to be ignited in a small waterproof frame or case of pasteboard, milled board, papier maché, or other suitable material, which can with the greatest facility be slidden or pressed into, or otherwise attached to a suitable holder contiguous to the touch hole of the gun, pistol, or other fire arms. In the accompanying Drawing, Fig. 1, represents the internal construction of my improved gun lock suited for a musket, the cock being brought down as after firing; Fig. 2, represents the same when at full cock. *a, a,* is the lock plate, the lower edge of which is formed more nearly in a straight line than usual; *b, b,* is the main spring, the end of which has a boss or knob *c,* acting upon the hollow part of the tumbler *d,* and the other end of the spring bears against a solid block or ledge *e.* The bridle piece *f,* which receives the axle of the cock *g,* and holds the scear *h,* I secure to the lock plate by the two screws *i, i,* as shewn. The striking plug *k,* of the cock is made hollow, and has perforations through its sides, for the purpose of letting out the smoke after the

A. D. 1841. December 14. No. 9177.
RICHARDS' Specification.

(1 SHEET)

FIG. 1.

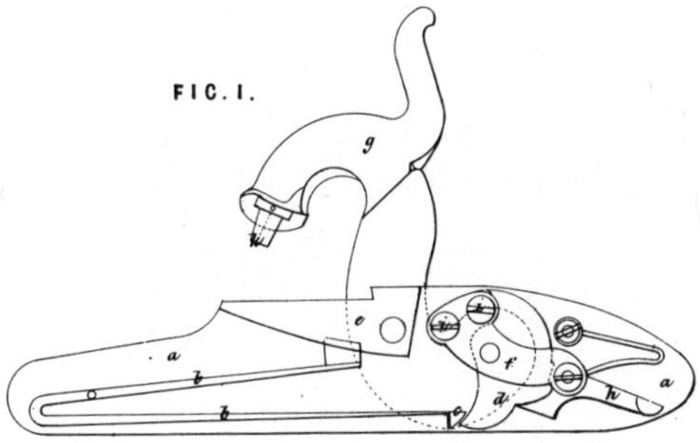

FIG. 2.

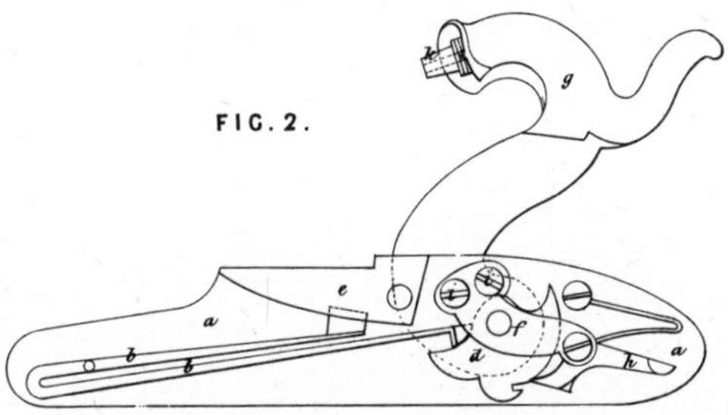

FIG. 3.

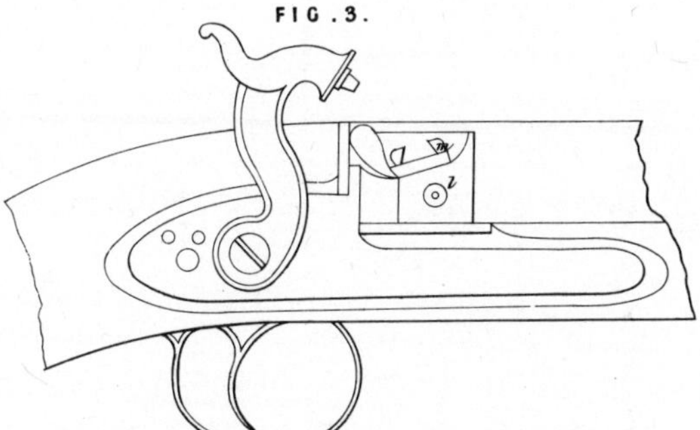

FIG. 4.

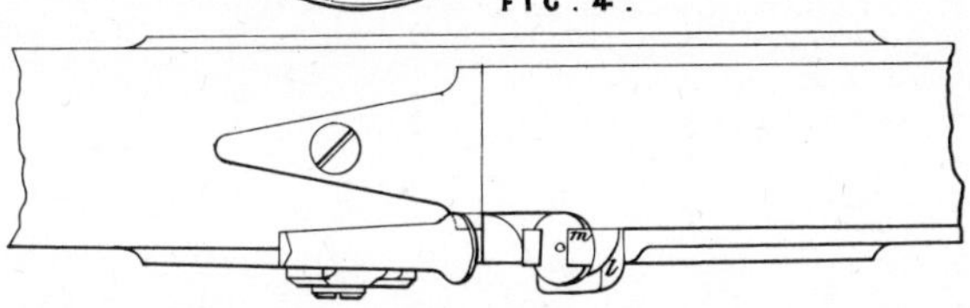

FIG. 6. FIGS. 5. FIG. 7. FIG. 11. FIG. 10. FIG. 8. FIG. 9. FIGS. 13. FIG. 12.

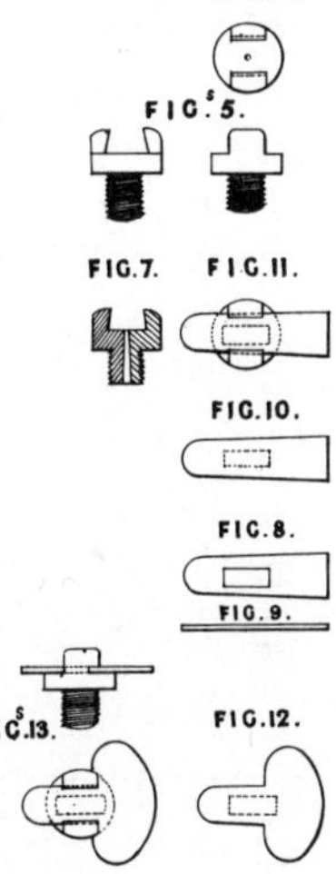

The enrolled drawing is colored.

Malby & Sons Lith.

London: Printed by George Edward Eyre and William Spottiswoode, Printers to the Queen's most Excellent Majesty. 1854.

ignition of the charge, and which assists in keeping the touch hole clean, and prevents the plug being injured by the repeated explosions. Fig. 3, represents the external appearance of a fowling piece, having my improved holder and touchhole adapted thereto. Fig. 4, is a top or horizontal view of the same; *l*, is the breach through which the touchhole passes, and *m*, is the holder to receive one of my improved primers, which holder may be made part of the touchhole piece, or detached therefrom if preferred. Figs. 5, represent the touchhole piece and priming holder separate from the gun; Fig. 6, the top view of the same; Fig. 7, is a vertical section taken through the holder and touchhole. This piece is screwed into the breach *l*, as shewn at *m*, in Figures 3, and 4. The form in which I propose to make my improved primers is shewn at Figures 8, 9, and 10. A piece of milled board, papier maché, or other light suitable material is cut out to the slightly wedge form shewn, about an inch long, and the sixteenth of an inch in thickness. The middle of this piece is perforated, which perforation or hole is to receive the percussion priming matter. The frame or case of the primer should be made firm, which may be done by rolling or pressing the millboard or other material. Fig. 8, represents the flat side of the case; Fig. 9, its edge view. One side of the case is to be first covered with tin foil, or other thin waterproof material, stuck on with shell lac or other adhesive matter not liable to be affected by water. Into the recess in the case the detonating priming composition is then to be introduced, and the other side of the case covered with tin foil, or other suitable thin waterproof material, as before, which will inclose the priming and keep it secure and dry, as at Fig. 10, or the priming may be made secure in the recess without a covering. In priming the piece one of these cases so charged is to be slidden sideways into the holder *m*, as at Figure 11, the folding lips being dovetailed to retain the priming case more securely. The plug of the cock then descending with great force upon the central part of the primer will explode the detonating composition, the fire from which descending through the touchhole will thereby discharge the fowling piece or other fire arms to which it is so applied. Fig. 12, is another form of case for holding the detonating composition, which is to be made and prepared in the same way as before described in reference to Figures 8, 9, and 10 This form of primer is to be inserted into the holder over the touchhole in the way shewn in two views at Figs. 13. Lastly, I desire it to be understood that, as respects my improvements in gun and pistol locks, I claim merely the construction of the main spring divested of the hooked or bent end, the extremity of which spring works upon the tumbler without a connecting link or swivell. And as respects the primer, I claim the manufacture of primers for guns, &c. by the insertion of the priming or detonating composition in a recess within a thin case of millboard, papier maché, or other suitable light material, so formed as to admit of its being securely attached to the touchhole of the piece, by sliding laterally or by pressing it into a suitable clip holder placed contiguous to the touchhole, thereby obviating the necessity of employing a spring or cover for keeping the primer securely in its position.

In witness whereof, I, the said William Westley Richards, have hereunto set my hand and seal, this Thirteenth day of June, in the year of our Lord One thousand eight hundred and forty-two.

W. WESTLEY (L.S.) RICHARDS.

W. RUSSELL.

AND BE IT REMEMBERED, that on the same Thirteenth day of June, in the year above mentioned, the aforesaid William Westley Richards came before our Lady the Queen, in Her Chancery, and acknowledged the Specification aforesaid, and all and every thing therein contained in form above written. And also the Specification aforesaid was stamped according to the tenor of the Statute in that case made and provided.

Inrolled the Fourteenth day of June, in the year above written.

LONDON:
Printed by GEORGE EDWARD EYRE and WILLIAM SPOTTISWOODE, Printers to the Queen's most Excellent Majesty. 1854.

A.D. 1849 . . . . N° 12,543.

Priming and Discharging Fire-arms.

MANTON & HARRINGTON'S SPECIFICATION.

TO ALL TO WHOM THESE PRESENTS SHALL COME, we, George Henry Manton, of Dover Street, Piccadilly, and Josiah Harrington, of Regent's Circus, in the County of Middlesex, send greeting.

WHEREAS Her present most Excellent Majesty Queen Victoria, by Her Royal Letters Patent under the Great Seal of the United Kingdom of Great Britain and Ireland, bearing date at Westminster, the Twenty-eighth day of March, One thousand eight hundred and forty-nine, in the twelfth year of Her reign, did, for Herself, Her heirs and successors, give and grant unto us, the said George Henry Manton and Josiah Harrington, our ex̃ors, adm̃ors, and assigns, Her especial licence, full power, sole privilege and authority, that we, the said George Henry Manton and Josiah Harrington, our ex̃ors, adm̃ors, and assigns, and such others as we, the said George Henry Manton and Josiah Harrington, our ex̃ors, adm̃ors, or assigns, should at any time agree with, and no others, from time to time and at all times during the term of years therein expressed, should and lawfully might make, use, exercise, and vend, within England, Wales, and the Town of Berwick-upon-

Tweed, our Invention of "IMPROVEMENTS IN PRIMING AND IN APPARATUS FOR DISCHARGING FIRE-ARMS," in which said Letters Patent is contained a proviso that we, the said George Henry Manton and Josiah Harrington, or one of us, should cause a particular description of the nature of our said Invention, and in what manner the same is to be performed, by an instrument in writing under our hands and seals, or under the hand and seal of one of us, to be inrolled in Her said Majesty's High Court of Chancery within six calendar months next and immediately after the date of the said in part recited Letters Patent, as in and by the same, reference being thereunto had, will more fully and at large appear.

NOW KNOW YE, that in compliance with the said proviso, we, the said George Henry Manton and Josiah Harrington, do hereby declare that the nature of our said Invention, and the manner in which the same is to be performed, are fully described and ascertained in and by the following statement thereof, reference being had to the Drawings hereunto annexed, and to the figures and letters marked thereon, that is to say):

Our Invention consists of so combining or arranging apparatus used for priming and discharging fire-arms, that the act of moving the ramrod shall cause a percussion cap or priming to be placed in the position to be discharged; such apparatus requiring that the hammer should be moved back after discharging the fire-arm, and before commencing to recharge. And in order that our Invention may be most fully understood, and readily carried into effect, we will proceed to describe the means pursued by us.

DESCRIPTION OF THE DRAWINGS.

Figure 1, shews a side view of a double barrel fowling piece having our improvements applied thereto. Figure 2, shews another side view of the fowling piece, the lock plate being removed in order to shew the interior combination, the parts being in the position immediately after returning back the ramrod; that is, the cap is on the nipple, and the apparatus which caused it to be placed there has gone out of the way by the act of returning the ramrod. Figure 3, shews another side view of the fowling piece (the interior parts, as in the previous Figure, being uncovered), the parts of the apparatus being shewn in the positions they would assume when the ramrod is withdrawn, and remains in that position until the time of returning the ramrod back into its place after ramming down the charge. Figure 4, shews another side view, wherein the parts are in the positions they will assume when the ramrod is in the act of being returned to its place, and has caused the apparatus to put on a cap to the nipple, and the further forcing in of the ramrod will cause the apparatus to move out of the way into a position to receive another cap, in readiness for the next time the ramrod is withdrawn, the withdrawing of the ramrod allowing the apparatus to move into position carrying a cap up so that the return of the ramrod into its place will cause such cap to be placed on the nipple, and then cause the apparatus to retire. Figure 5, shews another side view, partly in section the barrels being removed; the other views shewn by the Drawing being of detached parts, which are marked with the same letters of reference as those in the Figures above mentioned, and therefore will readily be understood.

a, is an axis placed in a recess below and at right angles to the groove *g*, formed in the front end of the stock, to receive the ramrod, and is carried by bearings formed in the pieces of metal *b*, let into the stock at each side of the gun. *c*, is a lever fixed on the axis *a*, the short arm c^1 being always pressed downwards by the spring *d*, by which means the arm c^2, is pressed upwards against the end of the ramrod *e* the arm c^2 carrying a friction roller c^3. On the outer ends of the axis *a* (when two barrels are used) are affixed the lever arms *f*, the ends of which are furnished with pins f^1, projecting inwards, which take into a slot or opening formed in the sliding piece *i*, attached by a pin joint to the instrument *j*, which receives the cap from the reservoir *g*, to be placed by it upon the nipple, the construction of which instrument we will now more particularly describe. The front part of the instrument *j* is so formed that the side pieces *j* act as springs to retain the cap within their curved ends, and the opening between them is slightly contracted near the curved parts, so that as the instrument slides back into its groove to receive a fresh cap a projection j^2, on the part b, enters between the side pieces j^1, and opens their ends to receive a cap from the reservoir, tube, or channel formed in the front end of the stock. The parts or spring holding pieces j^1 are attached to or may form part of the back or pressing portion j^3 of this instrument, which near its lower end is attached by a pin joint to the sliding piece i, and terminates with a forked opening upon the horns or fingers 1 and 2, of which the pin f^1, on the lever f, alternately acts.

We will now proceed to describe how a cap is received from the reservoir g, by the instrument j, and placed on the nipple by the aid of the parts previously described. When the ramrod is pushed quite down into its place, the parts are in the position shewn in Figure 2, and the arm f is pressing on the projection h^1, on the lower end of the bolt h, (which is pressed upwards by the spring k, when the ramrod is withdrawn, and projects a short distance into the end of the reservoir g, to prevent the succeeding cap from moving forward,) and the parts j^1 are slightly expanded by the projection j^2. The lock being placed at half cock, and the gun lowered, a cap falls into the instrument j, and on withdrawing the ramrod the pin f^1 on the arm f carries forward the parts i and j along the curved groove formed in the part b, pressing upon the finger 1; and the parts will assume the position shewn in Figure 3, and the upper end of the spring k, pressing against the sliding piece i, retains it during the movement of the pin f^1, to press down the finger 2, which takes place at the time the ramrod is in the act of being returned, and causes the part j to turn on its pin or axis and place the cap on the nipple, the parts assuming the position shewn in Figure 4. By returning the ramrod its lower end presses down the lever arm c^1, and arm f, which withdraws the parts i, j, and leaves the cap on the nipple, the spring-holding parts j^1 expanding when withdrawn. l is a cover to the opening where the instrument j protrudes, and is moved up and down by a projection l^1, taking into a slot in the arm f.

Having thus described the nature of our Invention, and the manner in which the same is to be performed, we would have it understood that we do not claim any of the parts separately; nor do we confine ourselves to the details as herein given, so long as the peculiar character of our Invention be retained. But what we claim is, the so combining or arranging apparatus for priming and discharging fire-arms, that the act of moving the ramrod shall cause a percussion cap or priming to be put in position to be discharged and the apparatus to retire.

In witness whereof, we, the said George Henry Manton and Josiah Harrington, have hereunto set our hands and seals, this Twenty-seventh day of September, One thousand eight hundred and forty-nine.

GEORGE HENRY (L.S.) MANTON.
JOSIAH (L.S.) HARRINGTON.

AND BE IT REMEMBERED, that on the Twenty-seventh day of September, in the year of our Lord 1849, the aforesaid George Henry Manton and Josiah Harrington came before our said Lady the Queen, in Her Chancery, and acknowledged the Specification aforesaid, and all and every thing therein contained and specified, in form above written. And also the Specification aforesaid was stamped according to the tenor of the Statute made for that purpose.

JEFFERSON.

Enrolled the Twenty-eighth day of September, in the year of our Lord One thousand eight hundred and forty-nine.

LONDON:
Printed by GEORGE EDWARD EYRE and WILLIAM SPOTTISWOODE,
Printers to the Queen's most Excellent Majesty. 1854.

A. D. 1849. March 28. Nº 12,543.

(2 SHEETS)

MANTON & HARRINGTON'S Specification.

FIG. 4.

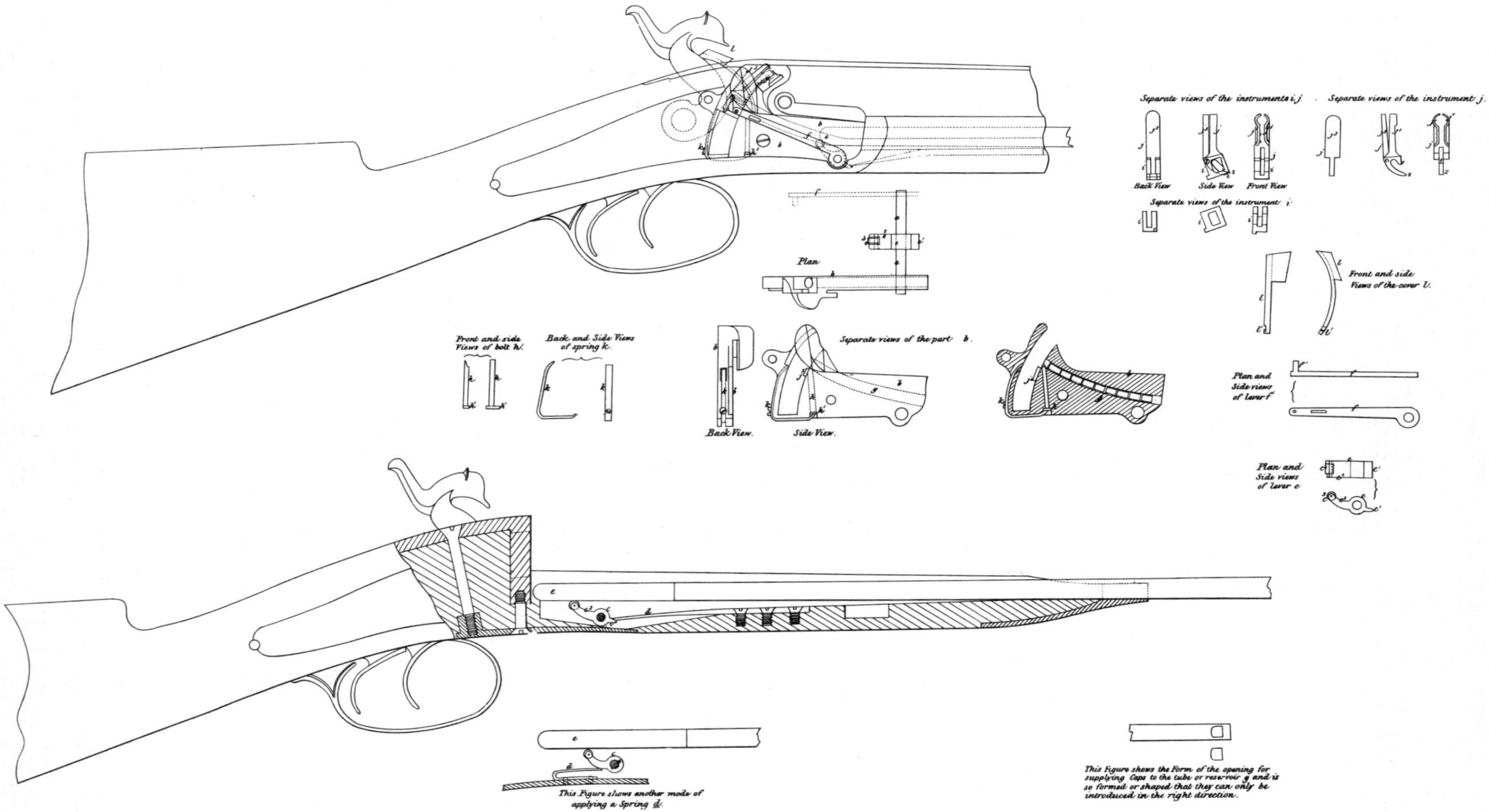

The enrolled drawing is partly colored.

Malby & Sons, lith.

London: Printed by George Edward Eyre and William Spottiswoode.

A.D. 1849 . . . . N° 12,668.

Revolving Fire-arms.

COLT'S SPECIFICATION.

TO ALL TO WHOM THESE PRESENTS SHALL COME I, Samuel Colt, of Trafalgar Square, in the County of Middlesex, Gentleman, send greeting.

WHEREAS Her present most Excellent Majesty Queen Victoria, by Her Royal Letters Patent under the Great Seal of Great Britain, bearing date at Westminster, the Twentieth day of June, in the twelfth year of Her reign, did, for Herself, Her heirs, and successors, give and grant unto me, the said Samuel Colt, Her especial licence, full power, sole privilege and authority, that I, the said Samuel Colt, my ex̃ors, adm̃ors, and assigns, and such others as I, the said Samuel Colt, my ex̃ors, adm̃ors, or assigns, should at any time agree with, and no others, from time to time and at all times during the term of years therein mentioned, should and lawfully might make, use, exercise, and vend, within England, Wales, the Town of Berwick upon Tweed, and in all Her Majesty's Colonies and Plantations abroad, and in the Islands of Jersey, Guernsey, Alderney, Sark, and Man, my Invention of "Improvements in Fire-arms," in which said Letters Patent is contained a proviso, obliging me, the said Samuel Colt, by an instrument in writing under my hand and seal, particularly to describe and ascertain the nature of my said Invention, and in what manner the same is to be performed, and to cause the same inrolled in Her Majesty's High Court of Chancery within six calendar months next and immediately after the date of the said in part recited Letters Patent, as in and by the same, reference being thereunto had, will more fully and at large appear.

NOW KNOW YE, that in compliance with the said proviso, I, the said Samuel Colt, do hereby declare that the nature of my said Invention, and the manner in which the same is to be performed, is particularly described and ascertained in and by the following description thereof, reference being had to the Drawings hereunto annexed, and to the letters and figures marked thereon (that is to say):

My Invention of "Improvements in fire-arms" consists in certain improvements upon that construction of guns and pistols which has a cylindrical revolving breech piece provided with a series of parallel chambers for containing a series of charges, which charges, by the revolution of the breech upon its shaft, may be successively brought into a line with the bore of the barrel and be severally discharged through the same. In the accompanying Drawing my improvements are shewn as applied to this construction of fire-arms, which formed the subject of a Patent granted to me in England on the Twenty-second day of October, One thousand eight hundred and thirty-five. Sheet I. of these Drawings represents my improvements as applied to pistols; and Sheet II. represents a rifle, carbine, musket, or shot gun with my improvements attached. Fig. 1, Sheet I., shews my improved pistol in side view; Fig. 2, is a side view, shewing the lock frame and recoil shield in section; and Fig. 3, is a longitudinal section taken through the middle of the breech and barrel. *a*, is the breech containing six chambers, more or less, as shewn in the end view Fig. 4; in the back of each of which a nipple is fitted (as shewn in end view at Fig. 5, and in section at Fig. 3,) to receive a percussion cap for firing the charge. This breech is supported by and is capable of turning on a spindle or arbor *b*, which is welded or fastened to and forms one piece with the recoil shield *c**, it being in a line parallel to the axis of the barrel. The shield is itself a continuation of the lock frame *c*, the whole being formed out of one solid piece of metal. By referring to the longitudinal section Figures 2, and 3 and to the cross section Fig. 6, the peculiar construction of this lock frame and shield will be readily understood. The shield c^*, stands up at right angles to the frame *c*, and forms a round head (somewhat like a bolt head) to its shaft *b*. The upper part of the shield is recessed to receive the hammer *d*, when it is thrown forward to effect the discharge of the pistol, and a recess is also made in the piece of metal which constitutes the lock frame and shield (see Fig. 2,) to receive the parts which respectively revolve the breech to bring round the charges in a line with the barrel *e*, and lock the breech to the frame for the purpose of insuring that the charges shall be in a line with the barrel before the firing takes place. When the pistol is on half cock or in the position shewn at Fig. 3, the breech is free to turn round on its arbor in the direction of the arrow, Figures 1 and 2. It may then be loaded and primed with facility without being removed from its place as was formerly requisite in charging this construction of revolving breech fire-arms, a free space being left in front of the mouths of the chambers, as will be seen by referring to the end view of this breech, Fig. 4, which shews the sectional area of the barre and its appendages in red lines. The barrel *e*, is supported in its place by the end of the spindle *b*, fitting into a socket in a bracket piece forming one with the barrel, as shewn best at Fig. 3; against the end of this socket the spindle is made to abut, and thus to determine the exact position of the barrel with respect to the face of the revolving breech. To keep the barrel secure in its place a key *f*, is introduced through slots in the bracket or projecting piece of the barrel, and through the spindle *b*, its upper edge acting on the forward end of the slot in the spindle, and its lower edge acting upon the lower end of the slots in the bracket of the barrel; the effect of which is to draw the barrel towards the cylinder breech and lock frame as the key is pressed in, and pins projecting from the end of the lock frame enter corresponding recesses in the bracket piece of the barrel. The key *f*, has a spring catch which rises when the key is forced "home," and by its turned up end coming in contact with the edge of the slot through which it has been passed it will prevent the key from getting loose and shaking out of its place by the concussion of the firing. This object is further ensured by the insertion in the barrel piece of the screw 1, the head of which would come in contact with the turned up end of the catch if it had escaped pass the edge of the slot, and prevent it from dropping or even being drawn out; this screw must therefore be removed before the key can be displaced. Jointed to the bracket part of the barrel by a pin 2, is a lever *g*, which is kept up in a position parallel to the barrel by a spring catch at its other end, taking into a catch on the lower side of the barrel. To this lever a plunger *h*, is connected by a pin 3, taking into a slot on the outer end of the plunger. The inner end of the plunger slides in a guide made for it in the bracket piece of the barrel. This plunger is intended to act as a ramrod, and to drive the bullets or cartridges into the several chambers of the breech consecutively as they are brought in a line with the plunger. To effect this the catch or the lever *g*, is disengaged, and the lever is brought into the position shown by dots in Fig. 2; the plunger is thereby driven forwards and made to ram the bullet (which has been previously inserted in the end of the chamber now brought in a line with the plunger) down into its proper place in the chamber. The plunger is then drawn back, and the next succeeding chamber brought in a line therewith, when the lever *g*, is again brought into the dotted position to thrust the plunger forward and ram down another charge and thus successively all the charges are acted upon. By referring to the Drawing Fig. 3, it will be seen that the mouths of the chambers and the inner end of the barrel are chamfered at their edges. This bevelling of the edges of the chambers is to prevent the lateral discharge between the breech and the barrel (which discharge is consequent upon the construction of fire-arms to which my improvements refer) from igniting the powder in the other loaded chambers; for the ignited matter, by coming into contact with the bevelled edge as it crosses the mouths of the chambers, will be deflected outward and effectually thrown off, and be prevented from reaching the powder in the chambers. The bevelling of the end of the barrel is intended to prevent its cutting the ball in its passage from the chamber. At Figures 6, and 7, it will be seen that a hollow is made in one side of the shield *c**; the object of this is to expose the ends of the nipples as the breech is revolved, and thus to allow of the percussion caps being readily placed thereon.

I will now describe the arrangements of the various parts for rotating and locking the breech, and for discharging the pistol, in order that the construction of the pistol may be the better understood. The hammer *d*, turns on a pin 4, in the lock frame *c*, and it is provided with stops or catches for the end of the trigger to abut against, as usual, and hold it at whole or half cock, as required. To the hammer is jointed a hand catch *i*, with the spring *k*, attached, (see Fig. 2,) which is pressed forward into contact with ratchet teeth formed on the end of the breech, and allows the hand to recede for passing below a second tooth of the ratchet. *l*, is a rocking lever, supported by a pin in the lock frame *c*, and carrying at one end a bolt which is intended to enter at certain times into the recesses 5, 5, in the periphery of the breech, a bearing spring *m*, (see Fig. 3,) giving it always a tendency to rise for that purpose. The other end of this lever is made thin so as to be capable of yielding laterally and recovering its position, thus allowing a stud 6, which projects from the hammer and has a chamfered or bevelled face, to pass the lever without disturbing the position of the bolt end when the hammer moves forward to fire the charge; and yet, when the hammer is drawn back, to present an obstruction to the stud 6, and be thereby tipped into the position shewn at Fig. 3, which movement will unlock the breech. As long as the hammer remains at half cock, the breech will be free to turn upon its spindle for the purpose of being loaded; but when the stud has passed the end of the lever, the spring *m* will again force the bolt of the lever into its original or locking position. The action of drawing back the hammer to its farthest extent (the bolt being first relieved) will raise the hand catch *i*, which being brought in contact with a ratchet tooth on the breech will turn the breech round in the direction of the arrow Fig. 2, to the extent of one tooth, and thus bring up a loaded chamber in a line with the barrel; and in succession the act of cocking will bring all the loaded chambers in like manner round in a line with the barrel and hammer to be discharged. In order to insure the insertion of the bolt of the lever *l*, into the recesses 5, 5, as they severally come round, and thereby to hold the breech firmly while the discharge takes place, I form a shallow channel or guide up to the edge of each, as shewn in the drawings, which will make the bolt feel its way to the recess and enter it more certainly than if it were required to fall into the recess suddenly. To prevent the fouling of the spindle and breech, I form upon the spindle a helical groove (as shewn at Fig. 3), the edges of which will more effectually prevent the smoke from passing between the breech and spindle than if the whole periphery of the spindle were in contact with the breech; and at the same time these edges will, as the breech is rotated, scrape off any matters that may have become deposited in the central bore of the breech, and deposit it in the grooves. By this means the contact surfaces will be kept clean, and the breech, which would otherwise foul after a few discharges and become fixed, will be free to turn on its spindle for a long period without requiring any cleaning.

In Sheet II., Fig. 1, is a side view; and Fig. 2, a plan view of a rifle, carbine, musket, or shot gun, with revolving breech. As the several improvements before described with reference to the pistol are equally applicable to guns for military and sporting purposes, I shall confine my further description to the modified arrangement of apparatus for ramming down the charges. In this instance, instead of the plunger being applied below the barrel it is attached to the side thereof. *a*, is a bracket projecting from the side of the barrel, and to it the lever *b*, is jointed by a pin 1; *c*, is the plunger, having at about the middle of its length a joint 2; it is forked at one end, and embraces the lever *b*, to which it is jointed by a pin 3: the cylindrical end of the plunger works between guide pins 4, 4, on the barrel; *d*, is a spring catch rivetted to the plunger *c*, and capable, when the lever *b* is laid parallel with the barrel, of taking in a notch cut in the bracket *a*. By means of this spring the ramming apparatus, when not required to be used, is retained in the position shewn at Fig. 1; but, by drawing the lever *b*, into the position shewn at Fig. 2, the catch will be immediately disengaged from the notch.

Having now described my several improvements upon fire-arms having breeches which revolve on a spindle parallel to the barrel, I wish it to be understood that I claim,—

First, making the lock frame and recoil shield in one and the same

A.D. 1849. June 20. No 12668.
COLT'S Specification.

(2 SHEETS.)
SHEET 2.

FIG. 2.

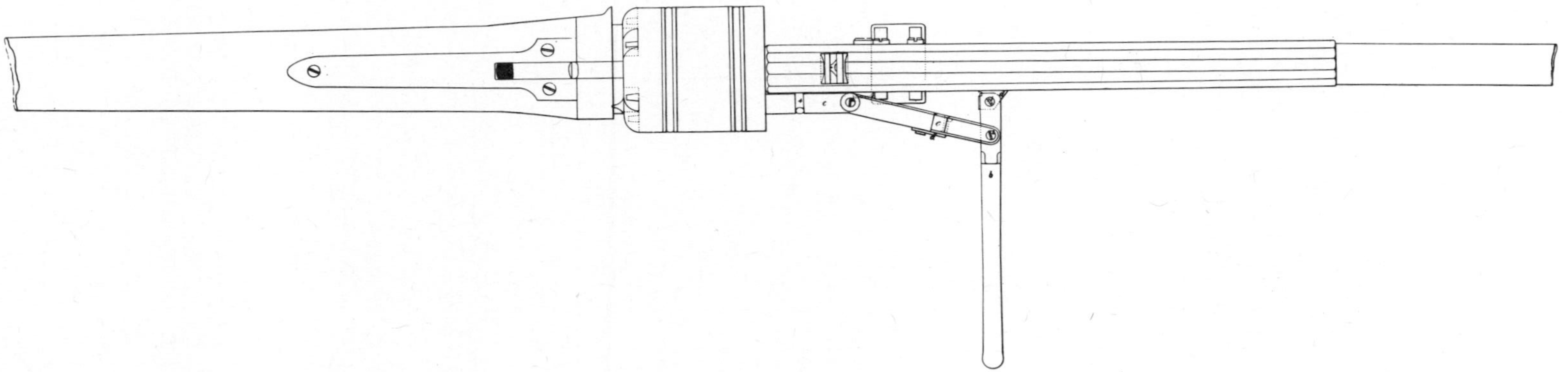

FIG. 1.

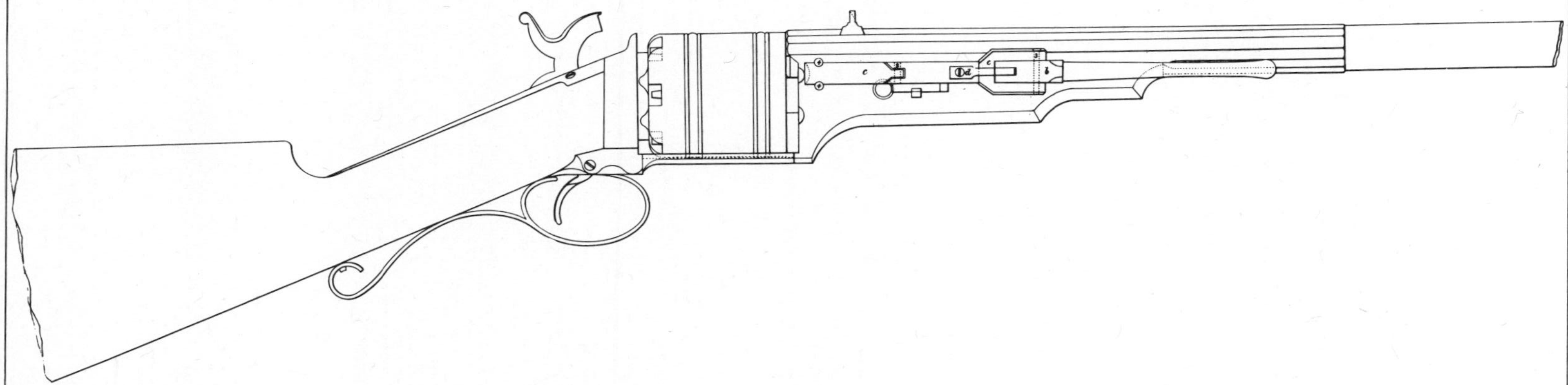

The enrolled drawing is colored.

Malby & Sons lith.

London: Printed by George Edward Eyre and William Spottiswoode, Printers to the Queen's most Excellent Majesty. 1854.

A.D. 1849. June 20. No 12,668.

(2 SHEETS)
SHEET 1.

COLT'S SPECIFICATION.

FIG. 1.

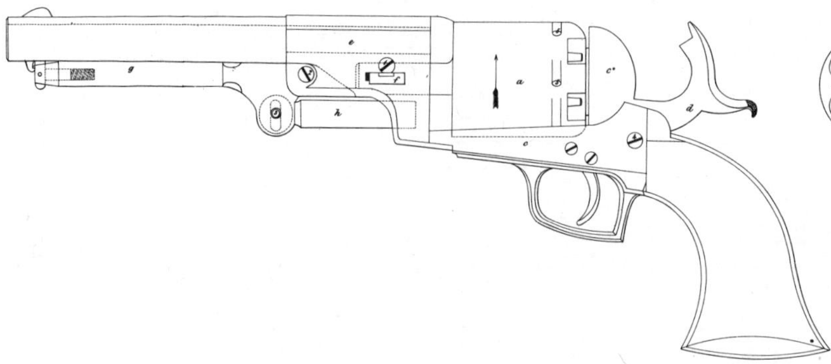

FIG. 4.

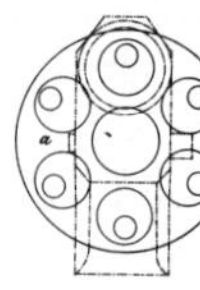

FIG. 2.

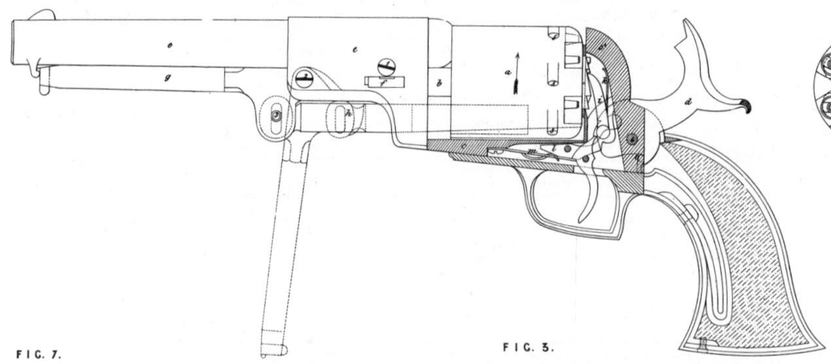

FIG. 5.

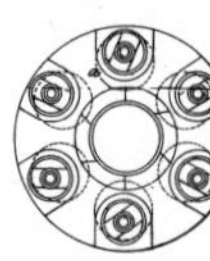

FIG. 7.

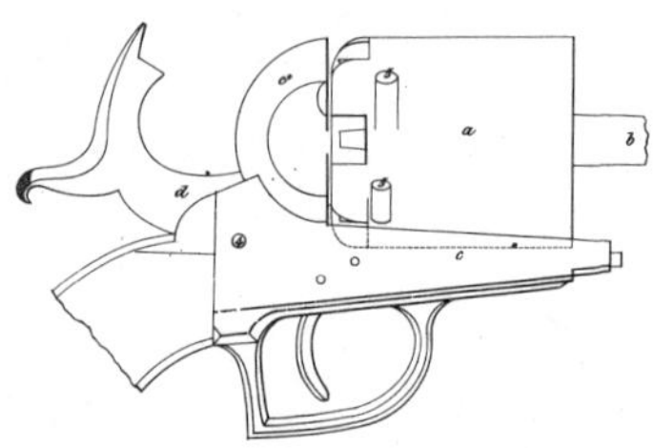

FIG. 3.

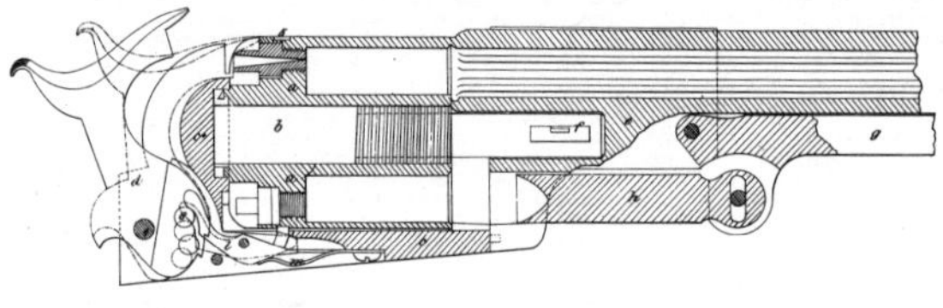

FIG. 6.

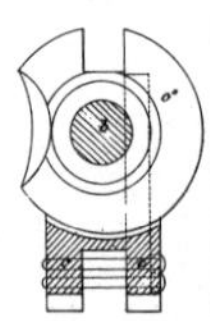

The Enrolled Drawing is colored.

Malby & Sons, Lith.

LONDON: Printed by GEORGE EDWARD EYRE and WILLIAM SPOTTISWOODE,
Printers to the Queen's most Excellent Majesty. 1854.

piece, whereby all possibility of the parts getting loose after repeated discharges is prevented.

Secondly, the general arrangement of the lock and apparatus for turning and locking the rotating cylinder breech.

Thirdly, the general arrangement of the parts whereby the operations of loading and priming may be effected without disconnecting the breech, as was heretofore requisite.

Fourthly, the application to guns and pistols of the lever apparatus for ramming down the charges into the several chambers with great expedition and effect, the same being a substitute for the loose ramrod.

Fifthly, the chamfering the mouths of the chambers and the inner end of the barrel; also the grooving of the periphery of the breech spindle; likewise the making sunk guides to the locking recesses on the periphery of the breech; and, further, the means of insuring the proper position of the barrel with respect to the face of the breech; all for the purposes herein-before described.

In witness whereof, I, the said Samuel Colt, have hereunto set my hand and seal, this Twenty-seventh day of October, in the year of our Lord One thousand eight hundred and forty-nine

SAMUEL (L.S.) COLT.

Taken and acknowledged by Samuel Colt (party hereto), at Paris, in the Republic of France, the Twenty-seventh day of October, One thousand eight hundred and forty-nine, before me

(L.S.) THOMAS PICKFORD,

Her Britannic Majesty's Consul at Paris.

Enrolled the Twentieth day of December, in the year of our Lord One thousand eight hundred and forty-nine.

LONDON:
Printed by GEORGE EDWARD EYRE and WILLIAM SPOTTISWOODE,
Printers to the Queen's most Excellent Majesty. 1854.

A.D. 1850 . . . . N° 12,920.

Breech-loading Fire-arms and Cartridges.

SEARS' SPECIFICATION.

TO ALL TO WHOM THESE PRESENTS SHALL COME, I, Matthew Urlwin Sears, of No. 36, Burton Crescent, in the Parish of St. Pancras, in the County of Middlesex, send greeting.

WHEREAS Her present most Excellent Majesty Queen Victoria, by Her Royal Letters Patent under the Great Seal of the United Kingdom of Great Britain and Ireland, bearing date at Westminster, the Eleventh day of January, in the thirteenth year of Her reign, did, for Herself, Her heirs and successors, give and grant unto me, the said Matthew Urlwin Sears, my exors, admors, and assigns, Her special licence, full power, sole privilege and authority, that I, the said Matthew Urlwin Sears, my exors, admors, and assigns, should at any time agree with, and no others, from time to time and at all times during the term of years therein expressed, should and lawfully might make, use, exercise, and vend, within England, Wales, and the Town of Berwick-upon-Tweed, and in the Islands of Guernsey, Jersey, Alderney, Sark, and Man, and also in all Her said Majesty's Colonies and Plantations abroad, my Invention for "**The Improved Construction of Guns and Cannons, and Manufacture of Cartridges for the Loading or Charging**

THEREOF;" in which said Letters Patent there is contained a proviso, that I, the said Matthew Urlwin Sears, shall cause a particular description of the nature of my said Invention, and in what manner the same is to be performed, by an instrument in writing under my hand and seal, to be enrolled in Her said Majesty's High Court of Chancery within six calendar months next and immediately after the date of the said in part recited Letters Patent, and by the same, reference being thereunto had, will more fully and at large appear.

NOW KNOW YE, that in compliance with the said proviso, I, the said Matthew Urlwin Sears, do hereby declare that the nature of my Invention, and the manner the same is to be performed, are particularly described and ascertained in and by the following statement thereof, reference being had to the Drawings hereunto annexed, and to the figures and letters thereon (that is to say):

My Invention consists, firstly, in a new and improved construction of guns that are loaded at the breech and fired by the protrusion of a needle or pin into a priming contained in the interior of the cartridge, wherewith such gun is charged, this improved construction being applicable to military muskets, rifles, carrabines, fowling pieces, and pistols, and, as far as concerns the loading at the breech, also to cannons; and, secondly, in an improved cartridge for charging or loading such guns.

Fig. 1, Sheet First of the Drawings, is a full sized horizontal section of a rifled gun at half cock, and with the cartridge put in. A, is the stock. The barrel B, B, ends in a cone at *a*, *a*; its bore is enlarged here so as to form a convenient lodging for the cartridge C^1. This part of the barrel is screwed at *b*, *b*, to a tube D, D, of which Fig. 2, 3, 4, give different crossections, and Fig. 5, a ground plan. This tube is truly bored inside, and has a longitudinall slit *c*, *c*, *c*, extending to the point *d*, where it is enlarged so as to form the oblong opening or hole *d*, *e*, *f*, *g*, just behind the screw threads; the tube D, has a small enlargement which serves for collecting powder, dust, and like impurities. E, is a cylinder accurately turned so as to move very truly, but with little friction, in the tube D. The different crossections of this sliding cylinder, which, in fact, forms the breech of the gun, and therefore is called the sliding breech, appear in Fig. 2, 3, 4, its upper part in Fig. 6, and its lower in Fig. 7. It is provided with a handle *h*, Fig. 1, 3, 6, by which it may be moved forward and backward in the tube D, or be pressed up against the conical part *a*, *a*, of the barrel in the position shown in Fig. 3. The lower part of the handle abuts with its back face against the shoulder *d*, *e*, of the tube D, which shoulder is made somewhat slanting in order to favor the locking up, while the valve-shaped fore part of the sliding breech E, lays firmly hold of the conical part of the barrel, and shuts the same; but when the handle is turned upwards to unlock it from the shoulder *d*, *e*, and then drawn backward to the point *c* of the slit *c*, *c*, the barrel will be opened, and the cartridge may be put into its chamber through the opening or loading hole *d*, *e*, *f*, *g*. The dotted lines in Fig. 3, show how far the handle *h* must be turned upwards in order to unlock it from the shoulder *d*, *e*. The fore part of the interior of the sliding breech E, is enlarged conically, and ground to the conical part of the barrel, so as to form a kind of valve shutting, and a small ring of copper or soft brass *i*, *i*, Fig. 1, is introduced to improve the shutting. Just behind this ring there is a space *k*, *k*, called the back chamber. Its use is not only to receive the residum of the burnt powder, but also to contain a quantity of air, which, by its elasticity, eases the recoil of the gun, and being strongly heated, and consequently dilated by the explosion of the powder, augments the force wherewith the bullet is driven out of the barrel. In the middle of the back chamber *k*, *k*, there stands out of the diaphragm l^1, l^1, the needle guide *l*, a small tube, whose office is shown by its name; it is square outside in order to be held by a key when it is to be screwed into the diaphragm. The back part of the sliding breech E, contains another sliding tube F, seen in longitudinal section in Fig. 1, in crossecsion in Fig. 3, 4, from above in Fig. 8, and from below in Fig. 9, and called the spring barrel. It is shut behind, where it bears the knob *w*, rising upwards, but open in front, and contains the main spiral spring *m*, the needle bar *n*, with the nut *o*, and the needle *p*. The spring barrel is flattened on the lower side, and has a slit there *q*, Fig. 1, 3, 9, for the teeth 1 and 2 of the nut *o* to work in. The greater part of the upper side is flattened also, as seen in

Fig. 1, 3, 4, 8. By this last flattening, and a little hollowing out of the sliding breech E, a space is formed for the reception of a stout spring *r*, Fig. 1, 3, 4. The spring *r* has two teeth, *s* and *t*, whereby it is held to the spring barrel F, and two catches which interlock with the sliding breech, thereby fastening the spring barrel to the latter. It has, moreover, a hook *v*, at the back end standing up above the tube D; by pressing down this hook so as to disengage the catches *u*, *u*, and using the knob *w* of the spring barrel as a handle, F may be moved forward and backward in the sliding breech, or be drawn entirely out of it and pushed in again. The spiral main spring is to be made to bear a weight of sixteen to eighteen pounds, and of a length somewhat greater than the space included between the narrow part of the spring barrel and the nut *o*, when this last is not held back by the detent lever *x*, but is in contact by its little copper ring *y*, with the diaphragm l^1, l^1, of the sliding breech. In other words, the spring *m* must be so long that it cannot be put into its chamber without using some force, and it is held there when the spring barrel is drawn out by the tooth *s* engaging with the upper tooth three of the nut *o*. The detent lever *x*, swings on a small pin or axle in a slit of the sliding breech; see Fig. 1, 4, 7. There is a spring *z*, riveted to *x*, abutting against the lower flat side of the spring barrel, and serving to engage the detent lever with either of the two lower teeth of the nut *o*. The trigger mechanism formed no part of this Invention, and may be varied according to convenience.

The movements for loading and firing this gun are as follows:— The lock being at rest, that is, the small copper ring *y* being in contact with the diaphragm l^1, l^1, of the sliding breech E, the needle or pin *p*, standing out forward of the needle guide *l*, and the handle *h* being locked to the shoulder *d*, *e*, hold the gun with the left hand, just so as if you were to put a percussion cap upon the nipple of a common gun, and then with the right hand, firstly, press down the spring *r* by means of the hook *v*, and draw out backwards the spring barrel F till the detent lever *x* catch the tooth 1, of the nut *o*; secondly, unlock the handle *h* of the sliding breech E, from the shoulder *d*, *e*, and draw it back to the point *c* of the slit in the tube D, whereby the loading hole *d, e, f, g,* will be opened; thirdly, put in the cartridge; fourthly, shut the loading hole by pushing forward the handle *h*, and locking it to the shoulder *d, e*; fifthly, puch the spring barrel home into the sliding breech; in this state the gun is at half cock, as shewn in Fig. 1, and may be handled and carried about with safety; sixthly, to bring the gun to full cock, repeat the movement N° 1, but draw back the spring barrel so far that the detent lever *x*, may engage with the tooth 2, of the nut *o*, and instantly repeat the movement No. 5. If, now, the trigger be drawn, the tooth 2 escapes from the detent lever, the needle bar and needle are forcibly pushed forward by the main spring, and the needle piercing the cartridge sets the priming on fire, which explodes the powder. It need scarcely be mentioned that, for quick firing, the bringing the gun to half cock may be omitted by drawing out the spring barrl so that the detent lever *x* may at once engage with the tooth 2 of the nut *o*.

The Drawings of Sheet Second show another arrangement for carrying out the Invention. Fig. 10, is a longitudinal section of a rifled gun half cocked with the cartridge put in. The barrel, with A, the stock. The barrel B, with its enlarged chamber for the reception of the cartridge C^1, is made conical at *a, a*, and screwed at *b, b*, into the tube D, D. Fig. 11, is a crossection of the barrel, to show in what manner I prefer to dispose the grooves internally, but without laying any claim to it. The tube D, is accurately bored, and there is also a small recess in it just behind the screwed part, for collecting any impurities that might obstruct the movements of the sliding breech E. The tube D, is shown in crossection in Fig. 12, and 13, and in a plan, but the under side turned upwards in Fig. 14. It appears from the Figures that the slit *c, c*, for the admission of the handle of the sliding breech, together with the loading hole *d, e, f, g*, are placed not at the upper but at the lower side, and that of course the cartridge must be put in from below the tube D, and sliding breech E. To make this putting in of the cartridge more easy there is a recess H, worked out in the stock. When the gun is loaded in this manner the loading hole *d, e, f, g*, need not be made so broad, consequently there is retained a greater quantity of iron in the tube D, and the fore part of the sliding breech is more fully encompassed by and more securely lodged in it. The construction of the sliding breech E, appears from Fig. 10, 12, 13, and 15, the latter Figure shewing its under side. The larger part is cylindrical, with the exception of a prismatical prominence or ridge on the lower side marked h^1, h^1, which in sliding the breech lengthwise through the tube D, enters the slit *c*, *c*. When the sliding breech is thrust forward as far as possible, and then turned sideways downwards, this ridge will stand in the middle of the loading hole *d*, *e*, *f*, *g*, as shewn by the dotted lines in Fig. 14. In this position the back face of the ridge h^1, h^1, abuts against the slanting shoulder *d*, *e*, of the tube D, while the valve-shaped forepart of the sliding breech with the metal ring *i*, *i*, are jammed fast to the conical part of the barrel conversely, the sliding breech being first turned upwards to unlock the ridge h^1, h^1, from the shoulder *d*, *e*, and then drawn back; the cartridge may be put into its chamber through the loading hole *d*, *e*, *f*, *g*. The handle *h*, for performing these movements of the sliding breech is riveted to the ridge h^1, h^1, or fastned to it by screws. It is seen in longitudinal section in Fig. 10, in crossection in Fig. 12, where it is shown going up at the right side of the tube in a recess of the stock; it is further seen in a horizontal projection of the under side in Fig. 16. The lower part of this handle is a curved plate, serving as an additional cover to the loading hole, and its upper part has the shape of on oblong ring. The dotted lines Fig. 12, show how far the handle *h*, is to be turned upwards in order to unlock the ridge h^1, from the shoulder *d*, *e*. There is a recess h^2, in the stock, serving for the reception of the handle when the sliding breech is drawn back to open the loading hole. The valve-shaped mouth piece of the sliding breech E, with its copper or brass ring *i*, *i*, back chamber K, needle guide *l*, and diaphragm l^1, are the same as those already described, but the after part of the sliding breech is of a different construction, there being particular spring barrel, and the large flat catch spring marked *r*, in the Drawings of Sheet First, with all its apendages, being ommitted. The construction hereby is much simplified, there being a single hole only to be bored through the back part of the sliding breech E, for the reception of the main spiral spring *m*, needle bar *n*, and *o*, and the needle or pin p. This hole is closed at the back by the screw 4, Fig. 10, and 17, it is perforated to allow the needle bar n, to pass through and between it and the nut o; the main spring m, may be compressed by drawing the needle bar backwards. The slit q, in which the teeth 1 and 2, of the nut o, and the detent lever x, are working, is continued through the back part of the sliding breech in such a manner that the needle bar, with the main spring, nut, and needle, may be drawn out or replaced after having removed the screw 4. The spring z, of the detent lever x, abuts against the lower side of the smaller part of the sliding breech E. The drawing back of the needle bar u, independently, whereby in this construction the gun is set to half or full cock, may be performed by different mechanical means, but the arrangement which I prefer, because it forms a convenient shutting for the after part of the tube D, is shown in a longitudinal section in Fig. 10, and in plan in Fig. 17. This cocking frame consists of a ring 5, screwed into a sheave 6, which by two bars 7, and 8, is connected to a short cylinder 9. The sheave 6, and the cylinder 9, are to be of such a diameter as to exactly fill the calibre of the tube D. The bottom of the cylinder 9, has a hole through which the needle bar passes, which is prevented from slipping of by the screw nut F. The upper connecting bar y, has a small recess in which the point of the screw 10, may slide, and the lower bar 8, bears small spring 11, by which means the cocking frame is kept to its place in the tube D.

Movements for loading and firing this gun :—Firstly, put the fore-finger of the right hand into the ring 5, and draw back the cocking frame and needle bar untill the detent lever x, catch the tooth 1, of the nut o, secondly, unlock the ridge h^1, of the handle of the sliding breech from the shoulder d, e, and draw it back to open the loading hole; thirdly, put in the cartridge; fourthly, shut the loading hole, by pushing forward the handle of the sliding breech and locking it to the shoulder d, e; the gun is now at half cock; and, fifthly, to bring the gun to full cock, draw back the needle bar by means of the cocking frame so far that the tooth 2, of the nut o, come to be engaged with the detent lever x, and instantly push the cocking frame into the tube D. The trigger mechanism is the same as that shewn in Fig. 1, Sheet first.

Fig. 18, Sheet Third, shows in what manner the shutting up of the sliding breech E, against the back part of the barrel is to be effected when the latter is not screwed into the barrel D, but both barrel and tube are in one piece. In this case, for the sake of convenience in workmanship, and also in cleaning the valve, face and ring *i*, are in the barrel, and the fore part of the sliding breach is made conical at the outside. Moreover, it appears from this Fig. and the crossections Fig. 19, and 20, that my improved construction is applicable to guns of comparatively small calibre, and by Fig. 21, that double barrelled guns may be made according to this principle, both loading holes in such a case being either at the top or at the bottom of the breech. Fig. 22, 23, and 24, show how, according to my Invention, a cannon may be made to load at the breech Fig. 22, is a longitudinal section. Fig. 23, a crossection; and Fig. 24, a plan of the back part of a cannon. D, D, that part of the barrel which serves as a tube to the sliding breech E; *c*, the slit; *d*, *e*, the shoulder; *d*, *e*, *f*, *g*, the loading hole; *i*, the valve ring, and *k*, the back chamber, and *l*, the handle of the sliding breech. It need scarcely to be mentioned that in large cannons the movement of the sliding breech cannot well be effected by hand only, but that some mechanical means to gain a purchase upon the handle, such as a rack and pinion, or worm screw and wheel, must be used. Fig. 25, Sheet Second, is a longitudinal section of a cartridge for loading or charging my improved guns. 1, is the powder; 2, a thin washer of pasteboard or millboard containing in the middle a percussion cap; 3, and 4, is the bullet. The powder, the washer, with its cap, and the bullet are put in the order here stated into a cylindrical case of fine paper, and, lastly, this is tied to the bullet with a cotton thread. The cartridge afterwards is to be smeared with tallow. The bullet has three rings round its cylindrical part, projecting as far as fully to fill the groves of the rifled barrel, while the cylindrical body of the bullet has a diameter fully equal to the calibre. By this means the bullet in its passage through the barrel not only prevents any loss of the gases evolved by the exploding powder, but acquires that spiral or spinning motion round its own axis which is necessary for guiding it in its flight. This arrangement of the cartridge has, moreover, this advantage, that the ignition of the powder is begun at its foremost part, and consequently its perfect combustion secured. As to the percussion caps to be used, they should be as flat as is consistent with their being well fastned in the middle part of the washer, and about a tenth of an inch in diameter; they should be primed with a well-made composition of fulminate of mercury, and the priming should be covered in with a thin piece of copper. Cartridges made with such percussion caps may be handled with as little danger as common soldiers' cartridges, and are not liable to miss fire or to corrode the interior of the barrel. If small shot is to be used, instead of a bullet, the barrel is not to be rifled, and a second washer, not perforated, is to be put in between that washer which contains the cap and the shot.

What I claim as my Invention under these Letters Patent is,—

Firstly, in regard to small fire-arms, the general arrangement of the mechanical means to be used and employed as herein-before set forth for loading or charging such guns at the breech, and firing them by means of a needle or pin protruding into the priming contained in the cartridge itself, and particularly *a*, the construction of the various sliding breeches drawn and described with valve shaped mouth pieces and rings, and detent levers attached to them; *b*, the shutting up of the back of the spring barrel marked F, in the Drawings of Sheet first; *c*, the construction of the needle bar with its nut, allowing the gun to be half cocked; *d*, the construction of the tube marked D, by which the gun may be loaded or charged from the lower side of the breech, together with the handle forming a cover to the loading hole.

Secondly, in regard to cannon, the loading them at the breech, and locking up of the sliding breech in the manner shown.

And thirdy, in regard to the cartridges, their arrangement in general, and the peculiar form of copper caps described to be used as a priming in the cartridges.

In witness whereof I, the said Matthew Urlwin Sears, have hereunto set my hand and seal, this Ninth day of July Eighteen hundred and fifty.

(L.S.) MATTHEW URLWIN SEARS.

A.D 1850. January 11. No. 12920.
SEAR'S' Specification.

(3 SHEETS.)
SHEET 1st

F I G. 1.

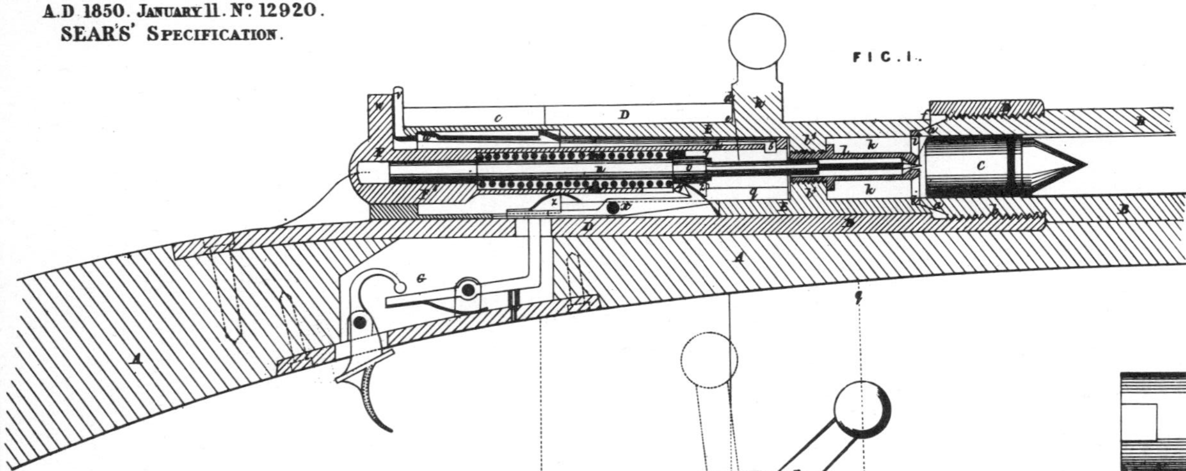

F I G. 3.

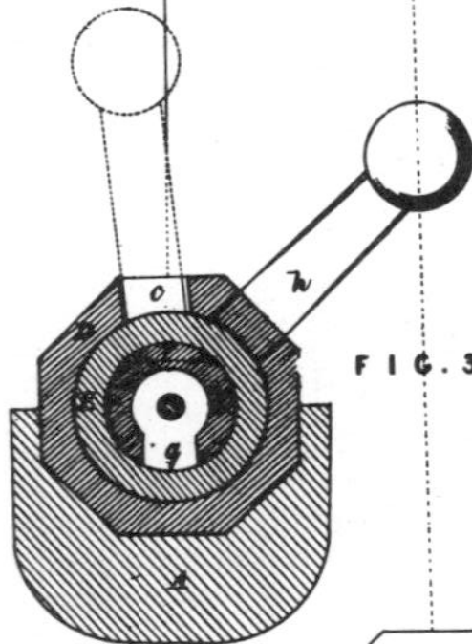

F I G. 4.

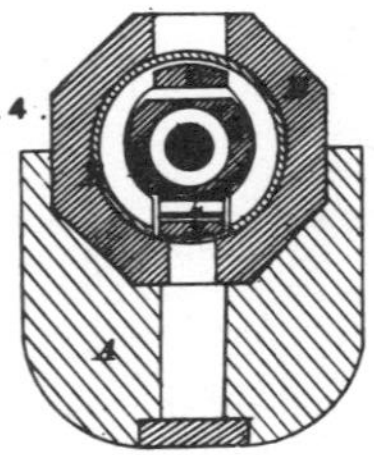

F I G. 2.

F I G. 5.

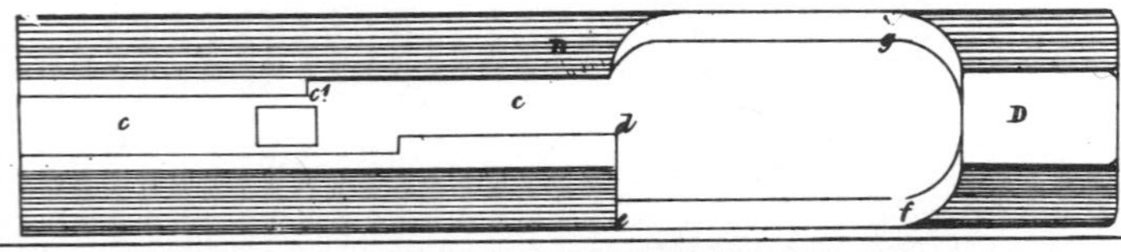

F I G. 6.

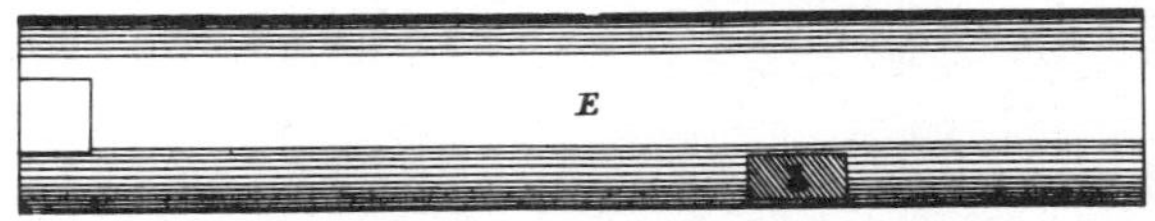

F I G. 7.

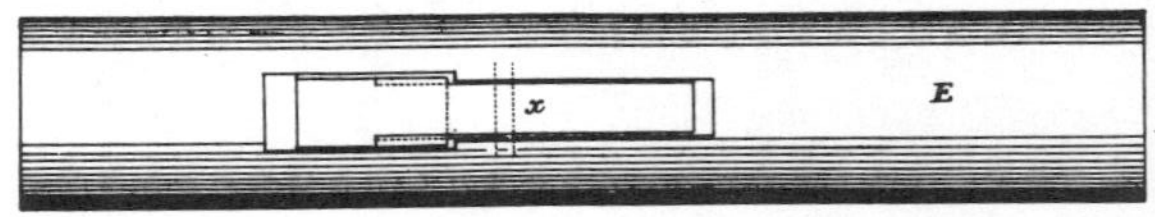

F I G. 8.

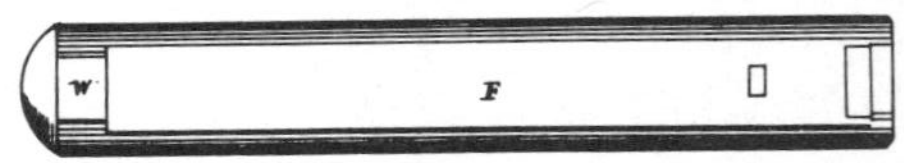

F I G. 9.

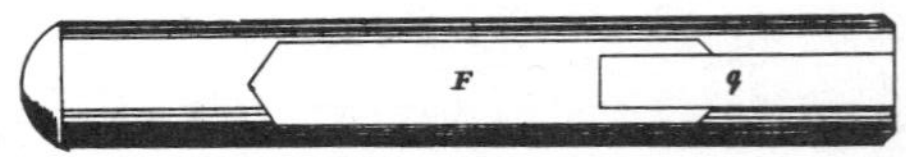

The enrolled drawing is not colored.

Malby & Sons lith.

London: Printed by George Edward Eyre and William Spottiswoode.

A.D. 1851 . . . . N° 13,527.

Rifled, Revolving, and other Fire-arms.

ADAMS' SPECIFICATION.

TO ALL TO WHOM THESE PRESENTS SHALL COME, I, Robert Adams, of King William Street, in the City of London, Gun Maker, send greeting.

WHEREAS Her present most Excellent Majesty Queen Victoria, by Her Royal Letters Patent under the Great Seal of the United Kingdom of Great Britain and Ireland, bearing date at Westminster, the Twenty-fourth day of February, (One thousand eight hundred and fifty-one,) in the fourteenth year of Her reign, did, for Herself, Her heirs and successors, give and grant unto me, the said Robert Adams, my ex̃ors, adm̃ors, and assigns, Her especial license, full power, sole privilege and authority, that I, the said Robert Adams, my ex̃ors, adm̃ors, and assigns, or such others as I, the said Robert Adams, my ex̃ors, adm̃ors, or assigns, should at any time agree with, and no others, from time to time and at all times during the term of years therein expressed, should and lawfully might make, use, exercise, and vend, within England, Wales, and the Town of Berwick upon Tweed, the Invention of "Improvements in Rifles and other Fire-arms," partly invented by me and partly communicated; in which said Letters Patent is contained a proviso that I, the said Robert Adams, shall cause a particular description of the nature of the said

Invention, and in what manner the same is to be performed, by an instrument in writing under my hand and seal, to be inrolled in Her said Majesty's High Court of Chancery within six calendar months next and immediately after the date of the said in part recited Letters Patent, as in and by the same, reference being thereunto had, will more fully and at large appear.

NOW KNOW YE, that in compliance with the said proviso I, the said Robert Adams, do hereby declare that the nature of my said Invention, and the manner in which the same is to be performed, are fully described and ascertained in and by the following statement thereof, reference being had to the Drawings hereunto annexed, and to the figures and letters marked thereon, that is to say:

The Invention consists,—First, of improvements in the construction of rifle barrels of fire-arms.

Secondly, the Invention consists of improvements in the construction of gun and pistol locks.

Thirdly, the Invention consists of improvements in the construction and arrangement of safety bolts or catches applied to fire-arms.

Fourthly, the Invention consists of improvements in the construction of guns or other fire-arms arranged to load at the breech.

Fifthly, the Invention consists of improvements in the construction of fire-arms having only one fixed barrel, with a series of loading chambers which revolve upon a centre, so as to bring each chamber alternately in a correct line with the barrel.

And in order that the Invention may be most fully understood, and readily carried into effect, I will proceed to describe the means pursued by me.

The improvements in the construction of rifle barrels of fire-arms consist in forming raised ridges on the inner surface of such barrels in place of grooves or the ploughed hollows commonly called "rifling;" and I prefer to form such ridges angular, as shewn by the section of a barrel at Figure 1, but the shape and number of such ridges may be varied, my Invention consisting in forming raised or projecting spiral ridges on the interior surface of the barrels of fire-arms. *a*, *a*, are the projecting ridges which are to be made in a spiral direction similar to that heretofore resorted to for the grooves. *b*, *b*, are the grooves, which have a considerable extent of width as compared with the more narrow ridges. Heretofore the hollows or grooves were narrow as compared with the projecting parts, and in fact as heretofore constructed the diameter of the bore of the barrel was taken across from the raised or internal projecting parts, the grooves being external of the measured diameter. According to my mode of construction the measured diameter of the bore of the barrel is from *b*, to *b*, the ridges being projecting interiorly. I prefer to employ three such ridges at equal distances apart, but this, as before stated, may be varied, and the balls used with such barrels should be cast with notches or grooves to fit the barrels, and I have found that by such means no patches are required to be used with the balls.

Another improvement in the construction of the barrels of fire-arms consists in forming them near the breech with an internal hollow projection, which I prefer to be coned on the outside as shewn at *a*, Figure 2, upon which the ball having a recess at one end corresponding in shape of the projection is placed. The passage or chamber from the end of the projection *a*, to the priming being of the proper dimension to receive the charge of powder suitable for such fire-arm, and the charge may be placed therein by inverting the barrel over the end of a tubular ramrod, into which the proper charge of powder has been first put, and then turning the barrel on end so as to discharge the contents of the ramrod into the breech; or the ramrod may be tubular, so as to allow the powder to be poured through it. The ball is next rammed down on to the projection *a*, thus effectually retaining the powder.

I will now describe the second part of the Invention. Figure 3, shows front and back views of a bar or front action lock constructed according to this part of the Invention; and Figures 4, 5, 6, and 7, show several views of the parts of the lock separately. This lock is composed of four separate parts only, and consists of a hammer or cock *a*, a plate *b*, a spring *c*, and a screw *d*. The cock *a*, works on a pivot or axle b^3, attached to and forming part of the plate *b*; and through the pivot the screw passes, which holds the cock firm in its place, and also attaches the lock to the stock of the gun in the back action lock, but in the bar lock by the usual tumbler pin or nut. The action of the lock is produced by the union of the main and sear springs in one piece which spring at the end c^1 is provided with a semi-circular piece, in plaçe of a tumbler, and the concave side of the part c^1 is notched or cut into notches to catch the tooth of a projecting sear a^1 fixed on to the inner side of the cock a, the sear a^1 passing through the curved slotted opening b^1 in the plate b, to take into the notched parts c^1 of the spring, when the hammer is placed at half or whole cock; and the part c^1 has a stud c^2, which projects at right angles from it into the stock to be struck or pressed by the action of the trigger to set the cock at liberty. The sear piece a^1 is rounded at its lower end, and traverses in the curved opening b^1 in the plate b, when placing the hammer on half or whole cock, and the curved ends of the opening b^1 limits the action of the cock in either direction. The stud c^3, fixed to and forming part of the spring c, passes through the slotted opening b^2 in the plate b, and takes into a recess a^2 on the inner surface of the cock, and when the cock is pulled back the recess a^2 acting on the stud a^3 compresses the spring. When the sear a^1 is released by the trigger depressing the stud c^2 a rapid motion will be given to the cock or hammer by the stud c^3.

I will now describe the third part of my Invention. Figure 8, shows a side view of apparatus commonly called a safety bolt or catch for bolting the cock or hammer, and preventing guns being discharged by accident; Figure 9, is a plan of the same parts; and Figures 10, and 11, show a side view and plan of the hammer and cock plate with the safety catch applied thereto. On the circular part a of the cock or hammer are formed notches or ratchet teeth d, into which the bent end of the lever or pall bit e takes and retains the hammer at half or whole cock until released in the following manner:—The lever e is attached to and carried by a pin or axis f, which passes through the lock plate into the bridle leg; on the axis f, is fixed a second or inner sear g, the bent arm g^1 of which is lifted or depressed by the lever h, carried by a projection h^1; on the inner surface of the trigger plate j is the grip lever, one end being attached by a pin joint to the projection h^1, the other resting on the end of the slot in the trigger plate. *i*, is a spring which is screwed to the inner surface of the trigger plate, one end of the spring always pressing the arm of the lever *h* in contact with the lever g^1 upwards, so as to keep the lever arm *e* in connection with the ratchet teeth on the hammer; the other end of the lever *h*, resting upon the grip lever *j*, which is constantly pressed outwards by the spring *i*, the other end of the spring *i*, acting on the trigger to keep it in position; the lever *h* has a spring h^2, affixed to it as shown, and when the grip lever is pressed upwards the end of the spring h^2 depresses the arm g^1, and removes the lever *e* from the notches on the hammer when at half or full cock; but when the cock is resting or depending on the safety catch any accidental blow or pressure on the grip lever is received on the spring h^2, thereby preventing any injury or derangement of the safety apparatus, and prevents the lock being disengaged until placed at half or whole cock.

I will now describe the fourth part of the Invention. Figure 12, shews part of a side view, and Figure 13, a plan of a gun arranged according to this part of the Invention. Near the breech end of the barrel a mortice or opening is formed, through which the chamber or moveable breech turns horizontally upon a vertical axis. By this arrangement or construction the upper part of the barrel is continued along the upper part of the stock, and forms a plate to which it is affixed, and the metal at the breech is cut away, as shewn, to receive the moveable breech, the barrel being enlarged for that purpose. *a*, is the fixed part of the barrel of the gun; and *b*, the moveable breech turning with the pin *e*, to which the trigger plate is attached, and is capable of turning at right angles with the gun, as shewn by red lines, so as to place the end of the moveable breech in a position to be loaded; *d*, is a bolt to retain the breech in a right line with the barrel when loaded.

I will now describe the fifth part of the Invention. Figure 14, shews a side view; Figure 15, a longitudinal section; Figure 16, a cross section; Figure 17, separate views of the revolving chambers; and Figures 18, 19, 20, and 21, shew some of the parts separately of a gun or pistol having the improvements applied thereto. *a*, is the tube or barrel of any suitable length, and open at each end; the upper part being continued at a^1 descends at right angles, returns, rises again, and joins the under side of the barrel, forming a rectangular opening, in which the series of chambers containing the several charges revolve to be successively brought opposite to the barrel a, and discharged as herein-after explained. The revolving chambers b are supported by and revolve on the axis b^1, which slides in a groove on the under side of the barrel through the rectangular frame a, the end being supported in the back part of the frame, which is solid, with the barrel as shewn, and the series of revolving chambers may be readily detached from the opening through the frame by drawing the axis b out towards the mouth of the barrel, or the series of revolving chambers may be supported on two conical centres, one or both of which may be capable of sliding out, so as to detach the series of chambers as shewn in section at Figure 22. The back of each chamber is recessed, as shewn, to receive the nipple b^2; at the end are formed ratchet teeth b^3, into which the perpendicular ratchet lever c^2 (shewn in red lines) works; the other end being connected to the trigger c by the pin c^3, which forms an axis for the sear c^1, which takes into a bent or recess e^1 formed in the front part of the cock or hammer e, and when the trigger is pulled back the sear c^1 lifts the hammer to a sufficient height for the action of the main spring to give it sufficient force to explode the cap, and the sear c^1 is thrown out of the recess e^1 by the projecting neb or tooth on the cock as shewn, when the cock is forced back to the fullest extent. The hammer strikes the cap in an horizontal direction, the striking part passing through a hole formed in the hinder part of the frame a^1; a spring bolt d, when pressed inwards, comes against a shoulder on the hammer head, and retains it back to prevent the hammer discharging the cap by an accidental blow, or for safety in carrying or when disengaging the part b; f^1 is a spring to return the trigger and parts attached to their original position; c^4 is a stud on the trigger which enters recesses on the outer circumference of the revolving chamber at the time each chamber is brought opposite to the barrel a, and retains it there whilst the discharge takes place.

A.D. 1851. Febʸ 24. Nº 13,527.
ADAMS' SPECIFICATION

(4 SHEETS)

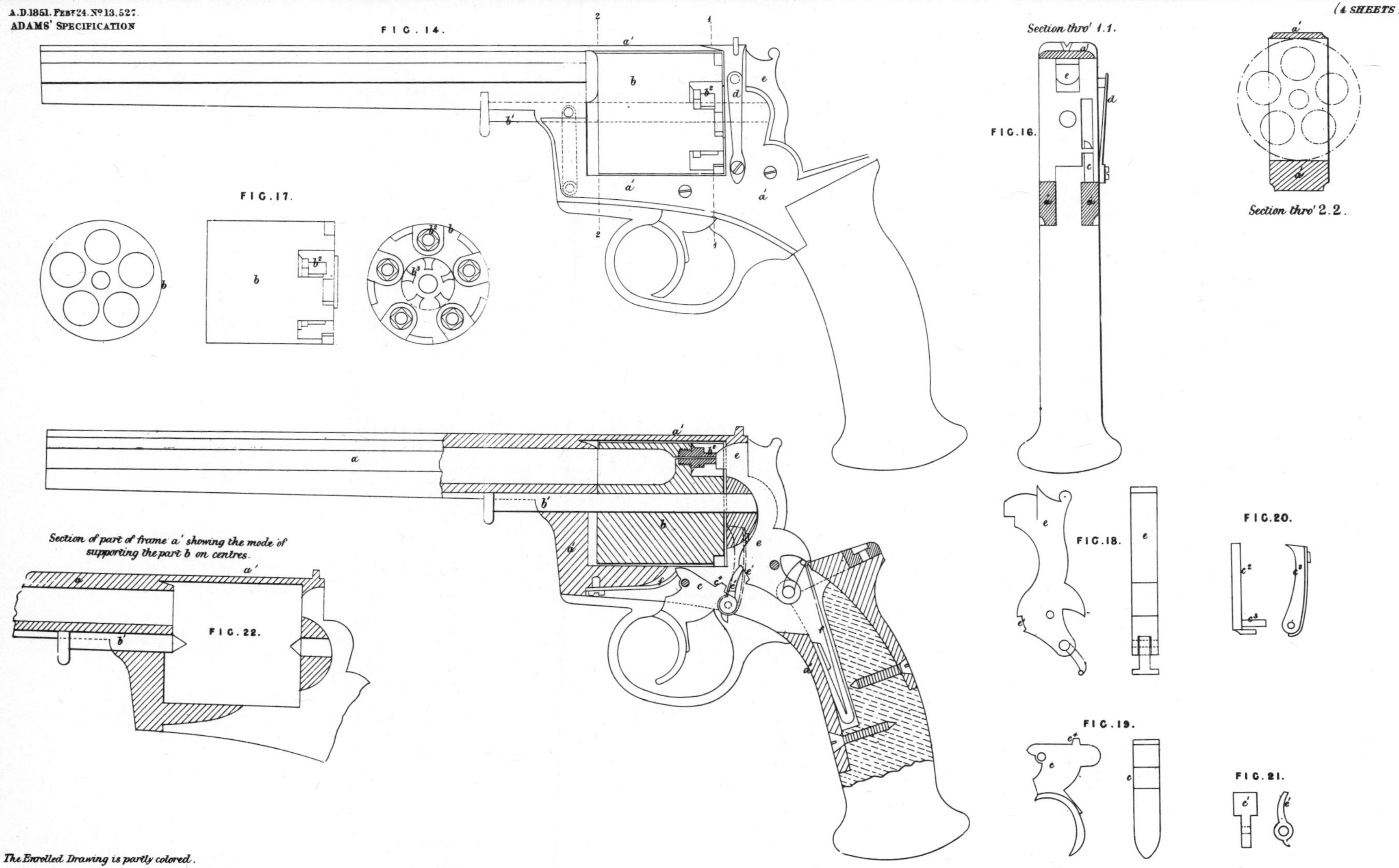

The Enrolled Drawing is partly colored.

Malby & Sons lith

LONDON: Printed by GEORGE EDWARD EYRE and WILLIAM SPOTTISWOODE. Printers to the Queen's most Excellent Majesty. 1854.

Having thus described the nature of the Invention, and the manner of performing the same, I would have it understood that I make no claim to the several mechanical parts separately, nor do I confine myself to the several details, so long as the peculiar character of any part of the Invention be retained. But what I claim is,—

First, the constructing the interior of rifle barrels with ridges. I also claim the manufacture of barrels with internal projections for the powder to pass into, and for the ball to fit on.

Secondly, I claim the improvements herein described of gun and pistol locks.

Thirdly, I claim the improvements herein described in safety bolts or catches of fire-arms.

Fourthly, I claim the improvements herein described in guns and other fire-arms arranged to load at the breeck.

And fifthly, I claim the mode of constructing the breech end of a barrel to receive a revolving series of chambers, and also the arrangement or combination of locks for such description of fire-arms.

In witness whereof, I, the said Robert Adams, have hereunto set my hand and seal, this Twenty-second day of August, in the year of our Lord One thousand eight hundred and fifty-one.

ROBERT (L.S.) ADAMS.

AND BE IT REMEMBERED, that on the Twenty-second day of August, in the year of our Lord 1851, the aforesaid Robert Adams came before our said Lady the Queen in Her Chancery, and acknowledged the Specification aforesaid, and all and every thing therein contained and specified, in form above written. And also the Specification aforesaid was stamped according to the tenor of the statute made for that purpose.

CROSBY.

Enrolled the Twenty-third day of August, in the year of our Lord One thousand eight hundred and fifty-one.

LONDON:
Printed by GEORGE EDWARD EYRE and WILLIAM SPOTTISWOODE,
Printers to the Queen's most Excellent Majesty. 1854.

A.D. 1852 . . . . No 13,994.

Needle Guns.

KUFAHL'S SPECIFICATION.

TO ALL TO WHOM THESE PRESENTS SHALL COME, I, George Leopold Ludwig Kufahl, of Christopher Street, Finsbury, London, Engineer, send greeting.

WHEREAS Her present most Excellent Majesty Queen Victoria, by Her Royal Letters Patent, under the Great Seal of the United Kingdom of Great Britain and Ireland, bearing date at Westminster the Third day of March, in the Fifteenth year of Her reign, did, for Herself, Her heirs and successors, give and grant unto me, the said George Leopold Ludwig Kufahl, Her especial licence, full power, sole privilege, and authority, that I, the said George Leopold Ludwig Kufahl, my exors, admors, and assigns, and such others as I, the said George Leopold Ludwig Kufahl, my exors, admors, or assigns, should at any time agree with, and no others, from time to time, and at all times during the term of years therein expressed, should and lawfully might make, use, exercise, and vend within England, Wales, and the Town of Berwick upon Tweed, within the Islands of Jersey, Guernsey, Alderney, Sark, and Man, and in all Her Majesty's Colonies and Plantations abroad, my Invention of "Improvements in Fire-arms," in which said Letters Patent is contained a proviso, requiring that I, the said George

Leopold Ludwig Kufahl, by an instrument in writing, under my hand and seal, shall particularly describe and ascertain the nature of my said Invention, and in what manner the same is to be performed, and shall cause the same to be enrolled in Her said Majesty's High Court of Chancery, within six calendar months next and immediately after the date of the said in part recited Letters Patent.

NOW KNOW YE, that in compliance with the said proviso, I, the sad George Leopold Ludwig Kufahl, do hereby declare the nature of my Invention, and in what manner the same is to be performed, to be particularly described and ascertained in and by this present instrument in writing, as follows, reference being had to the Drawings hereunto attached, and to the letters and figures marked thereon, that is to say:—

My said Invention relates to various novel and useful improvements in that class of fire-arms ordinarily known as needle guns. My previous improvements in the same class of fire-arms, under the title of "Improved Construction of Guns and Cannons, and Manufacture of Cartridges for the Loading or Charging thereof," have been protected by Letters Patent, bearing date the Eleventh day of January, One thousand eight hundred and fifty, and granted to Matthew Urlwin Sears, of Burton Crescent, London, as a communication from me. The objects of the present Letters Patent consist of various mechanical arrangements, apparatus, or means whereby the lock mechanism of fire-arms on the needle principle, whether shot guns, rifles, pistols, or repeaters, is made more durable, safe, and efficient than heretofore, and these improvements are entirely distinct from and unconnected with the subject of the said Letters Patent granted to the said Matthew Urlwin Sears. And in order that my said Invention may be readily understood, I shall now proceed to describe the various Figures on the Sheet of Drawings hereunto attached.

Figures 1, 2, 3, 4, 5, 6, and 7, on my Sheet of Drawings, illustrate the application of my improvements to an ordinary rifled needle gun. Figure 1, is an external side elevation of the lock portion of a needle rifle; and Figure 2, is a vertical longitudinal section of the same parts corresponding. Figure 3, is a transverse vertical section through the lock, at the line *a*, *b*, in Figures 1 and 2. Figure 4, is a similar transverse section, a little further forward at *c*, *d*, in Figure 2; and Figure 5, is a third section still further forward, at the line *e*, *f*, in Figure 2. Figure 6, is a horizontal longitudinal section, along the line of the bore of the barrel and the lock cylinder, as at *g*, *h*, in Figure 2; and Figure 7, is an external plan of the lock and its details.

In carrying out my several improvements I have merely retained the outside tube or socket A, screwed into or made in one piece with the barrel B, the needle C, with its guide D, and the lever handle L, as used in the common Prussian needle gun, such portions being common to all needle guns. The other details of the ordinary gun I have either laid wholly aside, or have so modified them as to suit the particular intention of my improvements. The igniting needle C, is screwed into a conductor E, carrying at its fore end a piston F, well packed with some elastic material, say stout leather, and the other or hind end is formed with a stud G, and disc H. The needle conductor E, is capable of actuation by the tumbler I, which has a rounded head for that purpose. This tumbler is set on a fixed stud centre K, near which it is actuated by the flat main spring J, and a sear, sear spring and trigger, and its axis projecting through the gun stock carries a cocking lever M. As the parts are delineated in the Drawings, they are in the position arising from the lock being at rest, or just discharged by the needle in passing into the priming. The head of the tumbler rests on the stud G, forcing the needle conductor and its piston up to the projecting hind end of the needle guide, so as to prevent the escape of the gaseous products of the explosion at this point, which escape is a very serious objection to the ordinary needle arms, especially with powerful charges. The head of the tumbler in this position, as marked 1, stands up through two corresponding slots, one in the external fixed tube or socket A, and the other in the inner moveable tube N; and hence the latter cannot now be moved round until the withdrawal of the tumbler head. At the same time the abutment O, of the inner tube is firmly interlocked with a shoulder of the external tube A, as represented in Figure 1.

Now in order to withdraw partially the tube N, from the outside one for loading or cleaning, the abutment O, is first to be brought opposite to a longitudinal slot P, on the left side of the external tube A, so that the breech of the gun can never be opened either wilfully or accidentally, either in the act of firing or for the purpose of introducing a charge, whilst the tumbler is in this position, and whilst the needle protrudes at all from its guide. This arrangement constitutes one of the means provided for insuring perfect safety in handling the gun, such safeguard being entirely wanting in all other needle guns which may be opened at the breach and charged whilst the needle stands out more or less from its guide, so that the charge is thus liable to accidental explosion. When the breech is to be opened for the introduction of a charge through the loading hole Q, the lock is, in the first instance, to be brought to half cock. This movement will carry the tumbler to the position indicated by the dotted line 2, in Figure 2, and the action causes the head of the tumbler to act on the face of the lower projecting portion of the disc H, of the needle conductor, and thus to draw back the needle to screen it inside the needle guide. At this position of half cock, but at no other throughout the entire range of action, the tumbler head is directly opposite a transverse slot R, formed in the left side of the inner moveable tube or chamber N, and arranged to receive the tumbler easily, so that the inner tube may then be turned partially round to the left by means of its lever handle L. When this movement has been accomplished the abutment of the lever stands opposite the longitudinal slot P, of the outside tube or socket A, as represented by the dotted lines in the plan Figure 7, the amount of turn being accurately determined by the bringing up or the bearing of the abutment O, against the left side of the longitudinal slot P. In this condition the slot P, allows of the inner tube N, being slided backwards until the abutment O, reaches the point S, so as to uncover the loading hole, the head of the tumbler meanwhile entering a second longitudinal slot T, of the inner tube, which latter slot communicates with or opens into the transverse slot R. The piece is now ready for loading, and while open at the breech for that purpose the tumbler, and consequently the main actuating spring J, are entirely disconnected from the needle conductor and needle, the stud G, being passed to the right side of the tumbler head. This contrivance is the second means of safety not to be found in any other class of needle gun. The cartridge having been introduced through the loading hole Q, into the space U, the inner moveable tube N, must again be pushed forward by means of its handle. This action covers up the charging hole, and the movement is rendered complete by turning the lever handle L, back again to the right, so as to interlock the abutment O, with the shoulder W^1, of the socket A, and all this must be done before the tumbler can act on the needle conductor. As the gun is now in a condition to be fired the lock may be brought to full cock, as indicated by the third radial dotted line 3, of the tumbler's position in Figure 2. In this position of the details, the head of the tumbler standing up at once in the lower longitudinal slots of both the outer and inner tubes, prevents the inner one from being turned, so that there is no possibility of the loading hole being opened. By this means it is that the arm is rendered perfectly safe in handling, for the breech cannot be opened for loading or any other purpose, nor can it fly open accidentally so long as the tumbler, main spring, and trigger have any communication with or power of action on the needle, which is the only means of firing the charge. To facilitate the periodical cleansing of the gun, a simple means is provided for the entire withdrawal of the inner tube N, from its socket, and its replacement without trouble or annoyance. This is done by first bringing the inner tube to its second position, or that which it occupies during loading, when the abutment of the lever handle L, is drawn back into the longitudinal slot P, until it reaches the point S, the lock being first brought to half cock to allow of this movement, as already described. The tumbler must now be drawn back to the extent of two notches, that is until the notch V, which is one notch beyond full cock, shall catch upon the detent, and this brings the tumbler to the position of the furthest dotted line marked 4. With the tumbler in this position its head is brought entirely beneath and clear of any part of the inner tube N, so that the latter may be at once drawn out of its socket by means of its handle L. When cleaned, the tumbler being in the position 4, the needle conductor is brought to the position in tube N, which it occupied when that tube was withdrawn. The tube is then reinserted at the open end of the socket, and is then pushed forward at once to the position for loading. The lock is then lowered to half cock and the loading hole is closed, as already described. By the contrivance of the position 4, of the lock an important advantage is gained, for without it the inner tube could never be wholly withdrawn from its socket, nor could the gun be effectually cleansed without removing it from the stock, and unscrewing the lock. After loading, if the piece is not to be fired at once, the lock is kept at half cock, or it may be let down to the position of rest. This latter position, however, is only to be arrived at when the charge is in by bringing the abutment O, of the lever handle opposite the longitudinal slot P, in the external tube. It is this feature of the engagement and disengagement of the needle with the tumbler and main spring which constitutes an essentially distinct feature of safety of action in my improved gun, for the ordinary needle guns must be kept either at half or full cock when loaded, otherwise the needle would enter and explode the charge. Of the minor details, with the assistance of which the accuracy of the various movements of the lock is insured, the mode in which the piston is caused to retain the position assigned to it while the lock is at half cock, and the loading hole opened for receiving a charge, or other purposes, is very important. To give this holding power a small flat spring W, is adjusted on the right side of the needle conductor, and a small projection on this spring, as represented in Figure 6, falls to a recess in the interior of the inner tube whenever the lock is brought to half cock. This gives the spring a slight hold upon the piston, sufficient to stay it until the latter is forcibly drawn along by the lock action. To prevent the escape of powder gas, or the gases resulting from the explosion of charge, at the joint between the termination of the barrel and the end of the inner tube N, the projecting open end of the latter may be turned down to a knife edge, so as to furnish a thin closely fitting and slightly elastic exterior bearing surface in contact with the interior of the barrel end, to form a tight joint when the loading hole is closed. But instead of this contrivance I prefer to insert into the front open end of the tube N, a short thin elastic tube X, of copper, or other suitable metal, which being fast in the tube N, moves with it, and it is left projecting a slight distance beyond the end of the tube, so as to enter a short distance into the end of the barrel, when the breech is closed for firing. This tube being very thin at the edge becomes expanded and pressed from within outwards, against the interior of the barrel by the effect of the explosion when the piece is fired, hence the joint is effectually closed and the gases are prevented from escaping, just as the cup leather of a hydrotatic press prevents the escape of water. For the best guns this thin joint tube is of platinum, so as to last for a very long period. Both these kinds of joints are further assisted in their accuracy of fitting by turning an external cone on the end of the inner tube N, to fit to a similar internal cone in the barrel end. The flat main spring J, is substituted for the helix ordinarily employed for actuating the lock mechanism of needle guns, as being much more durable and satisfactory in its action, and an important feature exists in the external cocking lever M, the position of which at any angle at once points out whether the lock is shut down at rest, or at full or half cock, or on the last notch. Although I have herein-before referred to the use of cartridges in my improved fire-arms, as affording the greatest facilities for speedy loading, I may remark that they are not by any means absolutely required for the effective service of the gun, for instead of cartridges, the powder may be easily measured into the breech, to lay round the needle guide, and the bullet, with the priming, may be thereafter inserted. And, although I have herein-before referred especially to the construction of needle rifles on my improved principle, it will be obvious to the practical gun smith that muskets, carabines, shot guns, pistols, and other arms may be so fitted and managed, as I have described. But the construction of needle revolvers, or repeating pistols, which has never hitherto been attempted, necessitates the introduction of certain modifications of the details, which I shall now proceed to describe. Figure 8, on my sheet of Drawings is a longitudinal side elevation of a repeating needle pistol, as fitted with my improvements. Figure 9, is a vertical longitudinal section of the same corresponding. Figure 10, is a transverse section at the line *a*, *b*, through the centre of the tumbler spindle in Figure 9, and Figure 11, is a second traverse section through the line *c*, *d*, in Figure 9. Figure 12, is a front end elevation of the revolving or repeating chamber, forming the magazine of charges, and,

Figure 13, is an end elevation of the hindmost open end of the barrel. The pistol is delineated in the Figures as unloaded, and having its lock put down or at rest. The fixed external tube or socket A, is formed at the upper side of its front end, next the barrel, with a longitudinal and a transverse slot meeting in cruciform fashion, for receiving and allowing of the requisite movements of the lever handle L, of the inner or moveable tube, containing the needle conductor E, with its packed piston F, the stud G, and disc H, as in the same details of the needle rifle. The lower part of the socket A, is extended downwards at its front end and on the left side, forming in combination with an adjustable plate on the right side and the trigger plate, a rectangular box for the reception of the tumbler I, the main spring J, the sear spring upper part of trigger, as well as those parts by which the revolving magazine B, containing the charges is made to revolve. When unloaded and at rest, with the tumbler in the position 1, as delineated in Figure 9, the tumbler head projects upwards through the corresponding lowest longitudinal slots in both tubes, so as to prevent the rotatory movement of the inner tube. At the same time the needle C, projects into the highest chamber (for the time being) of the magazine, thus also preventing the latter from revolving. A needle guide is in this arrangement quite unnecessary as the needle is so short that it may be used in the interior of the chambers. In commencing to load from the position delineated in my Drawings, Figure 9, the needle is first of all withdrawn from the most elevated chamber; this movement being accomplished as in my needle rifle, by carrying back the tumbler I, to half cock, that is to the position of the dotted line 2, by means of the external cocking lever M. The loading lever handle L, is then turned over to the left side as indicated by the dotted lines in Figure 10, whereby the head of the tumbler entering the transverse slot R, in the inner tube is disconnected from the needle conductor. The charges may then be most expeditiously entered into the ring of chambers by removing the barrel B^1, of the piece from the centre pin or arbor K^1, on which the magazine revolves. This gives free admission to the whole series of chambers and the charges consisting of powder and a bullet with suitable priming may be at once entered into the chambers successively, a washer of paper, thin felt, or other suitable material being first laid in the bottom or closed end of each chamber to cover up the small hole by which the igniting needle C, is to enter. When the chambers are all duly charged the barrel is replaced and held in position by a transverse pin O, and the lever L, being still on the left side, the tumbler may be let down to set the lock at rest. In this movement the head of the tumbler moves forward in the left hand longitudinal slot T, of the inner tube, which slot is now on the bottom side and is in accurate correspondence with the bottom slot of the external socket, and being disconnected from the needle conductor E, the tumbler leaves the latter behind to remain in the same position which it occupied on the occurrence of the half-cock movement. Here it is kept by a flat side spring W, precisely as in the rifle. In this state, as the needle cannot by any means reach the priming or detonating compound in the interior of the chambers of the magazine, the pistol cannot be discharged, and it may be carried about and subjected to the roughest usage without the slightest danger. When the arm is to be fired the tumbler is again moved to half-cock by the lever M, the lever handle L, being returned to the right hand position as in Figure 8, so that the tumbler head regains its position between the stud G, and disc H, of the needle conductor. On going on to full-cock or to the position of the dotted line of tumbler 3, the magazine B, will revolve one-sixth round, six chains being here used. This action arises from the effect of a small bell crank lever O', the short lower arm of which carries a pin fitting into a long slot in the link piece P', attached to the tumbler. The charge in the highest chamber may now be fired, and afterwards the remaining five in succession, by going at once to full-cock and drawing the trigger. When the lock is to be cleaned the inner tube N, is to be removed altogether from its outer socket, carrying with it the needle conductor, the barrel and magazine being first slipped off their arbor K'. But to permit of the removal of the inner tube with its internal mechanism, the tumbler must be brought to the position of the dotted line 4, when its head will be disengaged just as in the rifle. The means of preventing the escape of the gases of explosion, namely, the piston F, of the needle conductor shutting the needle hole in conjunction with the thin metal-tubes X, are the same as in the rifle, only that the elastic tubes X, are inserted into the chambers of the revolving magazine B, where the discharges take place. To bring the front part of the highest chamber tube into the open end of the barrel, and to hold it there during the discharge, a strong horizontal bolt or locking piece Q^1 is fitted into the lower part of the lock apparatus. This bolt works through a vertical guide S^1, screwed into the trigger plate, and its front end is kept constantly in contact with the inner or back face of the magazine, whilst its opposite end similarly bears by means of a small pulley upon a rounded prominence on the tumbler. Whilst the arm is at half or full cock, the magazine and the bolt Q^1, are retained in a back position and free from contact with the barrel end by a small blade spring U^1, situated between the front of the magazine and the barrel end, so that the thin tubes X, are all clear of the latter. But so soon as the trigger is drawn, the rounded prominence of the tumbler presses forward the bolt Q^1, and thus forces the magazine up against the barrel end, pushing the prominent end of the thin tube into the open end of the barrel, so as to close up the joint in a most efficient manner before the needle can reach the priming or detonator and explode the charge. The recoil of the magazine due to the explosion is not taken off by the main spring but by the strong axis K, of the tumbler, the bolt Q^1, being in a horizontal line with that centre. Besides these services the bolt Q^1, steadies the magazine as the latter revolves, and provides that a needle hole shall always be in its proper place for the entrance of the needle. It affects this with the assistance of the slight spring V^1, which acting on a pin in the bolt Q^1, constantly tends to force a small projection at the front end of the latter into the lowest needle hole after each revolution of one-sixth. To prevent any dislodgement of the charges in the chambers by the recoil, the end of the barrel is provided with an enlargement W^1, which all but touches the points of the bullets in the five chambers not being fired, and thus the charges are entirely prevented from moving forward by the concussion. Figure 14, on my Sheet of Drawings represents a vertical section of a metal disc pierced with six holes to correspond with the magazine chambers, and having a central hole to fit the arbor K^1. This disc is intended for screening the thin projecting edges of the elastic tubes X, from injury during loading by being placed over the face of the magazine. It will be obvious that the same general arrangements are applicable more or less in the construction of repeating needle rifles, carabines, and shot guns.

Having now described and particularly ascertained the nature of my Invention, and the manner in which the same is or may be used or carried into effect, I may observe in conclusion that I do not confine or restrict myself to the precise details or arrangements which I have had occasion to describe or refer to, as many variations may be made therefrom without deviating from the principles or main features of my Invention, but what I consider to be novel and original, and therefore claim as the Invention secured to me by the herein-before in part recited Letters Patent, is,

First, the application and use of the stud and disc, or details substantially of a similar nature, to the needle conductor for the purposes herein-before described.

Second, the system or mode of engaging and disengaging the needle with the tumbler or lock apparatus, as herein-before described.

Third, the application and use of a piston or traversing filling up piece, for the prevention of gaseous escape into the lock.

Fourth, the adaptation of the lock tumbler, for the prevention of the withdrawal of the inner tube or chamber from its socket, except at a certain position of the tumbler.

In witness whereof, I have hereunto set my hand and seal the third day of September, in the year of our Lord One thousand eight hundred and fifty two.

GEORGE LEOPOLD (L.S.) LUDWIG KUFAHL.

Signed, sealed, and delivered in the presence of

EDWIN P. ALEXANDER,

BRIS. HUNT,

Clerks to Mess[rs] W. & J. H. Johnson,

47, Lincoln's Inn Fields.

A.D. 1852, March 3. Nº 13,994.
KUFAHL'S Specification.

(1 Sheet.)

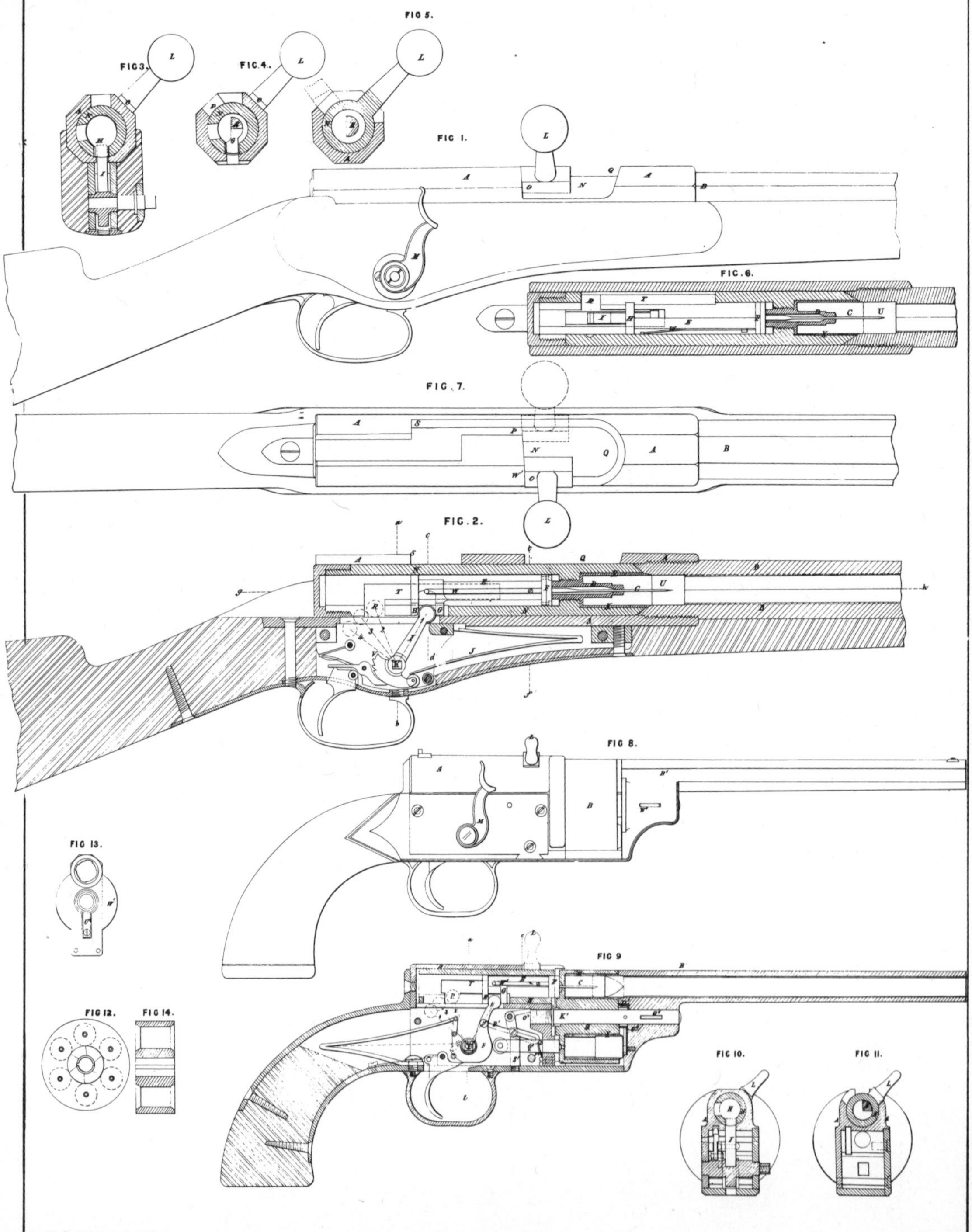

The Enrolled Drawing is not Colored

Malby & Sons Lith

London Printed by George Edward Eyre and William Spottiswoode.
Printers to the Queens most Excellent Majesty. 1854

A.D. 1852 . . . . N° 14,052.

Breech-loading Fire-arms.

POOLE'S SPECIFICATION.

TO ALL TO WHOM THESE PRESENTS SHALL COME, I, Moses Poole, of the Patent Office, London, Gentleman, send greeting.

WHEREAS Her present most Excellent Majesty Queen Victoria, by Her Royal Letters Patent under the Great Seal of the United Kingdom of Great Britain and Ireland, bearing date at Westminster, the Thirty-first day of March, One thousand eight hundred and fifty-two, in the fifteenth year of Her reign, did, for Herself, Her heirs and successors, give and grant unto me, the said Moses Poole, my ex̃ors, adm̃ors, and assigns, Her especial licence, full power, sole privilege and authority, that I, the said Moses Poole, my ex̃ors, adm̃ors, and assigns, or such others as I, the said Moses Poole, my ex̃ors, adm̃ors, or assigns, should at any time agree with, and no others, from time to time and at all times during the term of years therein expressed, should and lawfully might make, use, exercise, and vend, within England, Wales, and the Town of Berwick-upon-Tweed, the Invention of "Improvements in Fire-arms," communicated to me by a certain foreigner residing abroad, in which said Letters Patent is contained a proviso that I, the said Moses Poole, should cause a particular description of the nature of the said Invention, and in what manner the same is to be performed, by an instrument in writing under my hand and seal, to be inrolled in Her said Majesty's High Court of Chancery within six calendar months next and immediately after the date of the said in part recited Letters Patent, as in and by the same, reference being thereunto had, will more fully and at large appear.

NOW KNOW YE, that in compliance with the said proviso, I, the said Moses Poole, do hereby declare that the nature of the said Invention, and the manner in which the same is to be performed, are fully described and ascertained in and by the following statement thereof, reference being had to the Drawing hereunto annexed, and to the figures and letters marked thereon, that is to say:

The Invention relates to that class of fire-arms termed breech loading guns or fire-arms, or, in other words, those wherein the charge of powder and ball or shot, is inserted at or near the breech, instead of first being placed therein through the muzzle of the barrel. A gun made in accordance with the improvements herein-after specified admits of a cartridge being inserted in the rear part of the barrel, and on a line with the axis of the bore of the barrel. It also causes the cartridge to have its rear end cut off by the action of the sliding breech, so as to expose the powder to the fire of the priming or percussion powder. And furthermore, it secures the percussion nipple to be moved out of the way of the hammer or cock during the operation of loading the piece, thus preventing at such time a liability to explosion of the charge by any accidental depression, fall, or throw of the hammer down upon the nipple or percussion cap, or powder thereon.

Of the Drawings herewith presented Figure 1, denotes a side view of the said fire-arm; Figure 2, is a vertical central and longitudinal section of it; Figure 3, is a top view of it.

To enable others skilled in the fabrication or use of fire-arms to make and operate my Invention, I will proceed to explain it. In the said Drawings, A represents the barrel, and B the stock of the gun; the barrel being joined to the stock in any proper manner by a metallic connection, such as will allow of an aperture C made down through it, and next to the rear end or breech of the barrel, such aperture being of a size and shape to receive and hold a sliding breech or block of metal D, which is made to fit closely to it, and to slide up and down through or in it. The said slide or block D is connected with a lever E, by a stirrup or link F, jointed both to the lever and slide, so that when the lever (which turns on a fulcrum or pin H, arranged as seen in the

Drawings) is turned down, or in a direction away from the stock, as denoted by dotted lines in Figure 2, the block will be moved by it downwards a sufficient distance to cause it to freely and fully uncover the bore at the rear end of the barrel, and so as to allow a cartridge of powder to be inserted into such bore through such rear end of it. The lever, link, and block or slide should also be so made that when such lever is turned up close against the under side of the stock, the slide or block shall be carried up into the position denoted in Figure 2, so as to entirely cover over the rear end of the bore, and cause the block to be a breech or wall to resist the back force of the charge during explosion of it. The slide block should have a shearing or cutting edge on its top, as seen at *a*, so that when raised up against the rear part of a cartridge it shall cut off, shear, or sever such part of the cartridge from the rest or that portion of it which may be within the barrel. A front view of the slide block is given in Figure 4; a horizontal section of it in Figure 5; and a vertical section of it, through its nipple, in Figure 6.

The nipple for explosion of the percussion cap is seen at *b*, it being secured in or affixed to the said slide D, so as to move with it when it is moved either upwards or downwards. It may, however, be arranged or affixed to the stock or lock so as to be independent of the slide D; but this latter mode of applying it I do not consider so advantageous as that first described, as when it is attached to the slide it is carried out of reach of the cock I, when the slide is depressed for the purpose of enabling a person to load the gun.

The lever E, forms a guard for the trigger K, and may at its rear end be locked (whenever desirable) to the stock by means of a slide catch L, or any mechanical equivalent therefor.

The bore of the barrel is counterbored or properly enlarged at its rear part to receive the cartridge and shot or ball, such bore and cartridge being of such lengths as to cause the rear part of the cartridge when placed in it to project a short distance beyond it, so as to enable the slide D, when forced upwards, to shear or cut off the end of the cartridge, and allow the powder to rest against the slide and pass into a passage *c* thereof, made to communicate with the passage or bore of the nipple. The passage *c* is made with a shearing edge or inclined surface, as seen at *d*, whereby the powder is made to more readily descend or enter it than it would be otherwise likely to do.

A small spring *h*, is applied to the lever E, and the stock of the gun, and made to operate so as to hold or press the rear end of the lever up against the stock when it is in such position.

To use this gun, first prepare a cartridge by making a paper tube nearly of the size of the counterbore of the barrel. Into one end of such tube put the ball or balls, shot or shots, having a patch or piece of cloth or other suitable material properly drawn over the same, and secured to the tube. Next, charge the cartridge with gunpowder or other explosive material, and close it up of a length sufficient to cause it to project a little from the rear end of the barrel when put into place within the rear or counterbored part of it. The parts of the gun being in a situation for the insertion of the cartridge through the rear part of the bore of the barrel it is to be introduced therein, and the lever E turned up so as to cause the rise or full upward movement of the sliding breech shearing block and the sheering off of the back end or part of the cartridge. The percussion cap may next be put on the nipple, and the gun is then ready for being discharged in the ordinary way.

The advantages of this gun over other breech loading guns are, that the charge of the cartridge is never fully exposed to action of the fire of the priming or percussion cap. The slide is so little exposed to the explosion that it will not foul or expand to any material degree so as to have its motions seriously affected. Again, the movements of the slide breech being performed by or through a lever of high power, the slide can be very closely fitted to the rear end of the barrel. As the nipple is attached directly to the slide, and is moved with and by it out of the way of the cock, the cap does not reach the hammer or cap until the slide breech has effectually covered the bore of the barrel. Thus danger to the use of the gun from premature explosion of the cartridge rearward is prevented.

The main principle of the Invention, and what I claim, is the affixing to or combining with a breech loading barrel a contrivance or mechanism for clearing off the rear end of the cartridge, whether such contrivance be connected directly with a moveable slide, or with the barrel made moveable, or with any other part of the gun. And I further claim to combine the breech slide with the barrel and stock, or the supporter of the slide, in such manner substantially as specified, that when the said breech slide is depressed or moved in a proper direction the rear end of the bore of the barrel shall be exposed, so as to enable it to have a cartridge inserted at such end instead of through the muzzle, and on or about on a line with the bore; and also that when the said slide is elevated or moved in the opposite direction to a proper degree, it shall cover up the rear end of the bore, and constitute a contrivance for resisting the rearward force of the explosion of the charge. And I further claim the arrangement of the shearing contrivance, or that which severs off the rear end of the cartridge directly on or its direct combination with the breech slide, so as to operate during and by the upward or return movement of the slide as above explained. I also claim the improvement of combining the cap nipple with the breech slide, so as to be moved by it or with it and away from and towards the cock, substantially as specified.

In witness whereof, I, the said Moses Poole, have hereunto set my hand and seal, this Thirtieth day of September, in the year of our Lord One thousand eight hundred and fifty-two.

MOSES (L.S.) POOLE.

AND BE IT REMEMBERED, that on the Thirtieth day of September, in the year of our Lord 1852, the aforesaid Moses Poole came before our said Lady the Queen, in Her Chancery, and acknowledged the Specification aforesaid, and all and everything therein contained and specified, in form above written. And also the Specification aforesaid was stamped according to the tenor of the Statute made for that purpose.

CROSBY.

Enrolled the Thirtieth day of September, in the year of our Lord One thousand eight hundred and fifty-two.

LONDON:
Printed by GEORGE EDWARD EYRE and WILLIAM SPOTTISWOODE,
Printers to the Queen's most Excellent Majesty. 1854.

A.D. 1852. MARCH 31. No 14,052.

POOLE'S SPECIFICATION.

(1 SHEET.)

FIG. 3.

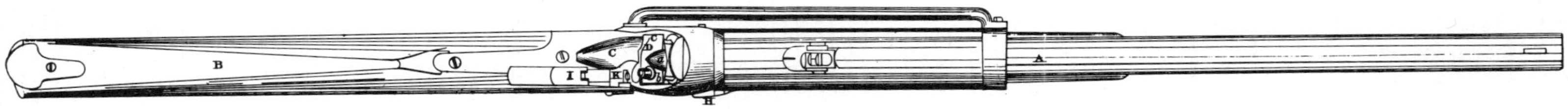

FIG. 4.

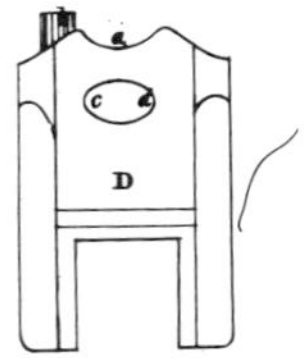

FIG. 2.

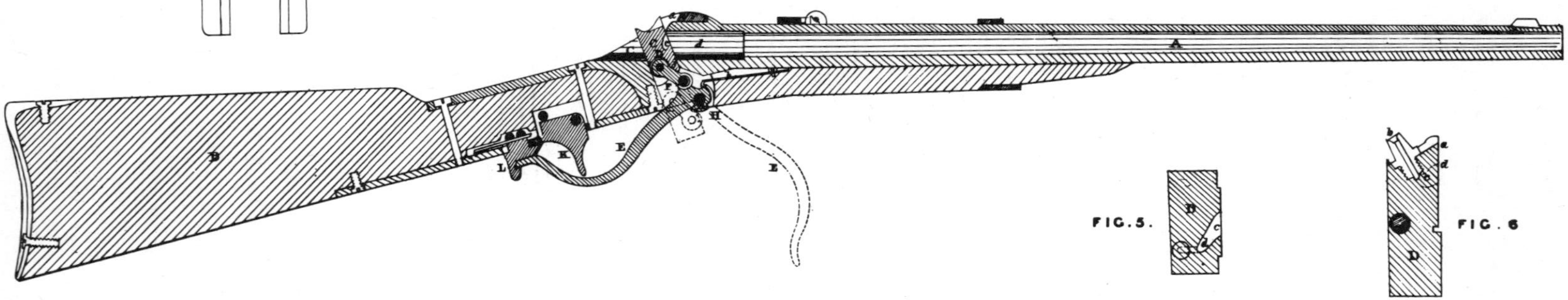

FIG. 5.

FIG. 6

FIG. 1.

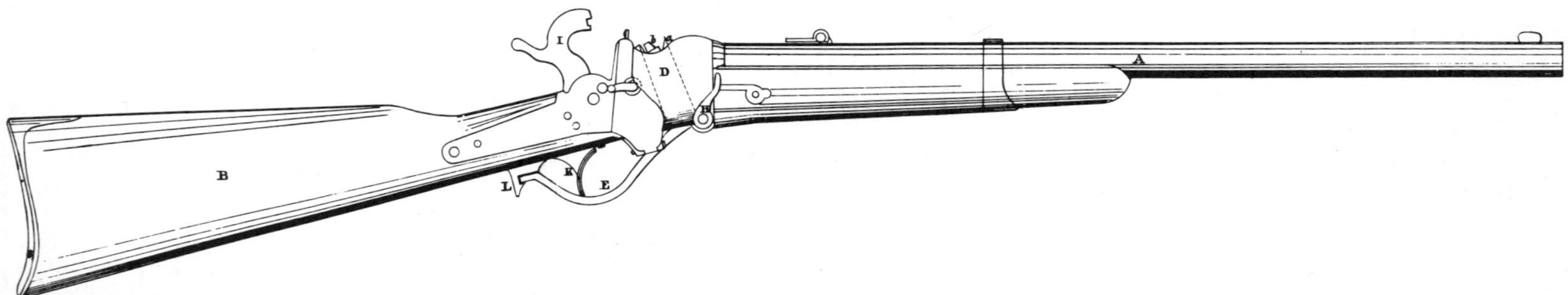

The Enrolled Drawing, is not Colored.

Malby & Sons, lith.

LONDON Printed by GEORGE EDWARD EYRE and WILLIAM SPOTTISWOODE

A.D. 1852 . . . . N° 14,066.

Fire-arms and Projectiles.

DELVIGNE'S SPECIFICATION.

TO ALL TO WHOM THESE PRESENTS SHALL COME, I, Henri Gustave Delvigne, of Brixton, in the County of Surrey, Gentleman, send greeting.

WHEREAS Her present most Excellent Majesty Queen Victoria, by Her Royal Letters Patent under the Great Seal of the United Kingdom of Great Britain and Ireland, bearing date at Westminster, the Seventeenth day of April, One thousand eight hundred and fifty-two, in the fifteenth year of Her reign, did, for Herself, Her heirs and successors, give and grant unto me, the said Henri Gustave Delvigne, my exors, admors, and assigns, Her especial licence, full power, sole privilege and authority, that I, the said Henri Gustave Delvigne, my exors, admors, and assigns, or such others I, the said Henri Gustave Delvigne, my exors, admors, or assigns, should at any time agree with, and no others, from time to time and at all times during the term of years therein expressed, should and lawfully might make, use, exercise, and vend, within England, Wales, and the Town of Berwick-upon-Tweed, my Invention of "**Certain Improvements in Fire-arms, and in the Methods of Discharging the same; also Improvements in Projectiles;**" in which said Letters Patent is contained a proviso, that I, the said

Henri Gustave Delvigne, shall cause a particular description of the nature of my said Invention, and in what manner the same is to be performed, by an instrument in writing under my hand and seal, to be enrolled in Her Majesty's High Court of Chancery within six calendar months next and immediately after the date of the said in part recited Letters Patent, as in and by the same, reference being thereunto had, will more fully and at large appear.

NOW KNOW YE, that in compliance with the said proviso, I, the said Henri Gustave Delvigne, do hereby declare that the nature of my said Invention, and the manner in which the same is to be performed, are fully described and ascertained in and by the following statement thereof, reference being had to the Drawings hereunto annexed, and to the figures and letters marked thereon, that is to say:

Figures 1, and 2, represent side elevations of one of my improvements; the sides of the body of the pistol being removed in Figure 1, in order that the parts in the interior may be seen. *a*, is the barrel which turns upon a joint *b*, connecting it with the body of the pistol *c*. *d*, is the nipple on which the percussion cap is put for the discharge of the pistol. The breech end a^1 of the barrel *a*, is open, and the loading of the said barrel is effected at the said open breech end. *e*, is a screw which passes through the part c^1 of the body of the pistol *c*; an arm or lever *f*, is fixed on the end of the screw *e*, for the purpose of giving facility in turning the said screw; the end of the screw *e* is made slightly conical, and fits accurately the end a^1 of the barrel *a*, into which it enters, when by the arm *f* the said screw is made to advance in the part c^1 of the body of the pistol. By the end of the said screw *e*, entering the open end of the barrel a^1, the said open end of the barrel is closed, and the said barrel fixed in its position, as represented in Figure 2. *g*, is the hammer, by which the percussion cap placed on the nipple *d* is struck, and the pistol discharged. The said hammer is concealed in the interior of the lock, but the arm *h* on the head of the said hammer projects through a slot in the lower part of the body of the pistol, and by pressing the thumb plate h^1 on the end of the said arm *h* in the direction indicated by the arrow in Figure 1, the said hammer is brought first into half and afterwards into full cock. *i*, is the main spring; the end i^1 of the said spring enters the recess k in the hammer, and presses the said hammer upwards. The other end of the said spring bears against the trigger l, and presses the end l^1 of the said trigger against the end g^1 of the hammer. Two teeth m, n, are formed on the end g^1 of the hammer, and the end of the trigger l^1, by engaging in the tooth m, holds the hammer g at half cock, and by entering the tooth n, holds the said hammer at full cock. When the pistol is to be discharged it is grasped by the right hand, in the manner indicated in Figure 2; and by the pressure of the second finger on the trigger l, the hammer g is disengaged from the said trigger, and falling upon the percussion cap on the nipple d, discharges the pistol. The barrel a may either be rifled or plain. By turning the arm f, the screw e may be withdrawn from the end a^1 of the barrel a, the said barrel raised into the position represented in Figure 1, the barrel loaded, and a fresh percussion cap put on the nipple d.

Figure 3, represents in side elevation another of my improvements. o, is the barrel, which is loaded in the ordinary manner. that is to say, from the end o^1. p, is the nipple on which the percussion cap is placed, by the detonation of which the pistol is discharged.

The said nipple being situated in a line with the axis of the barrel o, the ignition of the powder is effected in the centre of the same instead of at the side, as in ordinary fire-arms. q, r, t, is the hammer, by which the percussion cap on the nipple p is struck and detonated. The said hammer consists of a rod r, sliding in the tube s, in a direction parallel to the axis of the barrel o; one end of the said rod r is furnished with the cap or head q, which on the discharge of the pistol strikes the percussion cap on the nipple p; and the other end of the said rod is furnished with a thumb plate t, for the purpose of cocking the pistol. u, v, are two teeth, formed on the under side of the rod r. w, is a helical or coiled spring, which surrounds the rod r. The said spring is confined between the shoulder r^1 on the said rod and the plate s^1 screwed on the end of the barrel s; the said spring being in a state of compression, presses upon the shoulder r^1, and tends to force the cap q against the nipple p. x, is a spring which is attached at one end by the screw x^1 to the under side of the barrel s; an arm y, affixed to the said spring passes through a slot on the body of the pistol, and the external part of the said arm constitutes the trigger y. z is a tooth attached also to the spring x; the said tooth is pressed against the rod r by the said spring, and when by pressing with the thumb or finger upon the thumb plate t, the rod r is made to slide in the direction indicated by the arrow in Figure 3, the said tooth z engages in one or other of the said teeth on the rod r, and holds the hammer at half or full cock. By grasping the body of the pistol in the hand, in the manner herein-before described and represented in Figure 2, and pressing the second finger upon the trigger y^1, the tooth z becomes disengaged from the teeth on the rod r, and the spring w expanding, causes the cap or head of the hammer q to strike upon the percussion cap on the nipple p, and discharge the pistol.

The general arrangement and construction of the parts of pistols herein described, and illustrated in Figures 1, 2, and 3, of the accompanying Drawings, may be applied to guns and other fire-arms; but the said application will be so evident to a workman acquainted with the construction of fire-arms that I do not deem it necessary to enter into further details.

Figure 4, represents in elevation my improved projectile to be used with the pistol, represented in Figures 1 and 2, or any other rifled pistol or fire-arm, loaded at the breech end. It consists of a bullet of a conical form, in the base of which a wire or stem 1 is inserted, as shewn in the Figure. The use of the said wire or stem is to prevent the crushing and compressing of the powder with which the barrel is charged. The length of the said stem 1 is such that it projects above the charge of powder with which the pistol is charged; and when the end of the said barrel a^1, Figure 1, is closed by the plug on the end of the screw e, the said plug pressing against the end of the wire or stem 1, forces the projectile forward in the barrel, and thus enlarging the cavity in which the powder is contained prevents the crushing and wedging of the said powder. The said projectile may also be used with fire-arms loaded in the ordinary way.

Having now described the nature of my said Invention, and the manner in which the same is to be performed, I wish it to be understood that I do not limit myself to the precise methods herein described of carrying my Invention into effect, as the same may be varied without departing from the nature of my said Invention; but I claim,

First, the general arrangements of the parts of pistols herein-before described, and represented in Figures 1, 2, and 3, of the accompanying Drawing, and the application of the same arrangement to guns and other fire-arms to which they may be applicable.

Second, the construction of the projectile herein-before described, and represented in Figure 4, of the accompanying Drawing.

In witness whereof, I, the said Henri Gustave Delvigne, have hereunto set my hand and seal, this Eleventh day of October, in the year of our Lord One thousand eight hundred and fifty-two.

HENRI GUSTAVE (L.S.) DELVIGNE.

Taken and acknowledged by Henri Gustave Delvigne (party hereto), at Paris, in the Republic of France, this 11th day of October 1852, before me,

THOMAS PICKFORD,

(Consul's Seal.) Her Britannic Majesty's Consul at Paris.

A.

This is the Specification referred to in the affidavit of James Murdoch, sworn before me, this 14th day of October 1852,

HENRY MOWBRAY.

James Murdoch, of Staple Inn, Holborn, in the County of Middlesex, Patent Agent, maketh oath and saith, that he is acquainted with the manner and character of the handwriting of Thomas Pickford, Her Britannic Majesty's Consul at Paris, in the Republic of France; and this deponent further saith, that the signature "Thomas Pickford" set, subscribed, or written in the margin of the Specification hereunto annexed marked A, and dated the Eleventh day of October now instant, and purporting to be executed by Henri Gustave Delvinge, of Brixton, in the County of Surry, Gentleman, whereby it is stated that the Specification was acknowledged by him, the said Henri Gustave Delvinge, before the said Thomas Pickford, on the said Eleventh day of October, at Paris aforesaid, is of the proper handwriting of the said Thomas Pickford.

JAMES MURDOCH.

Sworn at the Chancery Affidavit Office, Southampton Buildings, Chancery Lane, in the County of Middlesex, this 14th day of October 1852, before me,

HENRY MOWBRAY.

By Affid[t].

AND BE IT REMEMBERED, that on the oath of James Murdoch the Specification aforesaid was enrolled word for word as above written. And also the Specification aforesaid was stamped according to the tenor of the Statute made for that purpose.

Enrolled the Fourteenth day of October, in the year of our Lord One thousand eight hundred and fifty-two.

LONDON:
Printed by GEORGE EDWARD EYRE and WILLIAM SPOTTISWOODE,
Printers to the Queen's most Excellent Majesty. 1854.

A.D. 1852. APRIL 17. Nº 14,066.
DELVIGNE'S SPECIFICATION

(1. SHEET.)

FIG. 1.

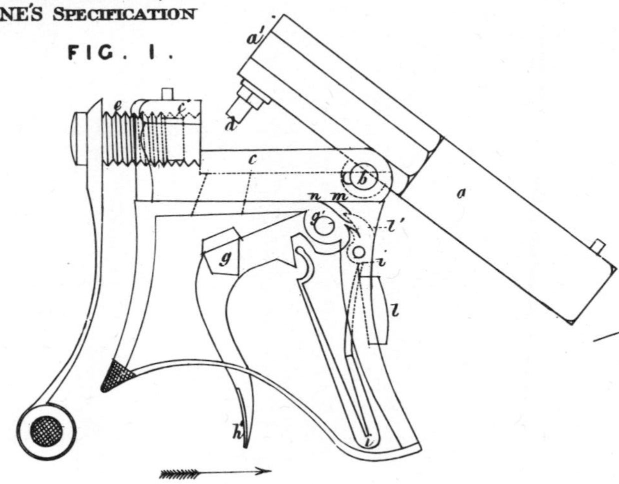

FIG. 2.

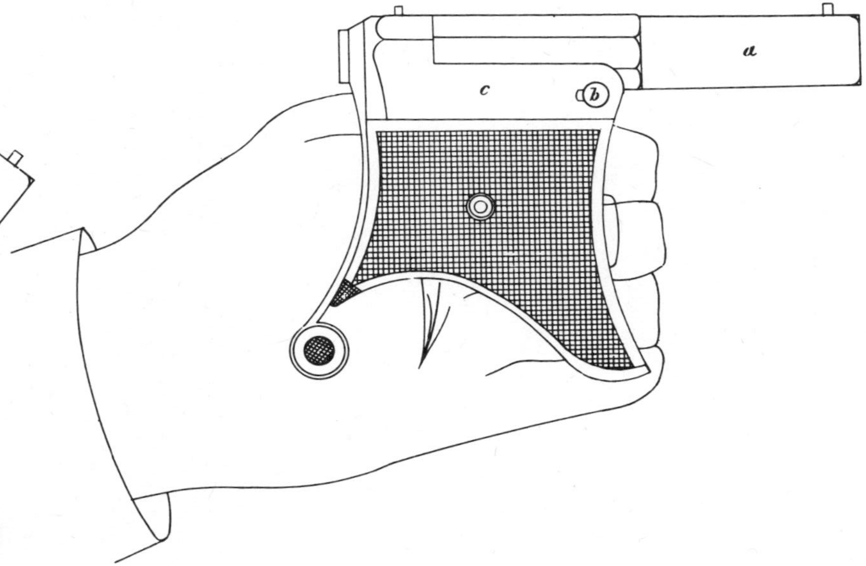

FIG. 3.

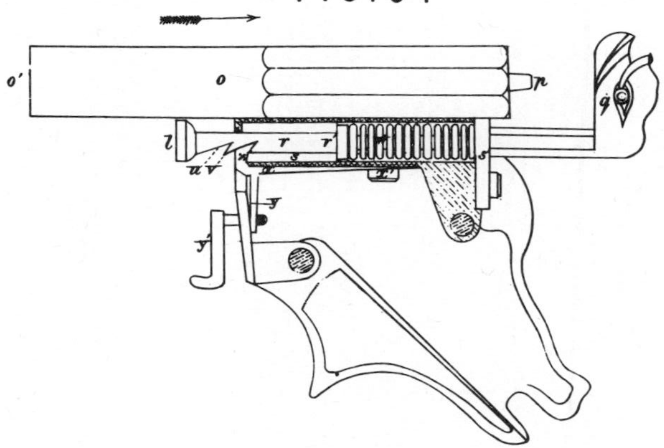

FIG. 4

The Enrolled Drawing, is partly Colored.

Malby & Sons, lith.

LONDON: Printed by GEORGE EDWARD EYRE and WILLIAM SPOTTISWOODE
Printers to the Queen's most Excellent Majesty. 1852.

A.D. 1852 . . . . N° 184.

Breech-loading Fire-arms.

LETTERS PATENT to Joseph Needham, of 26, Piccadilly, in the County of Middlesex, Gun Manufacturer, for the Invention of "IMPROVEMENTS IN BREECH-LOADING FIRE-ARMS, AND IN APPARATUS CONNECTED THEREWITH."

Sealed the 5th January 1853, and dated the 2nd October 1852.

PROVISIONAL SPECIFICATION left by the said Joseph Needham at the Office of the Commissioners of Patents, with his Petition, on the 2nd October 1852.

I, JOSEPH NEEDHAM, of 26, Piccadilly, in the County of Middlesex, Gun Manufacturer, do hereby declare the nature of the said Invention for "IMPROVEMENTS IN BREECH-LOADING FIRE-ARMS, AND IN APPARATUS CONNECTED THEREWITH," to be as follows, that is to say :—

This Invention relates to certain modifications and improvements on breech-loading fire-arms, whereby the action of loading and firing is rendered much more simple and rapid than by the arms upon that principle now in use. The improvements consist in enclosing a short piercer in a metal tube, which is hinged to the breech of the gun barrel. This piercer is actuated by a helical spring enclosed within the tube, which is fitted with a cap screwing on to its free end; the closed end of this cap has a small perforation in its centre to allow the point of the piercer to pass through on its way to the cartridge or capsule. The operation of loading the gun is effected by a partial revolution of the cap of the tube, which movement allows of the tube being turned back on its joint, thereby leaving an opening in the breech sufficiently large to admit the insertion of the cartridge or capsule. The lower portion of the cartridge, or a suitable capsule to contain the detonating powder, is so formed as to be a sufficiently tight packing of itself to prevent the escape of gasses at the junction of the free end of the tube and the breech of the barrel. Having inserted the cartridge or capsule, the tube is shut down again, and the cap again partially revolved, so as to take its original position, thereby forming an effective joint with the breech end of the barrel. By drawing the trigger the sear is relieved from a notch in the shank of the piercer, and the piercer is instantly acted upon by the helical spring, thus effecting the firing of the arm. To prevent accidents a safety bolt is attached to the tube, and is so arranged that by partially turning it the piercer may be locked or released, as the case may be.

In connexion with these arms a reservoir of cartridges may be formed under the barrel and in communication with the breech, and any suitable arrangement may be employed for supplying the cartridges in regular succession as required.

SPECIFICATION in pursuance of the conditions of the Letters Patent, filed by the said Joseph Needham in the Great Seal Patent Office, on the 2nd April 1853.

TO ALL TO WHOM THESE PRESENTS SHALL COME, I, Joseph Needham, of 26, Piccadilly, in the County of Middlesex, Gun Manufacturer, send greeting.

WHEREAS Her most Excellent Majesty Queen Victoria, by Her Letters Patent, bearing date the Second day of October, in the year of our Lord One thousand eight hundred and fifty-two, in the sixteenth year of Her reign, did, for Herself, Her heirs and successors, give and grant unto me, the said Joseph Needham, Her special license that I, the said Joseph Needham, my executors, administrators, and assigns, or such others as I, the said Joseph Needham, my executors, administrators, and assigns, should at any time agree with, and no others, from time to time and at all times thereafter during the term therein expressed, should and lawfully might make, use, exercise, and vend, within the United Kingdom of Great Britain and Ireland, the Channel Islands, and Isle of Man, an Invention for "IMPROVEMENTS IN BREECH-LOADING FIRE-ARMS, AND IN APPARATUS CONNECTED THEREWITH," upon the condition (amongst others) that I, the said Joseph Needham, by an instrument in writing under my hand and seal, should particularly describe and ascertain the nature of the said Invention, and in what manner the same was to be performed, and cause the same to be filed in the Great Seal Patent Office within six calendar months next and immediately after the date of the said Letters Patent.

NOW KNOW YE, that I, the said Joseph Needham, do hereby declare the nature of my said Invention, and in what manner the same is to be performed, to be particularly described and ascertained in and by the following statement, reference being had to the accompanying Drawings, and to the letters and figures marked thereon, that is to say:

My said Invention relates to various arrangements of mechanism, apparatus, or means applicable as well to the construction and action of fire-arms of the breech-loading class as to the cartridges or explosive charges used therein, for the purpose of simplifying and rendering more rapid the loading and firing actions, and increasing the general efficiency of the arms. And in order that my Invention may be properly understood I shall now proceed to describe the several Figures on the Sheet of Drawings hereunto attached. Figure 1, on my Sheet of Drawings is a side or longitudinal elevation of all the important portions of a rifle made according to my improvements. Figure 2, is a plan, and Figure 3, is a vertical longitudinal section of the same, corresponding. Figure 4, is a longitudinal section of the lock mechanism detached, as drawn on a larger scale. Figure 5, is an enlarged elevation of the needle actuating cam; and Figure 6, is a plan corresponding. Figure 7, is an enlarged side elevation of the cocking lever and shifting cylindrical cap; and Figure 8, is a corresponding elevation at right angles thereto. The barrel A, of the gun is fitted to the stock B, in the usual manner by means of the terminal projection C, screwed to the stock and the slotted eyes D. The breech end of the barrel is half cut away longitudinally for a short distance on its upper side, so as to leave a species of trough for the reception laterally of the short tubular piece E, hinged on the fixed joint pin F, passed transversely through the barrel end. The hind end of this tubular piece is solid and rounded off, and bears against the breech end of the trough, and its front end has an external square threaded screw cut upon it for screwing into a similar internal screw in the front tubular cap piece G, which cap has a short lever H, formed upon it to work it by, and a small double crown cam I, is screwed inside its front end. This duplex cam, when made to revolve, serves to draw back the ignition needle J, which is a short piece of steel wire fitted concentrically into the front end or socket K, of the catch spindle L, and screwed therein by a side pinching screw, whilst it projects through the cam I, and the front solid end of the cap G, as a guide to keep it in a line with the axis of the rifle or gun barrel. The spindle L, is entirely concealed in the tubular piece E, and G, of the lock mechanism, and its tapered detent end extends back for connection with the trigger action. It is surrounded by a helix M, which abuts in front against the socket collar K, and behind against an internal shoulder of the tube E, so that it constantly tends to press forward the needle to the position represented in Figure 4, which is that occupied at the moment of exploding the cartridge N. And to prevent the spindle or needle holder from turning round, it is formed with the diametrically opposite feathers O, fitting into corresponding grooves in the interior of the tube E. The "sear" P, and sear spring Q, are both attached to the tubular piece E, the sear being set on a transverse stud centre, whilst the spring, which is a flat blade, is screwed outside underneath, its loose free end having a constant tendency to turn the sear upwards on its centre, and bring its catch notch into gear with the terminal detent on the spindle L, when the latter is drawn back, retaining it in this position until the trigger is drawn. The trigger R, set on a stud centre in the usual way has an elevated notch to fit to the tail of the sear, so that when the trigger is drawn back on firing, it depresses the tail of the sear and releases the needle holder detent, when the spring M, darts the needle forward to give the igniting puncture. The cartridge, one arrangement of which is shewn in elevation at Figure 9, and vertical section at Figure 10, is fitted with a capsule T, of gutta-percha or other suitable material at its base, a short tail-piece U, being formed upon this capsule, the interior of which is fitted with a thin piece of metal slightly dished to contain the fulminate for the explosion. This fulminate is reached by the needle, when necessary, through a central aperture made in the base of the capsule, the object of which capsule is the prevention of the wasteful escape of the explosive gases on firing. It remains in the barrel after each explosion, and may be drawn out by the tail U, which projects outside. Figures 11, and 12, represent an elevation and section of another construction of cartridge. The capsule T, being in this arrangement dispensed with. As the enlarged portion A, of the cartridge is formed by simply doubling over the edge of the paper before rolling the cartridge, the tail B, may consist of a loop of tape attached to the cartridge. The piece of metal C, containing the explosive powder is attached by simply fitting it to the lipped end of the cartridge, and pasting a thin piece of paper over it, thereby effectually securing it to the cartridge. To guard against accidental discharges, a small safety bolt V, Figures 1, 2, 3, and 4, is screwed into the upper side of the tube E, its projecting inner end, one side of which is cut away, being made to gear with a notch on the upper side of the spindle L, when the safety apparatus is in action. When so placed, the gun cannot be discharged; but on drawing over the small arm V^1, fast on the upper end of the bolt from one notch to the other on the tube E, the spindle L, becomes unlocked from the bolt V, in readiness for firing. When the breach is closed for firing, the lever H, of the cap G, is turned downwards, and fits into a recess cut in the side of the gun-stock, being held in this position by a small spring detent W, fast in the stock, and projecting through a slot in the lever. When the gun is to be loaded, this lever H, is turned over upon the axis of the cap G, which turns with it, carrying round also the duplex cam I. In this way the two inclines of the cam press against the feathers O, on the socket K, and so force back the striker or needle, until the after end of the striker or needle spindle is caught by its detent, and held by the sear notch, as delineated in Figure 3. The same movement also withdraws the front end of the cap G, from its partial entrance into the gun barrel, by reason of the action of the screw connecting the cap with the tube E, and the latter, together with its cap G, may then be turned back upon its centre F, until it stands up at right angles from the barrel, as shewn in red lines in Figure 1. This leaves the trough or cut away portion of the breech quite open to receive the cartridge N, and the tube E, is then shut down, and the lever H, returned to the position shewn in Figure 1, reversing all the actions already explained, and forcing the end of the cap G, into the breech end of the barrel. The piece is now ready for firing, and when the trigger is drawn, the needle explodes the fulminate in the breech of the cartridge by its rapid puncturing and striking action, and the charge is exploded. The breech is then opened again, and the empty capsule is withdrawn by its projecting tail U. As the combined length of the tubular pieces G, and E, varies with the different stages of the action of the piece, the junction edges of the two pieces are cut obliquely to correspond with the angle of the screw threads, by which they are connected, in order that no space may be left between these edges at any time for the entrance of dirt. Figure 13, is a partial section of the breech and screw cap G, shewing the application of an ordinary percussion cap. The side of the screw cap G, is cut away at *a*, and an ordinary nipple *b*, is screwed into the cap, the perforation through the nipple corresponding with the needle hole in the end of the screw cap. A gun of this description may be used with or without the percussion cap, as the case may be. The needle J, shewn in dotted lines, is contained in the thick short spindle *c*, which is secured to the socket collar herein-before mentioned, and is actuated in a precisely similar manner to the needle in Figure 4. When a percussion cap is used, the needle J, is removed from the spindle *c*, and the striking of the end of the spindle *c*, against the cap on the nipple explodes it, and consequently fires the charge. But when the gun is wanted as a needle gun, then the needle J, is fitted into the spindle *c*, and on firing the gun it passes through the nipple and enters the cartridge, as herein-before described. A swelling *d*, and annular groove is formed upon the face of the screw cap, in order that the end of the cartridge may be broken or burst in when the cap is screwed up, thereby leaving the powder exposed to the igniting action of the percussion cap. By this arrangement the junction between the end of the screw cap and the breech will be on the *side* of the cartridge, in place of the end of it, as will be clearly seen on reference to Figure 13. In this case, of course, the ordinary cartridge may be used. Figure 14, is an arrangement wherein the cartridge or charge is placed within a suitable chamber in the front end of the screw cap G, in place of inserting it in the breech of the gun barrel. This chamber is of a larger bore than the barrel, as shewn by the Drawing. In this arrangement the escape of the gases at the junction is prevented by the enlarged rim *a*, on the front of the cartridge striking against the projecting edge *b*, of the end of the barrel when the charge is fired, and being also expanded laterally by the passage through it of the enlarged zone C, on the ball, shewn in Figures 15, and 16. To facilitate this mode of packing, the bore of the cartridge chamber is made larger than that of the barrel. A similar arrangement to the one last above described is also applicable to all breech-loading revolvers. Figures 15, and 16, are an elevation and section of the cartridge employed in this latter arrangement, the rim *a*, being formed by doubling the paper before rolling. This rim fits in front of the projecting zone *c*, of the bullet. In place of a zone or rim being formed upon the body of the cartridge, a separate ring, Figure 17, made of card board, felt, soft metal, or any other suitable material, may be employed. Figure 17^{a}, represents, in vertical section, the shell of the cartridge A, formed in one piece with the ball B. Cartridges of this description may be made by a suitable die, into which the cold metal for forming them is pressed by a plunger actuated by a fly press or other mechanical contrivance. The great pressure thus exerted on the soft metal squeezes up the sides of the cartridge from the solid piece in the die below. The shoulder C, answers the same purpose as the projecting ring in Figures 15, and 16, herein-before described. It will be obvious to the practical man that the various constructions of cartridges herein-before described, for preventing the escape of the explosive gases in breech-loading fire-arms, are equally applicable to cannon of the breech-loading class. The cartridge may be made to expand bodily by making it of a spiral coil of card board, as shewn at Figure 18, the coil being merely held together by a very slight lapping of thread. A cartridge of this construction is easily made, and requires no packing capsule, as the explosion of the powder causes the cartridge to expand along its entire length, and consequently prevent any escape of gases. Ordinary cartridges may be used with my improved gun by the employment of the expanding capsule, Figure 19, which is stamped out of a piece of millboard, or any other suitable material, and dished; this capsule is perforated at the centre, and has a small piece of metal containing fulminating powder, similar to the one herein-before described. The cartridge having been inserted into the breech, this expanding capsule is put behind it. The explosion of the charge expands the capsule, and forms the packing required. Figure 20, is a side elevation of a double tube for double-barreled guns. Figure 21, is a front view of the same; and Figure 22, is a plan of the same. The needles are actuated in a manner similar to the arrangement shewn at Figure 4; but one handle cocks both simultaneously. A, is the body of the moveable double tube which turns on a stud pin at B; this pin also secures it to the breech end of the barrel in a similar manner to the pin F, in Figure 4. The handle C, is formed in one piece with the hollow screw D; this screw works into an internal screw in the piece A. E, is a face piece, which is attached to the boss of the handle C, by a screw pin F, which passes through the centre of the hollow screw D. The screw F, is introduced from the back of the piece A, which is bored entirely through for that purpose. The face piece E, has two projecting bosses G, G, formed upon it to fit into the barrel. H, I, are upper and lower projecting lugs on the face piece E, which fit accurately over the end of the tube piece A, and preserve the face piece from derangement of position. The junction edges of the piece A, and boss of the handle are made slightly inclined, as in the cap G, Figures 7, and 8, to correspond with the pitch of the screw D, so that when the handle is turned there is no space left for dirt to enter between the junction edges. The boss of the handle C, has two inclines J, cut upon it, each incline acting on a projecting lug K, on the side of the spindles L, which contain the needles. In Figure 21, half of the face piece is cut away, to shew more clearly the boss of the handle C. As the face piece is attached to the boss of the handle C, it follows that when the handle is drawn up, as in Figure 20, to cock the gun, the projecting bosses on the face piece will be withdrawn from the breeches of the barrels, and the piece A, will thus be left free to be turned back for the insertion of fresh cartridges into the breeches of the gun. When the cartridges are in their places, the piece A, is shut down, and the handle C, depressed, thereby forcing the bosses G, G, of the face piece E, into the ends of the barrels. Safety bolts are also attached to this arrangement, similar to the one lettered V, in Figures 1, 2, 3, and 4.

Having now described and particularly ascertained the nature of my said Invention, and the manner in which the same is or may be used or carried into effect, I would observe, in conclusion, that I do not confine or restrict myself to the precise details or arrangements which I have had occasion to describe or refer to, as many variations may be made therefrom without deviating from the principles or main features of my Invention; but what I consider to be novel and original, and therefore claim as the Invention secured to me by the herein-before in part recited Letters Patent, is,—

First, the system or mode of preventing the escape of the gases of the explosion by the use of a capsule or ring, to be expanded by the gases themselves, or by the use of an expanding cartridge, as herein-before described.

Second, the system or mode of expanding the cartridge case, or a ring thereon, by means of a ring or zone or an enlarged portion upon the ball, as herein-before described.

Third, the system or mode of forming cartridges with an end ring, or thickened portion of the case, for preventing the gaseous escape, as herein-before described and shewn at Figures 11, 12, 15, and 16, of the Drawings.

Fourth, the system or mode of constructing cartridges, whereby they may expand bodily throughout their entire length, and so form a packing, as described and shewn at Figure 18. Also the constructing of the packing rings or zones on cartridges in the same manner.

Fifth, the system or mode of constructing cartridges, by forming the case in one piece with the ball, as herein-before described and referred to in Figure 17[a], of my Sheet of Drawings.

Sixth, the application and use of a projecting tag or tail for removing the exploded cartridge case, as herein-before described.

Seventh, the application and use of an expanding capsule in the manner and for the purpose herein-before described and illustrated in Figure 19, of the Sheet of Drawings.

Eighth, the application and use of a detached packing ring in the manner and for the purpose herein-before described and referred to in Figure 17, of the Drawing.

Ninth, the forming of that portion of the gun which holds the tubular pieces containing the action or mechanism in one piece, with the barrel itself, as herein-before described and shewn in Figures 1, 2, and 3, of my Drawings.

Tenth, the application and use of a revolving spiral cam for putting the needle or striker into firing position.

Eleventh, the system or mode of forming the junction with the breech by a screw connection with the after tubular piece.

Twelfth, the system or mode of keeping the exposed edges of the joint between the two sections of the tubular case of the lock mechanism in close contact by an external spiral or incline.

Thirteenth, the system or mode of simultaneously cocking and withdrawing the needle or needles or strikers by turning the cap piece G, as described.

Fourteenth, the system or mode of preventing accidental discharges by means of a rotatory detent catch to engage with a catch or stop on the needle or striker holder.

Fifteenth, the system or mode of constructing fire-arms, wherein the case of the lock mechanism is withdrawn at an angle with the axial line of the barrel for charging.

Sixteenth, the system or mode of bursting in the end of the cartridge, in manner and for the purpose herein-before described with reference to Figure 13, of the Drawings.

Seventeenth, the application and use of an ordinary nipple for percussion caps in guns of the needle class, as herein-before described and shewn at Figure 13.

Eighteenth, the application and use of a charge chamber distinct from the barrel of the fire-arm, such chamber being of a larger bore than the barrel itself, for the purpose herein-before described.

Nineteenth, the application and use of the peculiar junction edges, shewn at Figures 13, and 14, wherein the escape of the gases, if any, would be at a right angle with the axial line of the barrel.

Twentieth, the arrangement and construction of double barrelled guns of this class, as herein-before described and illustrated in Figures 20, 21, and 22, of the Drawings.

In witness whereof, I, the said Joseph Needham, have hereunto set my hand and seal, the Second day of April, One thousand eight hundred and fifty-three.

JOSEPH NEEDHAM. (L.S.)

LONDON:
Printed by GEORGE EDWARD EYRE and WILLIAM SPOTTISWOODE,
Printers to the Queen's most Excellent Majesty. 1854.

A. D. 1852. October 2. Nº 184.

NEEDHAM'S Specification.

(1 Sheet.)

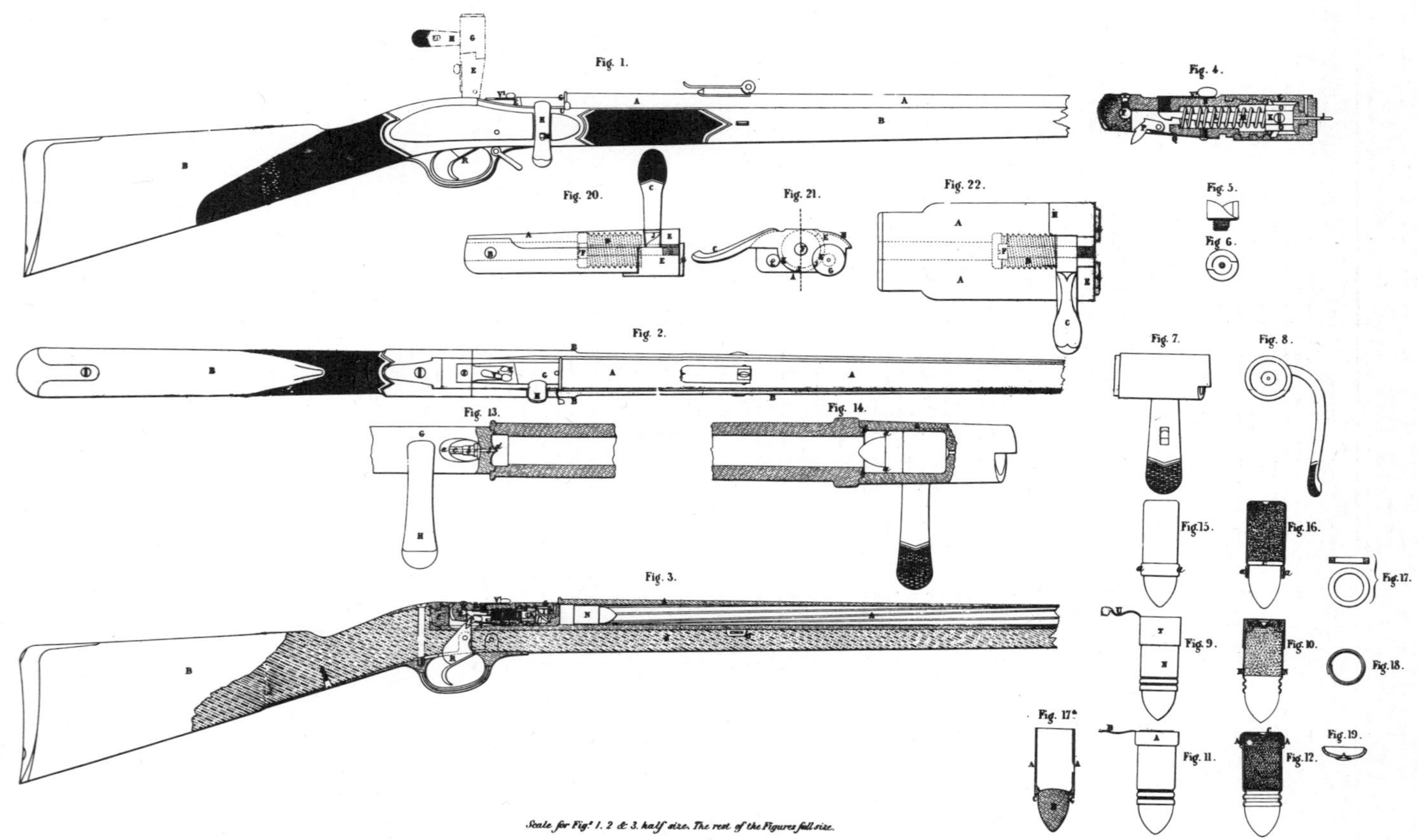

Scale for Figs. 1. 2 & 3. half size. The rest of the Figures full size.

The filed drawing is partly colored.

G. Fleiner Lith. Manchester.

A.D. 1852 . . . . N° 712.

Breech-loading Fire-arms.

LETTERS PATENT to Christian Sharps, of Hartford, in the State of Connecticut and United States of America, for the Invention of "IMPROVEMENTS IN BREECH-LOADING FIRE-ARMS."

Sealed the 21st January 1853, and dated the 11th November 1852.

PROVISIONAL SPECIFICATION left by the said Christian Sharps at the Office of the Commissioners of Patents, with his Petition, on the 11th November 1852.

I, CHRISTIAN SHARPS, of Hartford, in the State of Connecticut and United States of America, do hereby declare the nature of the said Invention for "IMPROVEMENTS IN BREECH-LOADING FIRE-ARMS" to be as follows:—

It consists, first, in providing the rear end of the barrel of a breech-loading gun with an adjustable bushing or short tube, that shall closely and tightly fit a recess in the rear end, contain or hold the charge either in part or in whole, and be capable of being pressed backwards against or toward the moveable breech, in order at any time to ensure a close or properly close joint between them.

Secondly, in providing the sliding or moveable breach, or the rear end of the barrel that abuts against said breech, with a facing or ring of platina, or material not easily oxidisable by any one or more of the volatile products of combustion of the charge or powder, the same being to prevent, in a great degree, the oxydation of those surfaces of the barrel and breech that are in contact at the time of an explosion of the charge or a discharge of the gun.

SPECIFICATION in pursuance of the conditions of the Letters Patent, filed by the said Christian Sharps in the Great Seal Patent Office on the 5th May 1853.

TO ALL TO WHOM THESE PRESENTS SHALL COME, I, CHRISTIAN SHARPS, of Hartford, in the State of Connecticut and United States of America, send greeting.

WHEREAS Her most Excellent Majesty Queen Victoria, by Her Letters Patent, bearing date the Eleventh day of November, in the year of our Lord One thousand eight hundred and fifty-two, in the sixteenth year of Her reign, did, for Herself, Her heirs and successors, give and grant unto me, the said Christian Sharps, Her special license that I, the said Christian Sharps, my executors, administrators, and assigns, or such others as I, the said Christian Sharps, my executors, administrators, and assigns, should at any time agree with, and no others, from time to time and at all times thereafter during the term therein expressed, should and lawfully might make, use, exercise, and vend, within the United Kingdom of Great Britain and Ireland, the Channel Islands, and Isle of Man, an Invention for "IMPROVEMENTS IN BREECH-LOADING FIRE-ARMS," upon the condition (amongst others) that I, the said Christian Sharps, by an instrument in writing under my hand and seal, should particularly describe and ascertain the nature of the said Invention, and in what manner the same was to be performed, and cause the same to be filed in the Great Seal Patent Office within six calender months next and immediately after the date of the said Letters Patent.

NOW KNOW YE, that I, the said Christian Sharps, do hereby declare that the following, taken in connection with the accompanying Drawings, is a Complete Specification of my said Invention, setting forth the nature thereof, and the manner in which the same is to be performed:

The nature of my Invention consists, first, in providing the rear end of the barrel of a breech-loading gun with an adjustable bushing or short tube that shall closely and tightly fit a recess in the rear end contain or hold the charge either in part or in whole, and be capable of being pressed backwards against or towards the moveable breech, in order at any time to ensure a close or properly close joint between them.

Secondly, in providing the sliding or moveable breech, or the rear end of the barrel that abuts against said breech, with a facing or ring of platina or metal not easily oxidisable by any one or more of the volatile products of combustion of the charge or powder, the same being to prevent, in a great degree, the oxydation of those surfaces of the barrel and breech that are in contact at the time of an explosion of the charge or a discharge of the gun.

Of the said Drawings, Figure 1, represents a side elevation of a cavalry carbine, constructed according to my Invention. Figure 2, is a central vertical and longitudinal section of the same. Figure 3, is a top view of it. Figure 4, is a front view of the slide block or sliding breech. Figure 5, is a rear end view of the barrel.

The carbine represented in such Figures is a breech-loading one, and is of the kind recently patented in England on the Thirty-first day of March, of the fifteenth year of Her present Majesty's reign, the Patent being taken out in the name of Moses Poole, and for an Invention which was a communication from a certain foreigner residing abroad.

In the Drawings above mentioned, A represents the barrel of the carbine or gun; B, the stock; D, the sliding breech; and E, the lever thereof; F, is the link by which the sliding breech is connected with the lever E, and made to slide in the passage C, made downwards against the rear end of the barrel. The barrel is counterbored at its rear end, or made larger in diameter there, and for a short distance from the extreme end, in order to receive an adjustable bushing or tube G, which is placed within in, and made to fit closely in it, and to slide rearwards towards and away from the sliding breech D. The tube G, is intended to receive the cartridge or charge, and to be forced back against the sliding breech, as occasion may require, in order to make a close joint between them. Were not a contrivance like the adjustable bushing used, or, in other words, were the rear end of the barrel made to rest in direct contact with the sliding breech, it would soon, by the friction of the breach against it, and by repeated explosions of the charge, become so worn as to make the joint more or less open, or too open.

By my improvement, that is to say, by the adjustable bushing, the joint may be kept sufficiently close, so far as wear on such bushing is concerned.

In regard to the second part of my Invention, I provide the moveable or sliding breech D, where it rests against the adjustable bushing, or the rear end of the barrel when no adjustable bushing is used, with a facing or ring of platina, which I insert in the surface and into the block, as seen at I, in Figures 2, and 4.

Under ordinary circumstances it is found that the volatile products of combustion of the charge are very apt to excoriate or otherwise injure the front surface of the sliding breech block; but by putting in said block, as described, a ring or facing of platina or other metal (capable of resisting the destructive effects arising from the explosion of the charge), the difficulty above mentioned is obviated.

What I claim as my Invention is as follows:—

Firstly, the adjustable bushing applied to the barrel, and a sliding or moveable breech, substantially in manner and for the purpose as set forth.

And secondly, the combining with the sliding breech and the barrel the ring or facing of platina, or other suitable material capable of resisting the destructive effects of the explosion of the charge as specified.

In witness whereof, I, the said Christian Sharps, have hereunto set my hand and seal, this Eighth day of March, in the year of our Lord One thousand eight hundred and fifty-three.

CHRISTIAN SHARPS. (L.S.)

Witness,

JOHN CLEVELAND PALMER.

GUY ROWLAND PHELPS.

LONDON:
Printed by GEORGE EDWARD EYRE and WILLIAM SPOTTISWOODE,
Printers to the Queen's most Excellent Majesty. 1854.

A.D. 1852. November 11. N° 712.
SHARPS' Specification.

(1 SHEET.)

FIG. 3.

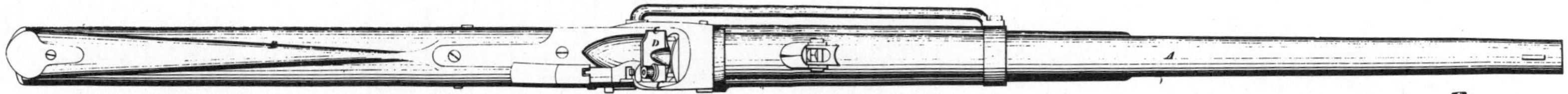

FIG. 4.

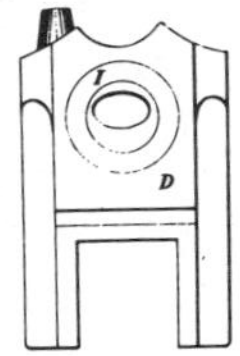

FIG. 2.

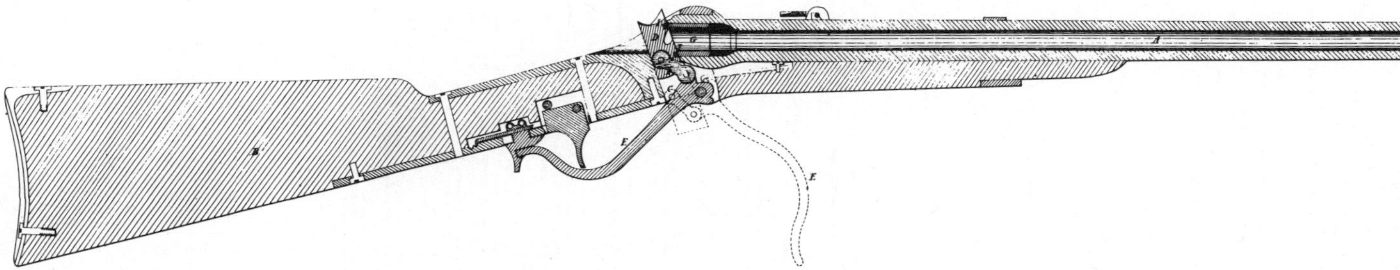

FIG. 5.

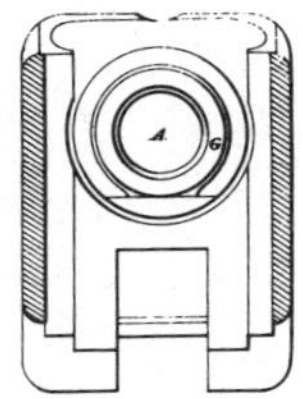

FIG. 1.

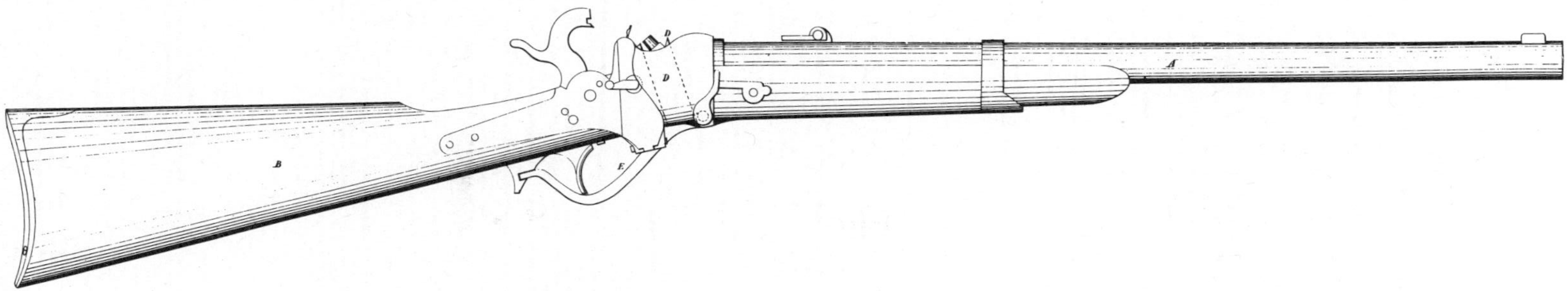

The filed drawing is not coloured.

Malby & Sons, Lithe Old Jewry

London: Printed by George Edward Eyre and William Spottiswoode.
Printers to the Queen's most Excellent Majesty. 1853.

A.D. 1853 . . . . N° 212.

Fire-arms.

LETTERS PATENT to William Tranter, of Birmingham, in the County of Warwick, Gun Maker, for the Invention of "**Certain Improvements in Fire-arms.**"

Sealed the 23rd March 1853, and dated the 28th January 1853.

PROVISIONAL SPECIFICATION left by the said William Tranter at the Office of the Commissioners of Patents, with his Petition, on the 28th January 1853.

I, William Tranter, of Birmingham, in the County of Warwick, Gun Maker, do hereby declare the nature of the said Invention for "Certain Improvements in Fire-arms" to be as follows:—

My Invention consists, in the first place, in adapting and applying to pistols and guns a double trigger or two triggers, one portion of the trigger or one of the triggers being used for raising the cock, and the other for discharging the piece. And with this trigger or triggers is combined a safety stop spring to prevent the cock falling, until the catch is withdrawn by means of the discharging trigger, or discharging portion of the double trigger. This double trigger or pair of triggers may be adapted in various forms to different constructions of fire-arms, so as to accomplish in each the double purpose of raising the cock, and at the same time withdrawing the safety stop spring, together with the discharging of the piece. The trigger or triggers may also have various forms of guards adapted and applied to them.

My Invention consists, secondly, in elongating the tube or socket of the revolving chamber in revolving pistols, through which tube or socket passes the short rod or axis on which the chamber revolves. The tube or socket may be slightly elongated, and the part of the frame into which it is required to fit may be countersunk and prepared to receive it in various ways, so as to increase the facility of applying the revolving chamber to its position in the frame of the pistol, and prevent the liability of the chamber to become inconveniently loose in its position.

My Invention consists, thirdly, in combining the aforesaid double trigger or pair of triggers with the chamber of a gun intended to be loaded at the breech. In this case a strong spring is used, both to hold the chamber in its place and to project the cock upon the nipple. The chamber may be turned round far enough to insert a loaded cartridge by pressing against a projecting lip and collapsing the spring, and the cock may be allowed to strike upon the nipple by withdrawing through the lifting trigger the means by which the projecting force of the spring upon it was counteracted. In this case, too, the parts may be retained on "half cock." The cartridge to be used in guns of this description is one so constructed as to dispense with the necessity of removing the closed end before placing it in the chamber by closing the end with a percussion cap.

My Invention consists, fourthly, in combining the double trigger or pair of triggers with the use of a needle or small rod to cause the explosion, in which case the said needle or small rod is projected by means of a spiral or other spring on the withdrawal of a bent lever by the lifting trigger.

And my Invention consists, lastly, in making gun barrels out of two or more spirals of iron or iron and steel wire platted or crossed round a mandril, the wire so platted being afterwards heated to a welding heat, and welded as in the ordinary manufacture of gun barrels, or it may be rolled in such state and with a mandril inside between two flat plates.

SPECIFICATION in pursuance of the conditions of the Letters Patent, filed by the said William Tranter in the Great Seal Patent Office, on the 28th July 1853.

TO ALL TO WHOM THESE PRESENTS SHALL COME, I, WILLIAM TRANTER, of Birmingham, in the County of Warwick, Gun Maker, send greeting.

WHEREAS Her most Excellent Majesty Queen Victoria, by Her Letters Patent, bearing date the Twenty-eighth day of January, in the year of our Lord One thousand eight hundred and fifty-three, in the sixteenth year of Her reign, did, for Herself, Her heirs and successors, give and grant unto me, the said William Tranter, Her special license that I, the said William Tranter, my executors, administrators, and assigns, or such others as I, the said William Tranter, my executors, administrators, and assigns, should at any time agree with, and no others, from time to time and at all times thereafter during the term therein expressed, should and lawfully might make, use, exercise, and vend, within the United Kingdom of Great Britain and Ireland, the Channel Islands, and Isle of Man, an Invention for "CERTAIN IMPROVEMENTS IN FIRE-ARMS," upon the condition (amongst others) that I, the said William Tranter, by an instrument in writing under my hand and seal, should particularly describe and ascertain the nature of the said Invention, and in what manner the same was to be performed, and cause the same to be filed in the Great Seal Patent Office within six calendar months next and immediately after the date of the said Letters Patent.

NOW KNOW YE, that I, the said William Tranter, do hereby declare the nature of my said Invention, and in what manner the same is to be performed, to be particularly described and ascertained in and by the following statement, reference being had to the Drawings hereunto annexed, and to the letters and figures marked thereon (that is to say):

My Invention of "Certain Improvements in Fire-arms" relates, in the first place, to the kind of fire-arms known as revolving pistols or pistols with revolving chambers; and consists, first, in adapting and applying to such fire-arms a double trigger or pair of triggers, combined with a safety stop spring, by means of which the piece is kept from being discharged until the spring is moved out of the way of the cock, which may be done simultaneously with the lifting of the cock and the turning of the chamber by drawing back the lifting trigger or lifting portion of the trigger with the second finger. The discharging trigger may then be moved with facility by the first finger, and the cock projected on the nipple to discharge the piece. This trigger or triggers may be used with various forms of guards. The lifting trigger or lifting portion of the trigger is slotted to hold the discharging trigger or discharging portion of the trigger. My Invention in relation to this kind of fire-arms consists, secondly, in elongating the tube or socket into which the hinder end of the revolving chamber is inserted, in order that the chamber may be thereby screwed forward when required, so as to keep the fore end always tightly fitting in the frame.

My Invention relates, in the second place, to guns intended to be loaded at the breech, and consists in adapting and applying thereto the double trigger or pair of triggers, somewhat modified in form as compared with that above referred to, and in the general arrangement of the parts of the gun, as herein-after more particularly described. In the present instance the trigger is so formed as to be capable of holding the gun at "half cock" by means of a tail piece on its end dropping into a notch or recess in the lower part of the cock, by which means the front or discharging part of this double trigger acts as a single trigger. In this case the revolving chamber may be turned round far enough to insert a loaded cartridge by pressing against a lip on the chamber, and collapsing the spring by which it is held in its place. The cartridge to be used in guns of this description is made so as to have a percussion cap inserted in the end opposite to the ball, by which the necessity for removing its ordinarily closed end may be obviated.

My Invention relates, in the third place, to guns, in which the arrangement and action of the revolving chamber and trigger or triggers are similar to those last referred to, but in which the gun is discharged by means of a needle or small rod projected by means of the spring which holds the chamber in its place, or by means of a spiral spring acting directly upon the rod carrying the needle.

And my Invention refers, lastly, to the mode of manufacturing gun barrels, and consists in making them out of two or more spirals of iron, or iron and steel wire platted or crossed round a mandril, the wire so platted being afterwards heated to a welding heat, and welded as in the ordinary manufacture of gun barrels.

In order, however, that my Invention may be fully understood I will proceed to describe it more particularly by the aid of the accompanying Drawings, in which are contained representations of the parts above referred to.

Figures 1, to 8, Sheet I., relate to revolving pistols. Fig. 1, is a side view of a pistol with my improved trigger or triggers and safety stop spring applied thereto, and shewing the cock projected on to the nipple, and the trigger or triggers drawn back. Fig. 2, is a transverse section of the same in the line *a*, *b*, of Fig. 1. Figures 3, and 4, are longitudinal sections, shewing the parts in different positions. Fig. 3, shewing the cock raised, and the lifting trigger drawn back; and Fig. 4, shewing the cock projected on to the nipple, having been liberated by means of the discharging trigger. Fig. 5, is an end view of the revolving chamber, with the nipples upon it. Fig. 6, is a longitudinal section of the same, shewing its elongated socket. Fig. 7, is a transverse vertical section of the frame within which the chamber turns, shewing the safety stop spring; and Fig. 8, represents another form of lifting and discharging trigger. In these several Figures the same letters of reference indicate corresponding parts. *a*, is the barrel of the pistol; *b*, the ordinary solid frame within which the revolving chamber *c*, is placed, so as to be capable of turning the required distance on the pivot or axle *d*, inserted in the frame *b*, in the usual manner; *e*, is the cock turning on a centre *f*, and projected on to the nipple by means of the spring *g*, and tail piece *h*. The lifting trigger is shewn at *i*, and the discharging trigger at *i**. These triggers turn on a centre *k*. The trigger *i*, is slotted to hold the trigger *i**. On the pivot or centre *l*, is fixed the spring pall *m*, which causes the chamber to revolve the required distance by acting against a tooth in the ratchet *n*, on the end of the chamber *c*, in the usual manner. On this pivot or centre *l*, is also fixed a tail piece or lever *o*, the upper end of which serves to raise the cock by pressing against a small recess in the same, as shewn at Figures 3, and 4, the lower end being acted on by the discharging trigger *i**, in order to withdraw the upper end from the said recess. The trigger *i*, is brought back to the position shewn in Fig. 4, after the discharge of the pistol by the pressure of the spring *p*. The safety stop spring is shewn at *q*, on the inner side of which is attached a projecting piece *r*, which projects inwards a sufficient distance (as shewn at Fig. 7,) to prevent the cock from going forward to the nipple until it is moved outwards into the position shewn in Fig. 2, by the pall or tail piece *m*, on the lifting trigger pressing against the curved end of the projecting piece *r*.

The action of this pistol is as follows:—The parts may be brought into the position shewn at Fig. 3, by drawing back the trigger by the second finger. The effect of which will be to raise the cock by means of the tail piece or lifter *o*, also to turn the chamber *c*, the required distance by means of the pall *m*, and also to remove the safety stop spring *q*, out of the position shewn at Fig. 7, into that shewn at Fig. 2, so that it may no longer present an obstacle to the projection of the cock on to the nipple. The pistol may then be discharged by a slight pressure of the first or forefinger against the trigger *i**, the end of which will thus be made to act upon the lower end of the tail piece *o*, and withdraw its upper end from the recess in the cock as shewn by dotted lines in Fig. 1. The removal of the pressure of the fingers from the trigger or triggers will cause the parts to assume the position shewn at Fig. 4, ready for another discharge.

It will be seen that at Fig. 8, the trigger or triggers are somewhat different in form from that just described, also that the tail piece or lifter *o*, is acted upon at the upper instead of the lower part. The action is otherwise the same as that above described, and I would here remark that my Invention is not confined to the exact forms thus represented, provided the action be in its general features the same as that above described.

I will now describe the elongated socket for the chamber *c*, which is represented at *s*, in Figures 3, and 6. It has a screw formed on it as shewn, and is prevented from turning by means of a set screw or pin *t*, inserted at the hollow part of the chamber *c*, (see Fig. 6.) This screwed socket may also be formed with a flange upon it, and may be used to hold the ratchet *n*, in the end of this chamber *c*, without screws; or the ratchet *n*, may be put on the flange of the socket *s*, and serve as a lock nut to prevent the socket from moving. The socket *s*, may also be made without a screw formed upon it, and in such case metallic washers may be used instead. When the ratchet *n*, is used to keep the socket *s*, in its place, screws must be used with the ratchet. The frame *b*, may be countersunk in order to receive the end of the socket *s*. By means of this elongated socket *s*, it will be seen that the chamber *c*, may be screwed forward or tightened against the fore part of the frame *b*, should it be required.

I proceed now to describe the second part of my Invention, which relates to guns intended to be loaded at the breech. Fig. 9, is a side view of such gun. Fig. 10, is a transverse section in the line *c*, *d*, of Fig. 9, looking towards the stock of the gun. Fig. 11, is a similar section in the line *e*, *f*, of the same Figure, looking towards the barrel. Fig. 12, is a longitudinal section, shewing the positions of the parts at the time when the cock is projected on to the nipple. A, is the loading chamber, which is kept in its position in the frame B, by means of a double acting spring C, pressing against the projection A*, and it is made to fit tightly against the end of the barrel by means of the bolt and screw D, and E. The chamber A, has on one side of it a lip F, by pressing upon which the chamber may be turned down to receive the charge; the spring C, also projects the cock. By means of this spring likewise the pivot or axle G, of the chamber C, is kept in its position; the end of the said spring, by bearing on one end of the arm or leaver H, causes a pin or stud on the same to pass into a hole in the axle G, and thereby hold it. I, is the lifting trigger, and I*, the discharging trigger, which are shewn in different positions in Figures 12, and 14; the former Figure shewing the trigger or triggers before they are drawn back, and the latter shewing the same with the parts as they would appear when the cock is lifted and ready to be projected on to the nipple.

The action of this gun is as follows:—The trigger I, is drawn back into the position shewn by dotted lines in Fig. 12, by which means the tail piece L, is made to move forward the bent arm or lever M, connected with the double spring C. In this manner the cock is lifted, and it is liberated by pressing back the trigger I*, which is thereby withdrawn from the notch in the bent arm or lever M, turning on a common centre with the cock, and the cock is projected on to the nipple by the tension of the spring C. It will be seen that in this case the lifting trigger only lifts the cock, which may be done by the hand without such trigger, so that, if preferred, a single trigger, as at I*, may be employed alone for this kind of gun. I would also remark, that the general arrangement of the parts of the gun just described may be adapted and applied to guns and pistols with revolving chambers. In the cartridge to be used with this gun the ball is elongated, with a shoulder extending into the body of the cartridge. At the reverse end the percussion cap is inserted in such a manner as to obviate the necessity for breaking off the end before use. Fig. 15, represents in longitudinal section the same kind of gun, but in which the explosion is to be produced by forcing a needle or small rod into the cartridge. It will be seen that the double acting spring C, which holds the chamber A, in its place is connected so as to act upon the rod to which the needle N, is attached by means of the bent arm or lever M, which in this case is lengthened upwards in order to adapt it for the purpose. In other respects the action of the gun is the same as that above described.

With reference to my improvements in the manufacture of gun barrels, above referred to, I would here remark, that the wire, when platted, is to be heated to a welding heat, and welded in the usual manner. The wire may be circular or of other form in the cross section.

Having thus described the nature of my Invention, and in what manner the same is to be performed, I would have it understood that I do not mean or intend to confine myself to the precise forms and arrangements of the parts shewn and described, as the same may be varied in some particulars without departing from the general features of my Invention. I would also here remark, that although I have described the double trigger or pair of triggers as applied to certain forms and constructions of fire-arms, yet my Invention is not confined thereto. And I would further remark, that the gun described under the second head of my Invention may be combined with the double trigger or pair of triggers as represented and described, or it may be combined with a trigger of another form.

I do not, however, claim as of my Invention the several well known parts of fire-arms which have been necessarily represented and referred to in order to make the object and essential character of my Invention evident. But what I claim as of my Invention is,—

First, the lifting and discharging trigger or triggers, combined with the safety stop spring, and capable of being acted upon in the manner and with the effect described.

Secondly, the mode described by which the revolving chamber may be made to fit with the required tightness in the frame which carries it.

Thirdly, the modified form of trigger or triggers applied to guns to be loaded at the breech, and the general arrangement and construction of such guns as described.

Fourthly, the same applied to guns to be fired by means of a needle.

Fifthly, the use of a cartridge of the kind described.

And lastly, the manufacture of gun barrels out of spirals of iron, or iron and steel wire, platted or crossed round a mandril as described.

In witness whereof, I, the said William Tranter, have hereunto set my hand and seal, this Twenty-sixth day of July, in the year of our Lord One thousand eight hundred and fifty-three.

WILLIAM TRANTER. (L.S.)

Witnessed by WILLIAM SPENCE,
Patent Agent.

LONDON:
Printed by GEORGE EDWARD EYRE and WILLIAM SPOTTISWOODE,
Printers to the Queen's most Excellent Majesty. 1854.

A.D. 1853. JANUARY 28. Nº 212.
TRANTER'S SPECIFICATION.

(2 SHEETS)
SHEET 1.

FIG. 1.

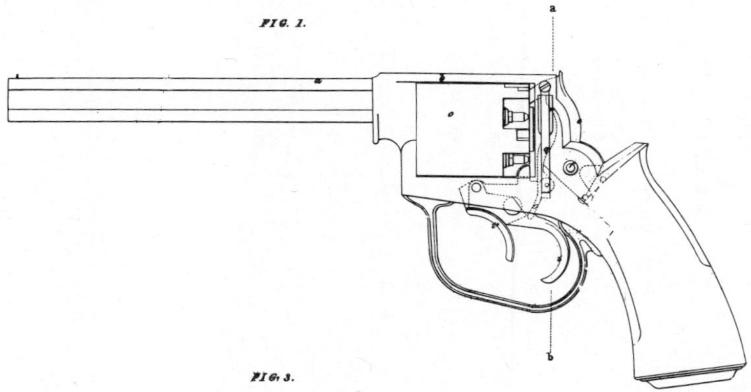

FIG. 2.

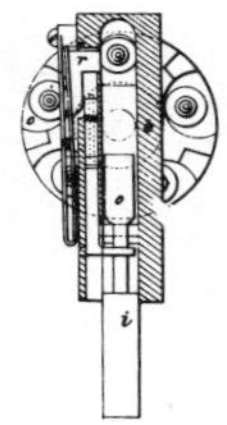

FIG. 5.

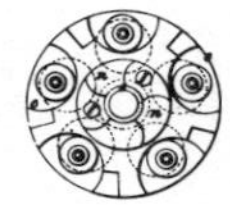

FIG. 6.

FIG. 7.

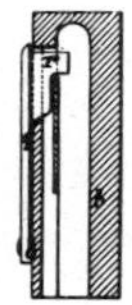

FIG. 3.

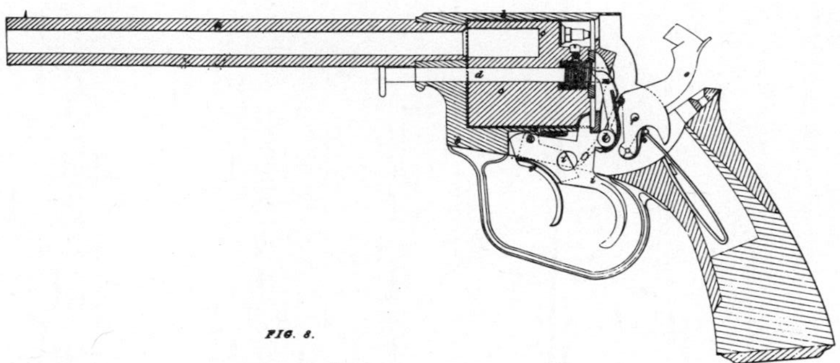

FIG. 4.

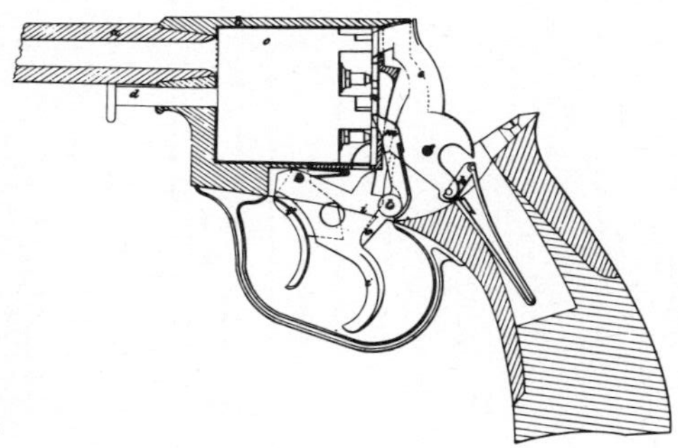

FIG. 8.

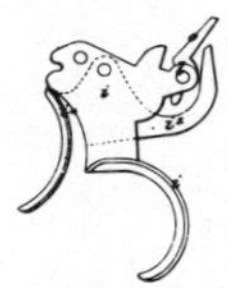

The Filed Drawing is partly coloured.

LONDON: Printed by GEORGE EDWARD EYRE and WILLIAM SPOTTISWOODE,
Printers to the Queen's most Excellent Majesty. 1853.

A.D. 1853 . . . . N° 305.

Repeating Pistols and other Fire-arms.

LETTERS PATENT to Philip Webley, of Birmingham, in the County of Warwick, Manufacturer, for the Invention of "**Improvements in Repeating Pistols and other Fire-arms.**"

Sealed the 16th April 1853, and dated the 4th February 1853

PROVISIONAL SPECIFICATION left by the said Philip Webley at the Office of the Commissioners of Patents, with his Petition, on the 4th February 1853.

I, Philip Webley, of Birmingham, in the County of Warwick, do hereby declare the nature of the said Invention for "**Improvements in Repeating Pistols and other Fire-arms**" to be as follows:—

The improvements relate to that description of fire-arm called "one barrel rotating chamber guns or pistols" for repeated firing. The Invention consists, first, in so supporting the barrel from the frame or body of the fire-arm that when it is required to remove or replace a cylinder (containing the charge chambers) the barrel may be partially turned round upon a pin as an axis placed parallel to the axis upon which the cylinder turns, and thus admit of its removal, and also in so arranging the framing as to obtain increased strength to resist the force of repeated fire.

The metal part of the frame passing under the cylinder is formed at its outer end into a pin or axis which passes through an arm projecting down from the inner end of the barrel; a stop prevents this arm being drawn off the pin or axis, but admits of its being with the barrel drawn slightly away from the cylinder, and from the outer end of the axis of the cylinder, in order to the barrel turning upon its pin or axis when the change of cylinder is required. This stop also (or another stop) serves to prevent the barrel at such time being turned too far round, and when the cylinder is in its place acts as a guide to the barrel in its again being brought to its central position for use, and when slided into contact with the cylinder a bolt or other stop secures the parts in position. The part of the frame upon which the pin or axis is formed to receive the arm from the barrel, and also that part of the arm which receives such pin or axis, are extended on the under side to obtain the increased strength for better resisting the strain of firing.

Secondly, the Invention relates to parts of the lock. The "sear" or instrument which acts between the trigger and the hammer I fix by a suitable joint to the hammer in place of to the trigger, and keep it in position by a spring. The driver, which acts to turn the cylinder so as to bring a fresh charge chamber after each fire, is kept in position by a spring affixed to the framing. A "single" in place of a "double" trigger spring is used. A lever stop is applied to the top strap to hold back the hammer from the nipples when required, such lever stop being so formed and placed as to be released by the hammer when fully drawn back by the trigger.

SPECIFICATION in pursuance of the conditions of the Letters Patent, filed by the said Philip Webley in the Great Seal Patent Office on the 4th August 1853.

TO ALL TO WHOM THESE PRESENTS SHALL COME, I, Philip Webley, of Birmingham, in the County of Warwick, Manufacturer, send greeting.

WHEREAS Her most Excellent Majesty Queen Victoria, by Her Letters Patent, bearing date the Fourth day of February, in the year of our Lord One thousand eight hundred and fifty-three, in the sixteenth year of Her reign, did, for Herself, Her heirs and successors, give and grant unto me, the said Philip Webley, Her special licence that I, the said Philip Webley, my executors, administrators, and assigns, or such others as I, the said Philip Webley, my executors, administrators, and assigns, should at any time agree with, and no others, from time to time and at all times thereafter during the term therein expressed, should and lawfully might make, use, exercise, and vend, within the United Kingdom of Great Britain and Ireland, the Channel Islands, and Isle of Man, an Invention for "IMPROVEMENTS IN REPEATING PISTOLS AND OTHER FIRE-ARMS," upon the condition (amongst others) that I, the said Philip Webley, by an instrument in writing under my hand and seal, should particularly describe and ascertain the nature of the said Invention, and in what manner the same was to be performed, and cause the same to be filed in the Great Seal Patent Office within six calendar months next and immediately after the date of the said Letters Patent.

NOW KNOW YE, that I, the said Philip Webley, do hereby declare the nature of the said Invention, and in what manner the same is to be performed, to be particularly described and ascertained in and by the following statement thereof (that is to say):

The improvements relate to that description of fire-arms for repeated firing called one barrel rotating chamber pistols or guns, and consist,—

First, in a novel mode of supporting the one barrel upon a pin from the frame or body of the fire-arm, so that when it is required to remove or replace a cylinder containing the charge chambers the one barrel may be partially turned round as upon an axis upon such pin placed parallel to the axis upon which the cylinder (containing the charge chambers) turns, and also to a mode of strengthening parts of such fire-arms to enable them better to resist the force of repeated firing; and,

Secondly, the improvements relate to the arrangement and combination of certain parts of the lock.

But in order to the Invention being fully understood I will proceed to describe the Drawing hereto annexed.

Description of the Drawing.

Figure 1, shews an external view; Figure 2, a plan; Figure 3, an end view; Figures 4, 5, 6, and 7, shew separate views of the metal frame *b*; and Figures 8, 9, 10, and 11, show various parts separately. *a*, is the rotating cylinder in which are formed the chambers to receive the separate charges, as is common in such like fire-arms. *b*, is the fixed metal part of the framing, which at its outer end at b^1, is formed into a pin or axis, upon which the barrel *c*, is supported, and is capable of turning by means of the projecting arm c^1, projecting from the barrel *c*, so that when a fresh cylinder *a*, has to be substituted for the one in use, or any change of cylinder has to be effected, the barrel *c*, may be turned upon the axis b^1, to admit of such removal or replacing.

And I would here state that I am aware that it has heretofore been proposed to arrange pistols of this description, that is, with one fixed barrel, and a series of charge chambers in a rotating cylinder, so that the barrel may be turned as upon a hinge in order to the withdrawal and replacing of a cylinder; but I believe in every case the part supporting the barrel has been caused to turn upon an axis lying across that of the axis of the rotating cylinder. And I mention this in order to state that one part of my improvements consists in forming an axis upon the framing so that the main barrel may be permitted to turn thereon in a plane parallel to that of the axis of the rotating cylinder, as shewn by the Drawing, by which arrangement I have been enabled to obtain great firmness of the parts during firing, combined with great facility in shifting the charge cylinders. When the parts are in position for firing, as shewn by Figure 1, the barrel is held in position by the pin or bolt *d*, which passing through the parts b^1, c^1, holds them correctly together. The pin or bolt *d*, has affixed to it a spring d^1, which gives elastic pressure to retain it in position. When a charge or removal of a cylinder is required, this pin *d* has to be withdrawn, then the barrel *c* may be slided slightly along the axis b^1, but is prevented coming entirely off the axis b^1, by the pin c^2 coming against the shoulder b^4; and the axis b^1, is cut away, as shown at b, to admit of this sliding movement, whilst it also serves to guide the barrel correctly to and from its correct position in relation to the other parts of the fire-arm. The axis b^1, is also cut away at b^3, so that when the barrel c is drawn along the axis b^1 to its full extent, the part b^3 will admit of the pin c^2 turning with the parts c, c^1, into the position indicated by dotted lines in Figure 3. When the replacement of a barrel has been effected, and the parts are again turned into position for firing, they are to be secured in that position by the sliding up of the pin or bolt d; and it will be seen that the parts b, and c^1, are extended on the under side at b^5, and c^3, so as to give additional strength for better resisting the strain in firing.

I would here state that although I have only shewn and described my improvements under this part of the Invention as applied to a pistol, a workman will readily make such modification in the arrangement of the parts by which the same may be used with guns or rifles having a rotating cylinder containing a series of chambers for separate charges.

I will now describe the second part of the Invention, which consists of modifications of certain of the parts by which simplicity thereof is effected. e, is the "sear" or instrument which acts between the trigger f, and the hammer g; and the improvement consists in affixing the "sear" by a suitable hinge joint to the hammer, in place of to the trigger, as usual in other like fire-arms where a "sear" has been used. The driver h, which acts to turn the cylinder a, partially round so as to bring up a fresh charge chamber after each fire, is borne upon by a spring i, affixed to the framing, as shown; and a single spring j, in place of a double spring, is used to act upon the trigger. l, is a lever stop, which turns upon the screw pin l^1, and at its other end l^2, is turned inwards, in order that when required that end may, by being forced inwards, rest within the hammer to hold it back from striking the nipples, but in the act of moving the hammer farther back to "full lock," the part l^3 of the lever l, is acted upon by the hammer to move the part l^2, out of the way.

Having thus described the nature of my said Invention, I would have it understood that I do not confine myself to the precise details herein shewn and described, so long as the peculiar character of either part thereof be retained.

In witness whereof, I, the said Philip Webley, have hereunto set my hand and seal, this Second day of August, in the year of our Lord One thousand eight hundred and fifty-three.

PHILIP WEBLEY. (L.S.)

Witness,

WILLIAM BROOKES,

73, Chancery Lane.

LONDON:
Printed by GEORGE EDWARD EYRE and WILLIAM SPOTTISWOODE,
Printers to the Queen's most Excellent Majesty. 1854.

A.D. 1853, Feb. 4. N° 305.
WEBLEY'S Specification.

(1 SHEET)

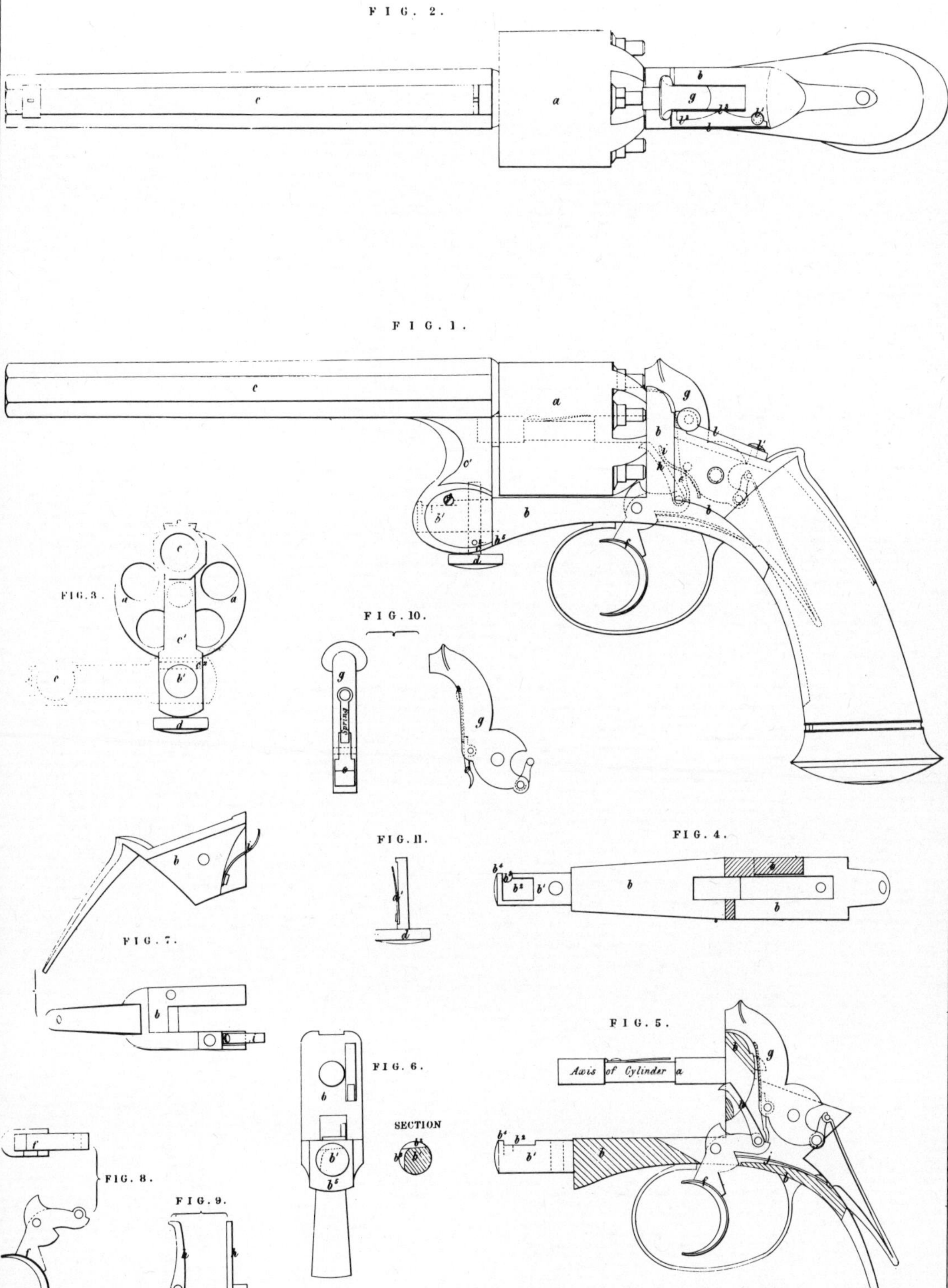

The Filed Drawing is not colored.

London. Printed by George Edward Eyre and William Spottiswoode.
Printers to the Queen's most Excellent Majesty. 1854.

J.R. Jobbins.